LEARNING FROM ICARUS

Aviation Safety Lessons from the Past and Present

Wisdom from great pilots and philosophers
to improve flight safety in the future

Bruce Landsberg

Vice-Chairman, National Transportation Safety Board – retired
Past President, AOPA Air Safety Institute

Disclaimer

The thoughts and opinions expressed are the author's and do not necessarily reflect the position or views of any organization.

Some images from official reports and other sources are of limited quality but are the best available. Where instrument approaches have not changed in a material way, a more current and higher-quality image may be substituted and identified accordingly.

Recommended Reading:

- *A Gift of Wings* – Richard Bach: Essays on why we fly and many engaging stories.

- *Weather Flying* – Rob Buck & Bob Buck: A great book on practical weather flying.

- *Engines* – Mike Busch: An excellent guide to understanding and operating piston engines.

- *Stick & Rudder* – Wolfgang Langwiesche: The most detailed explanation of angle of attack and stalls.

- *Stall/Spin Awareness* – Rich Stowell: An outstanding resource on what too many pilots fail to understand.

*"**Those who do not learn from the past are condemned to repeat it.**"*

George Santayana, 1905

Comments and questions?

✉ learningfromicarus@gmail.com

Table of Contents _Toc200369824

Dedication

Dedicated to Jan, Matt, and Neil, who entrusted their lives to me every time we went aloft.

To my father, Helmut, who started me on this odyssey, and my mother, Frances, who was my first passenger in a J-3 Cub.

Acknowledgments

Cover Photo Credits: NTSB, Vans AirForce Net, FAA, Author

There are many to thank—or blame—for this effort. No one is successful without the help and guidance of others. The leaders, pilots, groups, and authors listed below have all served extensively in the cause of aviation safety and have helped me and countless others along the way. A brief description is provided for each of these remarkable individuals and organizations, who deserve much more recognition than can be given here. **THANK YOU.** Profound apologies to those I may have missed.

Individuals

- **Col. Al Bonadies** – USAF pilot, consummate flight instructor, FAA Designated Pilot Examiner, and mentor from whom I learned instrument flying, procedure, technique, and the importance of professionalism. An unsung hero to the hundreds of pilots he taught to fly.

- **Phil Boyer** – President, Aircraft Owners and Pilots Association. A great leader, mentor, and pilot who flew across the North Atlantic, built his own aircraft, and flew everything from biplanes to light jets. An early adopter of all things electronic and one of the best communicators and businessmen in aviation.

- **Capt. Bob Buck** – Airline pilot and author of *Weather Flying* and *North Star Over My Shoulder.* A true aviation pioneer and a friend. He started with TWA in DC-2s and concluded a storied career in Boeing 747s. During WWII, he flew a B-17 all over the world to study weather that everyone else avoided.

- **Richard L. Collins** – Editor, *Flying Magazine* and *AOPA Pilot Magazine.* One of aviation's great authors, from whom I learned much about flying around thunderstorms and how dangerous flying can be if not approached with respect and ongoing study.

- **William Kershner** – Military pilot, aviation author, and factory test pilot. A great instructor, author, and mentor who put complex aviation ideas into plain language. Former owner of the world's smallest aerobatic flight school, consisting of one Cessna 150 Aerobat and one sick sack.

- **Russell W. Meyer, Jr.** – President and CEO, Cessna Aircraft Company. A great leader and pilot who demonstrated the highest corporate and personal integrity. When a brand-new Conquest II suffered a fatal accident due to a design issue, he immediately grounded the fleet and provided owners with other aircraft until the problem was fixed—textbook perfect.

- **Robert L. Sumwalt III** – Airline pilot, longtime member and chairman of the NTSB, and Executive Director of Embry-Riddle's Boeing Center for Aviation and Aerospace Safety. Responsible for supporting my appointment to the NTSB, cat lover, and a gentleman of high integrity. We even agreed sometimes.

- **Al Ueltschi** – President, FlightSafety International. An aviation pioneer who built the world's largest simulator training organization. He flew with Pan Am in its early years and became Juan Trippe's (Pan Am's CEO) personal pilot. A brilliant businessman, humanitarian, and mentor.

- **Jim Waugh** – Executive Vice President, FlightSafety International. A phenomenal leader who guided one of aviation's best marketing organizations with compassion, business savvy, and great integrity. The business world needs more like Jim.

Organizations

The groups below have played a critical role in the operation of the world's largest, safest, and most advanced aviation system. It was a privilege to work with all of them.

- AOPA Foundation's Board of Advisors and the President's Council
- AOPA Air Safety Institute and the tens of thousands of pilots who support it
- EAA (formerly the Experimental Aircraft Association)
- Federal Aviation Administration
- General Aviation Manufacturers Association
- National Aeronautics and Space Administration
- National Business Aviation Association
- National Air Traffic Controllers Association
- National Weather Service – Aviation Weather Center

Finally, many thanks to my friends and colleagues at the National Transportation Safety Board, who doggedly pursue and methodically document the probable cause of many crashes—and a few accidents (that will be explained later)—under some of the most trying conditions imaginable. They provided much of the content for this effort and the data upon which we all depend to fly safely.

Special thanks to *AOPA Pilot*, where some of this content originally appeared and has been re-edited. Deep appreciation to AOPA's former Senior VP of Communications and Editor-in-Chief of AOPA Pilot, Tom Haines, and William Garvey**,** former editor of *Flying Magazine* and *Business and Commercial Aviation.*

"If I have seen farther, it is by standing on the shoulders of giants."

Sir Isaac Newton, English mathematician, scientist, and physicist

The Goal

The goal is simple—**learn enough not to crash.** Looking back over half a century of flying. In some cases, I was good; in many cases, I was lucky; and in some, both attributes played a part.

You may learn something new here or simply be reminded that anything moving through the air (or on the surface) should be treated with great respect. These concepts and stories reflect the wisdom of great philosophers, authors, and pilots who came before us. By standing on their shoulders, we will all see farther and fly more safely.

Bruce Landsberg

Mount Pleasant, SC

Chapter 1
WHY ARE WE HERE?

The aviation safety book will never be written—there will always be more to learn. The basic art and science of flight aren't that difficult, but if one's only skill is regurgitating information on a written test or flying only occasionally, that does not serve well when faced with a nasty crosswind, a convective weather system, or a missed approach in low weather. Doing just "good enough" isn't the path to mastery.

I'll try to avoid obvious repetition, but reviewing essential truths helps new pilots flatten their learning curve. More experienced fliers may be reminded of how they gained that experience—and lived to talk about it.

The hits just keep on coming for the same reasons, and review helps keep us out of the rough. The scheduled airlines in some parts of the world have a remarkable safety record, but they follow stringent protocols for crash prevention. However, many of the same mistakes that are prevalent in general aviation (GA) still occur in the air carrier and corporate world— just not nearly as often.

As discussed later, comparing light GA aircraft safety to airline safety is a false equivalence. That doesn't mean we should stop trying.

The legend of Icarus is the first aviation crash that never happened—long before powered flight was possible. Daedalus and his son, Icarus, were imprisoned in a tower on an island by an evil Greek king. Daedalus planned an aerial escape and built two sets of wings by collecting feathers from birds that landed on the tower window and using candle wax from their prison cell. He warned Icarus not to fly too high, as the sun would melt the wax holding the feathers in place.

Icarus defeathered, Author

But Icarus, certain that he knew better and enthralled by the excitement of flight, flew much higher than his father recommended. The wax melted, and he crashed into the sea. His father knew from experience the properties of wax when subjected to heat and understood the limitations of the equipment, but his son misjudged the risk. How prophetic. We can learn a lot from Icarus.

"All limits are self-imposed."

Icarus, the first "aviator" who mythically crashed before airplanes existed

Some limits are circumstantial, while others are absolute. In too many of life's pursuits, we self-limit. The very act of becoming a pilot requires pushing beyond our limits. However, flying safely demands wisdom and knowing when to say "when." After every flight, take a moment to reflect on what went well and what could have been done better. That's how athletes, musicians, and professionals in every field reach the top of their game. Minor improvements add up.

As a new instructor, my students taught me more in the first six months than I ever taught them. The learning curve was steep for both student and teacher. I was teaching at the Part 141 school where I had trained, working alongside experienced CFIs who reviewed my students during the required phase checks. There was constant "feedback" on my failings—plenty of them. It was hard on the ego, but what an education.

For those of us who've been flying for a while, it's important to remember how we got this far. Self-accountability and humility go a long way. We've all heard of "pros" and high-time pilots who landed downwind, landed hot, flew into terrain, tangled with weather, or made basic errors that everyone knows they shouldn't. Some survived—many did not. No matter how full the logbook, the next flight is always the most critical.

Quotes

There are quotes from brilliant people throughout this book to help convey key points. In some cases, they are my own fabrication.

"Quote me as saying I was misquoted."

Groucho Marx, Comedian

Fix It?

The "silver bullet" solution—like the free lunch—doesn't exist because the opportunity to make bad decisions arises on every flight. Safety is always a journey, never a destination. Perfect safety in any mode of transportation doesn't exist—not yet—but we can get closer. The airlines have proven that.

Learning to fly is a growth process. Almost everyone who goes through it is significantly changed. The discipline, science, decision-making, and physical eye-hand-foot coordination required to fly are not to be dismissed lightly in this increasingly virtual, keyboard-driven world. Fortunately, these skills can be learned by most people. While new technology has the potential to make aviation easier, mastery still requires study, practice, and introspection.

Part 91 (Personal flight) pilots have the freedom to decide how much risk they take on, but they will be held accountable—whether by the authorities, the laws of physics (gravity), or their insurance companies— for bad decisions. However, non-participants (people on the ground or in other aircraft) and our passengers have an absolute right to life, regardless of how much risk we may personally accept.

Safety Evolution

Throughout my journey in both automobiles and aviation, I've seen significant technological advancements and an increased emphasis on safety. The aviation risk tolerance mindset—or culture—of what is considered acceptable has improved markedly.

Cars have evolved from having no seat belts to three-point harnesses, from steel dashboards to padded ones, and from basic safety features to crumple zones, collision avoidance systems, adaptive speed control, automatic emergency braking, blind spot warnings, backup cameras, and multiple airbags.

1940s J-3 Cub, AOPA

2000s Cirrus, AOPA

Airplanes have progressed from two-point seat belts to multi-point harnesses, airbags, collision and terrain avoidance systems, airframe parachutes, GPS, autopilots with envelope protection and emergency landing capability, weather-in-the-cockpit, noise-canceling headsets, and phenomenal instrumentation. The hardware has improved significantly, but when pilots push beyond their skill or the aircraft's capability, tragedy results.

We are rapidly moving into envelope protection and Simplified Vehicle Operations, as discussed in the Automation Chapter. With some help from our electronic friends, there may be breakthroughs that reduce complexity. But until that happens—and until the aircraft you're flying is equipped with such advancements—there will still be a need for some of that good old-fashioned pilot stuff.

"Aviation rules have expanded in all areas. In some cases, they were absolutely needed. In others, we got bureaucracy at its finest."

Author

There is misguided thinking among some in government that a rule alone will prevent a bad outcome. The unfortunate reason behind many ridiculous regulations is that someone, at some point, did something stupid. As a mostly reformed bureaucrat, I believe pilots still need to do some thinking. If they don't, the authorities will be more than happy to add to the paperwork and aggravation. We can help prevent that.

Yet, the NTSB never seems to run out of crashes to investigate. It's impossible to completely shield yourself from all hazards, but the vaccine of continuous improvement and recurrent training throughout your flying career is remarkably effective.

"Two unalterable rules govern all modes of transportation—gravity and Pauli's Exclusion Principle."

Author

Pauli-Illustrated, FAA

The assumption is that you understand gravity, having lived with it all your life. But Pauli is equally important. Wolfgang Pauli, an Austrian physicist, introduced his concept in 1925. You'll recognize it as the principle that no two objects can occupy the same place at the same time. This applies to all modes of transportation. Pilots, drivers, mariners, bicyclists—everyone has tested Pauli's principle at some point. He hasn't lost one yet.

Not Me

The typical pilot-observer reaction after a crash is, "I knew that," but the pilot involved certainly didn't. That's known as attribution bias—"I'm smarter than that." Maybe whatever was determined to be the probable cause always worked before, but situations that appear the same are often subtly different. The NTSB has a massive database of crashes to prove the point.

Hindsight bias and complacency affect all of us. The ability to rationalize that we'd never be *that* stupid is endless. But a solid understanding of risk—and, more importantly, ourselves—can make all the difference.

Some of the "don'ts" discussed here are hard-wired into us. It takes ongoing effort to overcome the idea that we're somehow better, that the weather really isn't *that* bad, that this maintenance item can be deferred, etc.

> *"The price one pays for pursuing any profession or calling is an intimate knowledge of its ugly side."*
>
> *James Baldwin, Author*

The following is an ugly paragraph. One thing you'll never see in an aircraft sales brochure is a discussion or picture of how, if the tool is misused, catastrophe lies ahead. It dampens sales somewhat.

Learn to fly? Maybe not today, FAA

Perhaps what's required is a bit more pilot imagination. Picture yourself in a bad situation created by your decision-making. Envision convective turbulence so severe that you can't read the instruments—then there's a loud bang as the airframe fails. Imagine trees looming during takeoff with an overloaded aircraft, approaching the departure end of a short runway. Listen as tree trunks begin ripping into the wings and feel the massive G-forces. Finally, picture yourself becoming so totally disoriented in clouds that, when trying to follow the instruments, your head violently disagrees as the aircraft hurtles earthward for a final, instantaneous stop. You get the idea.

Rapture of the Sky

My apologies for being so graphic, but on a day or in a place where we shouldn't fly, these images may help us make the right decision. This desire to fly—despite the serious prospects of crashing—could be called "Rapture of the Sky."

Flying is wonderful, uplifting, and even very practical. But the joy or desire to get somewhere or do something—when conditions, whether personal, environmental, or mechanical, are ill-advised—allows us to rationalize anything. Every new pilot, and some older ones, have periodically succumbed to this powerful aphrodisiac for any number of reasons.

Don't ask me how I know.

Most of the time, we live to fly another day. Some learn the lesson, and others… well, it always worked before.

"Safety isn't everything. It's the only thing."

What NFL Coach Vince Lombardi would have said about flying (he said it about winning).

What Lombardi said about winning applies here because, after a crash, nothing else matters. Once lives have been lost or injured, or an aircraft damaged, the **mission mentality**—the business deal, getting the boss or client to the resort, the trip to anywhere, the pleasure flight, the aerobatic sortie, the desire to get home—becomes a hollow concept.

There is no "mission" that is worth that.

But it won't happen to me—because that invisible, nonexistent protective shield has just been activated.

Wanna bet?

Unless you're in the military and at war, the term **"mission"** does not apply—as in *"We're on a mission from God"* from *The Blues Brothers* movie.

"A child's education should begin at least 100 years before he was born."

Supreme Court Justice Oliver Wendell Holmes Jr.

The Icarus "crash" never actually occurred, but the hard-earned wisdom of the Daedalians is partially lost as each new generation climbs into the cockpit. We all start out as Icarus, and Lady Luck has flown with every one of us. The best survival strategy is to get aeronautically educated before the luck runs out. I've made many mistakes and been guilty of some of the bad judgment you'll read about here. However, we can all learn a lot from the past 100-plus years of flight—if we're willing.

Everyone is an expert on preventing crashes – afterwards

"Lucky" Lindy understood risk

"The readiness to blame a dead pilot for an accident is nauseating, but it has been the tendency ever since I can remember. What pilot has not been in positions where he was in danger and where perfect judgment would have advised against going? But when a man is caught in such a position, he is there due to his error and is seldom given credit for the times he has extricated himself from worse situations. Worst of all, blame is heaped upon him by other pilots, all of whom have been in parallel situations themselves—but without being caught in them. If one took no chances, one would not fly at all. Safety lies in the judgment of the chances one takes."

Charles Lindbergh, first pilot to cross the Atlantic Ocean solo

Pull off a great crosswind landing, and you're the "Ace of the Base." But get blown into the ditch by an inopportune gust, and you're a dummy. Should've known better. It's the marginal cases where we are the most judgmental—where the outcome could go either way. It's hard to know where the edge is, but it sure is clear in hindsight. You only need to look at today's chat boards and TV to see that Lindbergh's observation is still practiced.

Humility was mentioned earlier. Unfortunately, that's not in the nature of many pilots, although they might have much to be humble about. The pros are quite risk-averse. They've seen what happens when carelessness, enthusiasm, or ignorance overcome the realities of a particular flight, and they understand the consequences.

Take your time, and don't try to absorb all of this at once. That avoids the TL;DR (*Too Long, Didn't Read*) temptation. There are decades of flight education here from master pilots. My role was merely to chronicle and provide some examples of hard-learned experience. Read the introductions before going to the case studies. The case studies were selected based on their experience potential. Some date back decades, but their lessons are just as pertinent today as they were when the crash happened. The same types of crashes still occur, or they played a significant role in shaping our regulatory and airspace systems.

Some of the communication transcripts are detailed, but take the time to understand how things got so balled up. The message will be repeated because that's how it sticks. The cost of learning all this was astronomical in terms of human suffering and dollars. Pilots better skilled and smarter than most of us have come to grief—but we don't have to.

There are air carrier, high-end general aviation, and light aircraft crashes—enough learning opportunities for everyone. When you're inevitably faced with a flight decision, remember the catastrophes cataloged here. Perhaps your random-access memory will kick in to advise caution—a small voice that says, *"I've seen this somewhere before."*

My goal is not to be preachy but to pass along the wisdom that was given to me, in some cases before I was ready to receive it. I hope you're smarter.

"Ships in harbor are safe, but that is not what ships are built for."

John Shedd, author and professor

Coach class, not always pleasant but they get where they're going.

Pilots and passengers enjoy the essence of flight – with care, Author

So it is with flying. No mode of transportation is perfectly safe. Understand the danger zones, and remember that they're not always obvious. Leave a margin for error—one that either you or somebody else just might make.

We who engage in accident analysis have the benefit of hindsight. Accident investigators, regulatory authorities, judges, juries, attorneys, other pilots, the media, and armchair analysts may debate for hours, days, or longer about what a pilot should or shouldn't have done—when, in reality, there may have been only a few seconds or minutes to act.

In some cases, disaster was caused by unbridled enthusiasm or ignorance. In others, well, it just seemed like a good idea at the time—but wasn't. Occasionally, there was an honest and momentary lapse in skill or judgment. Those sneak up on us and are the hardest to defend against. And sometimes—rarely—something broke, and there just wasn't a good solution.

CHAPTER 2
HUMAN FACTORS OR WHY IS IT ALWAYS US?

"Flying is one of the few popular sports in which the penalty for a bad mistake is death.... Such are the terms that flying lays down for pilots: Love me and know me and you shall be blessed with great joy. Love me not, know me not, and you are asking for real trouble."

Richard Bach, pilot, writer, A Gift of Wings

Aviation problems are generally self-inflicted. Aircraft-related failures resulting in accidents are the exception, typically accounting for around 10 to 15 percent. Even then, the "failure" is often due to deferred or poor maintenance. Once in a great while, a design problem goes undetected or unaddressed.

It may not be obvious why some of the crashes in this section are not listed elsewhere. In my opinion—yours may differ—all these events had their genesis in human factors. (We could probably say that about all crashes, but that would make categorization difficult.)

"Crash" and "accident" are not synonymous in NTSB vocabulary. An accident is an unforeseen, rare event that has not been well studied. A crash, however, is a common event where what happened is clear, why it happened is well understood, and the means to prevent it are well known. The vast majority of NTSB investigations, across all modes of transportation, involve crashes. However, investigations are conducted to uncover the few true accidents within the larger pool of crashes. These investigations also help reinforce what strategies and actions do not work and where efforts should be focused to improve safety.

Pilots (and other humanoids) are the probable cause in 65 to 95 percent of vehicle mishaps—whether involving aircraft, cars, tractors, trains, tugboats, or skateboards. We design, build, and maintain aircraft; develop and program flight management systems; operate the air traffic control system; conduct training; construct instrument procedures; fuel aircraft; and create performance documents and maintenance manuals. The aircraft and operational systems don't invent or operate themselves—the human touch is always there.

"If something can go wrong, it will."

Murphy's Law

Is it possible that, in a pilot's mind before a crash, there is an inability to contemplate a disastrous outcome? Pilots (we) often fail to consider what happens if the plan, skill level, or weather doesn't cooperate. It's uncomfortable to think about our own injury or demise, so we avoid it. That's optimism bias.

Humans are an unreliable bunch—we make mistakes, get tired, bored, ill, lazy, and distracted. Sometimes we're foolhardy, ignorant, or careless. Often, economic motives push us to make ill-advised decisions. In personal flying, it takes training, commitment, and a willingness to admit mistakes, identify weaknesses, and learn from them.

"Murphy was an optimist"

O'Toole's Commentary

Minuteman Launch- no room for error, USAF

I learned a lot about human reliability and systems design when the U.S. Air Force decided I should become a launch officer for the Minuteman II/III ICBM. The Air Force was selective, and "Oops" was not in the vocabulary.

We spent a lot of time learning (and relearning) every procedure and decision point. Single-point failures were largely eliminated, but in single-pilot, single-engine flight, that's not possible. So, we need to fortify the primary weak point—ourselves.

We'll discuss engines later.

Human factors textbooks discuss the likelihood that individuals will perform tasks effectively and accurately without errors. Competence, attentiveness, training, experience, and adherence to procedures all play a role.

My cockpit: Ten missiles to oversee and make sure none gets out of the barn by mistake, USAF

In aviation, medicine, and nuclear operations of any kind, human reliability is crucial for safety, efficiency, and overall execution. Performance improves through training programs, procedural enhancements, automation, and system designs that minimize human error.

In the launch control center, the system design was excellent—except for the air conditioning. The massive brine chiller units kept all the electronics of the day comfortably at about 65 degrees. It was a little brisk for the crew, but it did help keep us awake. That was by design.

> *"There are 3 kinds of men. The ones that learn by readin'. A few learn by observation. The rest have to pee on the electric fence for themselves."*
>
> *Will Rogers, 20th Century Humorist*

Will didn't mention women, who, as a group, seem to manage risk better than men. That conversation is well above my pay grade.

Unfortunately, many single-faceted solutions—whether focused on technique, procedures, new technology, or rulemaking—are not as effective as their passionate proponents suggest. Most of the low-hanging safety improvements were identified decades ago, but absorbing that hard-earned wisdom is more challenging.

Sometimes, it's easier to design the human out of the system. So far, that hasn't been completely successful—but it's improving.

> *"Hold my beer and watch this."*
>
> *Famous last words*

Risk Management

The aviation industry has discussed risk management for decades, with considerable success in some areas. There are at least three types of risk-taking behaviors:

A bit Risky – Organ Donor?

1. **Deliberate risk-taking** – When someone knowingly proceeds despite strong evidence that what

they're about to do isn't smart. *"Hold my beer and watch this."*

2. **Risk from ignorance or complacency** – When a pilot doesn't recognize the danger. *"I think this will work."*

3. **Calculated risk** – When a pilot fully understands the odds of success and the consequences of failure. *Lindbergh's approach.*

Some pilots and motorcyclists share similar personalities. The "organ donor" crowd rides and flies with wild abandon, convinced they can manage the danger. Meanwhile, the more cautious, staid type is typically more experienced. Survivors study their mistakes—and those of others. But crash-prone pilots and cyclists may not be so introspective until a millisecond before eternity, when the surprise is complete.

The willingness to learn applies to all higher-order activities, such as martial arts, sailing, writing, business, yoga, music, leadership, politics, and golf (as much as golf and politics could be considered higher-order activities). The guidance here is validated by hundreds of millions of flight hours.

New pilots usually fall into the ignorance category—through no fault of their own. There's a lot to learn about aviation, especially beyond the local traffic pattern and practice area. I speak from experience—it takes time to know what to look for and to develop discernment.

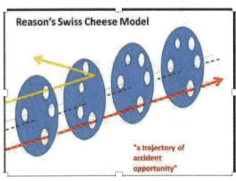

A model of crash causation, Dr. James Reason

The "Swiss Cheese Model developed by Dr. James Reason, conceptualizes how various layers within an organization can help prevent a crash. Typically, a bad outcome results from a series of misjudgments and contributing factors rather than a single mistake.

There are several ways to visualize this. Each layer represents a potential crash factor, such as risk tolerance (individual or organizational), pilot decision-making, maintenance, weather, and so on. Despite our best efforts, there are holes in every layer.

To complicate matters, imagine each layer as a spinning disk, reflecting the dynamic nature of daily operations. When all the holes align—it's a bad day.

A simpler concept is the **Accident Chain**. In chain theory, each link represents an opportunity to either cause or prevent an event. Break any one link in the chain, and the crash does not occur.

For example, a pilot decides to fly to an important family gathering. The weather is hazy, and nightfall is approaching. The pilot is delayed due to rush hour traffic, so the trip will now take place after dark. He is not instrument-rated but has completed some IFR training. The flight is over open water to some islands.

Chain Theory

On descent, he turns off the autopilot and becomes spatially disoriented. Sound familiar? This was John F. Kennedy Jr.'s last flight into Nantucket (see *Vineyard Spiral*). Change almost any one of these factors, and the sequence is disrupted.

But some accident chains are brutally short—selecting the wrong fuel tank before takeoff, leaving a control lock in place, or descending below landing minimums in instrument conditions can all lead to disaster.

The more sophisticated Swiss Cheese Model applies to larger operations and theorizes that many factors contribute to a disaster, including an organization's culture and management.

For example, an avgas-powered aircraft is misfueled with Jet A. Several failure "opportunities" occur along the way: not having the "duck-billed" nozzle permanently attached to the jet fuel hose (which prevents it from being inserted into an avgas fuel tank), failing to provide proper training for line personnel, inadequate supervision of line personnel, and management assuming that everything will be done correctly. The last person—and the easiest to blame—is the line worker who serviced the aircraft, but that was merely the final link in a chain of preceding errors.

Human Frailty in Aviation

See and Avoid

The "see and avoid" principle works about half the time—maybe. It underpins VFR flight operations, assuming that pilots flying in visual conditions will maintain a constant lookout and avoid close encounters. Most of the time, this works—not necessarily due to diligent scanning, but because the sky is vast and aircraft occupy only a small part of it. However, in confined airspace—near airports, on final approach, at airshows, in high-traffic practice areas, or dense sightseeing zones—encounters become more likely.

Several factors work against "see and avoid." Human vision and brain processing allow us to focus on only one area at a time. Anything on a collision course will have no relative movement—it remains a pinpoint in the windshield until it suddenly "blossoms" at the last moment, often too late. Sun glare and ground clutter further obscure targets.

Distractions from other cockpit duties also play a role. In some cases, an approaching aircraft is hidden behind a structural element, such as a windshield post, door frame, wing, or even the aircraft's nose. A passenger might also obstruct the view. The target may remain behind this blocking structure, only briefly visible during maneuvering—or never seen at all.

Note: A yellow circle superimposed on the left windshield represents the DHC-2, which is mostly hidden by the left windshield post.

Target hidden behind door post, NTSB

After the Cerritos, California, crash—where a Piper Archer brought down an AeroMexico DC-9—the entire air traffic control (ATC) system changed. Airliners were mandated to fly in largely protected airspace and were required to have a sophisticated Traffic Collision Avoidance System (TCAS), which detects potential threats and directs pilots to take evasive action. General aviation was only allowed into high-density airspace if equipped with Mode C transponders to broadcast altitude and presence, along with proper training and either clearance or established communication.

This concept was further expanded with the advent of ADS-B, significantly improving traffic awareness in the cockpit.

For nearly 40 years, this system worked flawlessly—until the collision at Reagan National Airport (DCA). Just before the publication of this book, a U.S. Army Black Hawk helicopter collided with a Bombardier regional jet on short final approach into DCA. At this writing, the accident is still in the preliminary investigation stages, so any analysis remains subject to change.

The details, briefly: It was a windy, clear VFR night. An American Airlines regional jet was cleared to land on crosswind Runway 33. Below its final approach path, at about 200 feet AGL, ran a high-density helicopter route along the east side of the Potomac River. Normally, a dedicated controller manages helicopter traffic while the local tower controller handles fixed-wing arrivals. However, in this case, the two roles were combined—possibly due to a lull in activity or staffing shortages.

Several factors will likely come under scrutiny, including the Army helicopter's adherence to route parameters and the overall risk posed by the low-altitude helicopter corridor. Prior reports had already documented close calls and missed approaches involving fixed-wing aircraft and helicopters in the area. This will likely prompt a review of why existing safety reporting systems failed to address the issue and whether corrective actions should have been taken earlier.

In response to the crash, the FAA immediately shut down the helicopter route.

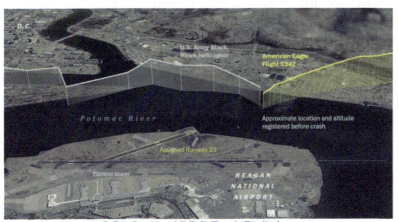

DCA Crash - ADS-B Track Preliminary

The Flight Data Recorders and Cockpit Voice Recorders were recovered from both aircraft, which will make assessing this tragedy much easier. Were the altimeters on the Black Hawk set properly and reading correctly? There are plenty of possibilities. Was the crew wearing night vision goggles, which can make spotting traffic more difficult against background lights? TCAS did not alert because, by design, it is disabled close to the ground.

My speculation is that this will likely be classified as a human factors event, with contributing factors including the Army pilots, FAA ATC staffing, the proximity of the helicopter route to DCA, and the failure to apply safety management systems despite previous close-call reports. There may also be an issue with altimeter accuracy, but it's too soon to tell. This is all purely speculative at this point, but if we meet on the ramp somewhere, I'll buy the coffee if I'm wrong. Of course, none of this erases the tragedy of the lives lost.

Another major issue with *see and avoid* is distraction. Programming navigation equipment, checking weather, consulting charts, communicating with ATC, and even simple complacency all take time away from scanning outside. Humans are poor monitors—we have short attention spans and struggle with multitasking. Unless an automated system or another person interrupts whatever task we're focused on, we miss a lot in a dynamic environment.

The science supports this. As noted in the Cerritos crash, a detailed Massachusetts Institute of Technology study found that pilots spotted other aircraft only about half the time, often at less than a mile. The Australian Transportation Safety Board also conducted a study that highlighted the significant limitations of *see and avoid.* However, when traffic avoidance systems like ADS-B are used properly, the success rate jumps to about 85%. I'm a strong believer in ADS-B as an awareness tool—the more aircraft equipped with it, the better.

That said, ADS-B isn't required in all airspace or on all aircraft, so it's not a perfect solution. But it's a huge improvement over flying "barefoot." When operating in high-density airspace, pilots must carefully split their attention between inside and outside the cockpit. (There's more on this in Chapter 3, *Takeoff and Landing.*)

> **"Human beings are flawed individuals. The Cosmic Baker took us out of the oven a little too early."**
>
> *Jimmy Buffett – singer, songwriter*

Risk Tolerance

Technology can reduce risk, but if the pilot simply increases their level of risk-taking, the overall danger remains the same. Some pilots feel braver flying parachute-equipped aircraft, rationalizing that they always have a bailout option. While these chutes have saved many lives, their safety envelope is limited, requiring intelligent use. A parachute won't help when descending below minimums, colliding with mountains or towers, or flying into a thunderstorm that tears the aircraft apart.

It's similar to anti-lock brakes on cars—if the reckless driver behind you insists on tailgating, the risk of a bad outcome remains. Technology evolves much faster than human behavior and may eventually begin to temper poor judgment. Maybe.

Automation that limits our ability to make errors is helpful, but how far can we engineer systems to counteract human incompetence or deliberate recklessness? Flight envelope protection continues to improve and may significantly reduce loss-of-control crashes. Stay tuned.

Basic synthetic vision is excellent for avoiding obstacles, but manuals specifically advise against using it for low-level, low-visibility flight. Anecdotally, one pilot attempted to fly VFR up the Columbia River Gorge at night in IMC using an early glass-equipped airplane under visual flight rules. He crashed into the same rock outcropping that has collected aircraft debris for decades.

Fortunately, most pilots use these technological advances as intended, improving overall safety. Weather-in-the-cockpit (datalink) is a great enhancement to situational awareness, but it doesn't change the weather. Like anti-lock brakes, its effectiveness depends on the pilot's willingness

Columbia River Gorge –VFR Only!, Google Earth

to use the information intelligently.

> *"Five hundred years ago, everybody knew the Earth was flat, and fifteen minutes ago, you knew that humans were alone on this planet. Imagine what you'll know tomorrow."*
>
> Agent K, The Movie, Men in Black

We're smarter now—maybe. The age of powered flight is well over 100 years old, yet early pilots were our intellectual equals. Human evolution occurs over tens of thousands of years, so our intelligence hasn't significantly improved since those early days. But something is different.

> ***"I wish to avail myself of all that is already known."***
>
> *Wilbur Wright – Letter to Smithsonian, 1894*

Wilbur had it right—there's a lot more knowledge about flying now. Regardless of what we think we know today, it's amazing how much more there is to discover.

Judgment

Formal judgment training helps, but it's well documented that humans behave differently when being monitored. Put a speed camera or a cop with a radar gun on the side of the road, and watch how traffic slows down.

Airline safety improved dramatically with the advent of cockpit voice recorders (CVRs) and flight data recorders (FDRs), which led to flight operation quality assurance (FOQA). With FOQA, everything is reviewed regularly, mostly by computers that look for significant deviations from prescribed procedures. It's embarrassing to admit we don't always perform to specification, but this process allows errors to be identified before a crash occurs.

Pilot unions have periodically fought against the full disclosure of what happens in the cockpit. Their stated concern is that recordings could be released to the media for tabloid exploitation or used for criminal prosecution. Some countries are quick to criminalize mistakes, but technology exists to manage data release and prevent inappropriate use. Not unreasonably, there is also concern that the FAA may resort to enforcement when additional training might be a better solution. However, there is much to be gained from firsthand reports—without punishing those who admit mistakes. Programs like the Aviation Safety Awareness Program (ASAP) and NASA's Aviation Safety Reporting System (ASRS) exist for this purpose.

By federal statute, the NTSB cannot release actual recordings (video or audio from crash aircraft) but can release transcripts. The only live recordings that typically emerge immediately after a crash come from the ATC system, including aircraft radio transmissions and ADS-B tracks, which are already in the public domain. From a crash investigation perspective, audio, video, and system recordings make the job much easier. But the real benefit is their preventive potential—trapping errors and correcting them before an accident occurs. The key is to routinely analyze the data.

On a check ride or with an instructor, pilots tend to fly conservatively. But the solo mindset is different when there's a trip we really want to take. Non-refundable hotel rooms, important meetings, not wanting to disappoint passengers, or simply wanting to sleep in our own bed don't change weather or mechanical problems.

For decades, the FAA has educated pilots about the Five Hazardous Attitudes:

1. **Anti-Authority** – "Don't tell me what to do."

2. **Impulsivity** – "Do something quickly."

3. **Invulnerability** – "It won't happen to me."

4. **Macho** – "I can do it."

5. **Resignation** – "What's the use?"

Has it worked? It's hard to say. Our statistical measuring tools are crude, but it certainly doesn't hurt. The crash rate is gradually decreasing, though that may be as much due to improved technology as to our collective learning.

Lack-of-Judgment Attributes

This is slightly different from the FAA's Hazardous Attitudes list. It's an unverified hypothesis, but in nearly every crash where the pilot was at fault, at least one of these five traits seems to be present:

1. **Ignorance** – Lack of knowledge, education, or awareness. This is the easiest to address; pilots can learn what they don't know—if they are willing.

2. **Arrogance** – An attitude of superiority, overconfidence, or presumption. An ignorant and arrogant pilot is the most dangerous creature aloft. Unfortunately, this trait is often hardwired into an individual.

3. **Complacency** – Self-satisfaction accompanied by an unawareness of danger or deficiency. Pilots can fly for thousands of hours in a routine environment and then face a critical situation in a moment. This is one of the hardest risks to guard against. Recurrent training helps.

4. **Distraction** – Failing to focus on the most important task. This factor is present in almost every crash. Humans like to think they

can multitask, but research has conclusively proven that we don't. Our brains process information sequentially, switching back and forth between tasks rather than truly handling multiple things at once. Think about gear-up landings. Pilots who forget to extend the landing gear typically say they were distracted, and the gear warning horn somehow blended into the background. It doesn't take a full-blown emergency to derail thought processes—another aircraft in the traffic pattern or a simple radio call can do it. The mantra is **Aviate, Navigate, Communicate**, yet, as the saying goes, *"When you're up to your neck in alligators, it's hard to remember that the original objective was to drain the swamp."* It's easy to recognize in hindsight but much harder to manage in real-time. Procedures and checklists help—if we follow them.

5. **Lack of Skill** – No matter how much aircraft and avionics improve, pilots will always need fundamental aircraft handling skills and sound judgment. Until flight systems become so fault-tolerant that they prevent critical errors entirely, human skill will remain essential. As we'll see in the section on automation, that's easier said than done.

"Well, that's the news from Lake Wobegon, where ALL the women are strong, ALL the men are good looking, and ALL the children are above average."

Garrison Keillor, writer, and radio show producer/performer

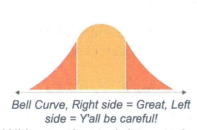

Bell Curve, Right side = Great, Left side = Y'all be careful!

No population is like the mythical residents of Lake WBG. We suffer under the tyranny of the bell curve shown here. Most pilots are average, a few are better, and some are worse. Equipment and procedures should be designed for the bottom half of the curve— the more capable among us will do just fine. With ongoing training, study, and the right kind of practice, higher performance isn't so far away, but it takes effort.

I'm better than those crash pilots.

Perhaps you are, but perhaps not. That's known as attribution bias, and pilots take comfort in it regularly. However, in aggregate, most of us fall somewhere close to the middle. This applies to doctors, CEOs, mechanics, plumbers, accountants, NTSB Board members—everyone.

Our ability to accurately self-assess is consistently lacking. That's known as the Dunning-Kruger effect (look it up). One curve does not define all of a person's attributes. Someone may be a whiz at making money but a lousy pilot. Or, as it is more elegantly stated:

> *"Financially gifted but aeronautically challenged."*
>
> Author, but somebody else has likely said it too

At FlightSafety International, while providing training for high-performance singles, twins, and turboprops, we saw many highly successful individuals. They excelled in their careers and could afford first-class aircraft, but some were in desperate need of professional aviation training to safely operate them.

Most improved significantly, some did not complete the training, and a few of those later crashed. There are people who simply shouldn't fly, and while the system does a fairly good job of filtering them out before accidents occur, it's not perfect.

> *"Ignorance more frequently begets confidence than does knowledge."*
>
> Charles Darwin, Scientist, author of On the Origin of Species

The freedom to pursue flight must always be balanced against the safety of the non-participating public and our passengers. There is an irrational fear among the public and some legislators that light aircraft pose a significant risk. They don't. In a typical year, there are only a handful of non-participating fatalities. Compared to other everyday risks—such as highway accidents—the chances of an aircraft falling on someone are exceedingly low.

However, when the rare crash does occur, it is sometimes used as justification to restrict flight. Passengers entrust their lives to us and deserve our utmost consideration.

Here are some terms used by NTSB human factors specialists:

- **Optimism Bias** – *It won't happen to me.* This attitude affects all modes of transportation. If anyone truly believed they were going to crash on a particular day, they wouldn't fly.

- **Overconfidence Bias** – *I can handle this—what could possibly go wrong?*

- **Fundamental Attribution Error** – *I'm better than you.* If a crash happens to me, it was unavoidable—a gust of gravity or wind at the worst possible time. But if someone else crashes, they were reckless, and it was obvious this would end badly.

- **Continuation Bias** – *Plan the flight, fly the plan—even when things are going wrong.* The ground speed is low, but we should have enough fuel. Approach minimums always provide a margin. I can handle these clouds VFR. I can get around the thunderstorms or icing. It's only another twenty miles to the destination. We can make this landing work, even though we're fast and high.

These biases are deeply ingrained, and our ability to rationalize is immense. Easy to fix? It is not.

> ***"Rapture of the sky enthralls many new pilots."***
>
> *Author*

It's cool to be a pilot and to do what only a tiny percentage of the population can. We naturally want to show off our new skills to family and friends—to impress them and share the thrill and beauty of flight. (I may have been guilty of this on occasion.)

However, there is a significant difference between how seasoned aviators assess flight risk and how a newbie does. Experienced pilots have seen what happens when unbridled enthusiasm takes over. It all comes back to that experience factor mentioned earlier.

> ***"Mix ignorance with arrogance and the results are guaranteed to be spectacular."***
>
> *Author*

Ignorance can be fixed, but arrogance is a personality trait—and much harder to adapt to the aviation safety mindset.

Hangar Story

"Bob" had just earned his private pilot certificate. It didn't come easily, but he managed to squeak through. He ran his own company and was in his late 20s. After purchasing a used Cessna Cardinal, he began working on an instrument rating. However, his attitude quickly alienated all the CFIIs at our airport, so he moved to a nearby airport—and burned through the instructors there as well.

One very hazy summer evening, several of us were on the taxiway performing pre-takeoff checks with students for night pattern work (there was no separate run-up pad). Suddenly, Bob veered off-road into the grass. He bucked and bounced his way to the head of the line, somehow avoiding the runway lights, then departed—without bothering to do an engine run-up or even make a radio call.

22

As the airport's Accident Prevention Counselor, as we were called in those days, I spoke with him the next day and noted that he seemed to be in a big hurry. Bob replied that he was late for an instrument lesson on the other side of the Chesapeake Bay. We discussed etiquette, patience in the cockpit, the dangers of uneven ground off the taxiway (including the risk of prop strikes), checklist discipline, and the wisdom of a VFR flight across the bay in three-mile visibility at night. "No worries," he said. "I can handle it."

Several months later, while "giving instruction" to his wife in the left seat, Bob somehow flew the Cardinal through a visual approach system about 75 feet off the edge of the runway. He couldn't remember which seat he was in, and besides, there's nothing illegal about acting as PIC from the right seat—but now the FAA took notice. They gave him a re-inspection "courtesy ride," which he passed.

Convinced Bob was a hazard, the airport manager at our privately owned, public-use airport invited him to go elsewhere. He never got that instrument rating. About a year later, a friend sent me a newspaper clipping: Bob had ended his flight career—and everything else—by flying VFR into IMC. Sadly predictable.

VFR into IMC

How many Visual Flight Rules (VFR) pilots inadvertently enter Instrument Meteorological Conditions (IMC)? We don't know. The failures are obvious and brutal, but what about the successes? What are the odds that an untrained pilot can survive?

Experimental ethics prevent us from conducting the kind of test needed to determine the chances of a successful VFR-to-IMC encounter. It might be interesting to try it in a simulator—comparing state-of-the-art technology against basic instruments with no autopilot. However, that could embolden some pilots to make a habit of flying in IMC without a clearance.

> *"If at first you don't succeed, then skydiving definitely isn't for you."*
>
> Stephen Wright, Author, humorist

Occasionally, you'll read someone's breathless account of how flying into IMC without an instrument rating really wasn't so hard, and that the fear surrounding it is overrated. Once again, we're only looking at one side of the picture. The caveat is that IMC comes in both gentle and not-so-gentle forms. It could be easy, or it could be life-ending.

Additionally, while the odds are pretty good that you won't collide with another aircraft, the irresponsibility of deliberately entering the clouds without informing ATC of your location is, in my view, a hanging offense.

Before and during my time at the NTSB, there were crashes involving airline pilots who had lied about their backgrounds to get hired. Their training records were poor, with multiple failures. Having a bad day or failing once isn't necessarily a red flag—it's the repetitive failures that indicate a serious problem. At the time, the pilot records system wasn't robust enough to catch these issues before they led to the destruction of aircraft, passengers, and sometimes even people on the ground.

Now, commercial operators are required to maintain training records for all their pilots and enter them into a central FAA database. The hope is that this system will prevent tragedies like the Colgan crash (cited later in this chapter).

Despite Lindbergh's quote in the introduction about how the system is too eager to devour anyone, aviation is a performance-based activity. When lives are at stake, things must be done properly. While there is some flexibility for minor mistakes, too many errors—or a single critical mistake—can lead to disaster.

However, it's not always the pilot at fault. There are systemic failures. Management often shares a major part of the blame, as demonstrated by Dr. Reason's Swiss cheese model—presuming there is a management structure in place at all. The NTSB strongly supports the Safety Management System (SMS), a core component of safe flight operations. The idea is that safety must be ingrained in an operation's culture, with everyone actively thinking about it, discussing it, and feeling empowered to raise concerns without fear of punishment.

While the concept sounds simple, keeping SMS effective day in and day out requires continuous effort from leadership down to frontline personnel. When a crash occurs, investigations often reveal cracks in the system that people had taken for granted.

For small Part 91 (non-commercial) and Part 135 (air taxi) operators, you are the management. At this writing, the FAA hasn't quite figured out how small operators can effectively implement SMS. The concept is solid, but long-term execution remains a challenge.

Success Is Overrated in Business—But Not in Aviation

Dozens of brilliant quotes highlight the human ability to make mistakes, and in some circles, failure is even considered a good thing. This approach works when lives aren't on the line. Some of the greatest commercial and scientific breakthroughs came after repeated failures.

- Formula 409® only worked after 408 failed attempts.
- Thomas Edison reportedly tried 10,000 times before perfecting the light bulb.
- Post-it Notes® were discovered due to persistent trial and error.
- In the stock market, you only need to be right 51% of the time to make a fortune.

Aviation, however, is different. We don't have to reinvent flight operations, aircraft, or the ATC system—what works and what doesn't has already been explored. That said, the rapid rise of Simplified Vehicle Operation (SVO) is pushing the boundaries of automation. But until these systems are truly foolproof, reckless experimentation carries Darwin Award-level risks.

Of course, serious failures still happen, like the Boeing 737 Max fiasco (see the Automation chapter), along with periodic Airworthiness Directives addressing lesser issues.

Mindset

The NTSB doesn't typically speculate on a pilot's mindset unless there is compelling evidence to do so. That's a tall order. Sometimes circumstantial evidence suggests that mental pressure or distraction played a role, explaining the "what" but rarely the "why."

Here's an oversimplified, unproven hypothesis on the fundamental causes of aviation accidents:

Aircraft Utility – Trying to extract too much performance from the aircraft. Examples include overloading, taking off or landing on a runway that's too short, running out of fuel, flying a non-deiced aircraft into icing conditions, or flying with a known mechanical issue.

Pilot Utility – Trying to extract too much performance from oneself. This includes VFR into IMC, attempting a challenging crosswind landing without recent practice, flying in low IMC without proper proficiency, or flying while fatigued or ill.

Too Much "Fun" – Engaging in reckless flying, such as buzzing, river-running, or improper aerobatics. How could anyone have too much fun flying? It's not the flying part—it's the crashing part. Usually, these incidents involve sensible regulations being ignored.

If you ever feel the urge to push your limits, ask yourself:

"How would I explain this to an FAA inspector, an NTSB administrative law judge, or worse—the families of my passengers who were injured or killed?"

That's assuming you survive.

Understand the power of the Dark Side.

"Baseball is 90 percent mental—the other half is physical."

Yogi Berra, Hall-of-Fame baseball player, manager, and master of the malapropism

With flying, it's about 58% mental, 30% physical, and 73.7% judgment. The math doesn't quite add up, but you get the idea. Rarely do crashes occur due to a single error—or even two.

New pilots, after completing primary and instrument training, are still learning where the risk areas exist. When we have only a little experience, our mental toolbox is limited. The saying, "When the only tool you have is a hammer, everything looks like a nail," applies. With limited experience, it's easy to miscategorize where problems are likely to develop. Reviewing the case studies cataloged here adds to the toolkit. The learning process should continue for as long as you fly.

Hot wings, AOPA

Docile Wings, AOPA

Transitions to Other Aircraft

Mishaps abound in a new-to-us environment. Changes in switch locations, control forces, or avionics configurations all impact situational awareness. What was once automatic in a familiar aircraft now requires more thought and new skills. Unlike the early days of airline aviation, standardization is now the mantra. Pilots once flew several different models regularly, but today, consistency is prioritized.

In general aviation (GA), our aircraft are usually less complex, but differences still exist and can occasionally trap the unwary. Fuel systems and performance data tend to be the most problematic. With GPS and Flight Management Systems (FMS), there are numerous opportunities for errors. Some amateur-built aircraft and warbirds can be particularly demanding—their performance is impressive, but their flight characteristics can be far less forgiving than those of certificated aircraft. Type clubs, knowledgeable instructors, and regular practice are essential.

The first 100 hours in a new aircraft type is an arbitrary benchmark for learning its nuances, assuming you're flying regularly—at least an hour per week. Additionally, pattern and practice area familiarization are very different from long cross-country flights in varying weather conditions. It's best to avoid carrying passengers for the first five to ten hours to devote full attention to the aircraft.

When I first flew corporate twins for Cessna Aircraft Company, that was their rule. For instrument flight, my assigned minimums were an 800-foot ceiling and two miles visibility. After 10 solo hours and another check ride, the restriction was lifted. We also had six-month proficiency checks. Airlines have similar policies—captains with fewer than 100 hours in a particular aircraft type, despite thousands of hours of jet flight experience, must raise instrument landing minimums by 100 feet and one-half mile above published minimums. That makes sense.

Flying in instrument meteorological conditions (IMC) is a rite of passage, but too many new instrument pilots experience it for the first time with passengers on board. It's far better to gain experience before bringing others along for the ride. An instrument rating or aircraft checkout is just the beginning.

Hangar Story

I had just earned my instrument rating along with another newly rated instrument pilot at our flight school. We both had the same highly experienced instructor, but because our airport lacked instrument approaches, our actual time in IMC was, charitably speaking, "nil."

My friend had offered to fly Sister Mary Catherine in his Piper Comanche to an airport about 150 miles away. The weather was benign IFR—fog and rain. At the destination, conditions were about 900 feet overcast with three miles visibility, but we'd be in the soup the entire way. My friend asked if I'd come along to assist.

Off we went, with Sister in the back, seemingly unconcerned, while we sweated bullets up front. I watched him, he watched me, and we both kept the Comanche aloft by sheer willpower. The trip went off without a hitch, and we both felt like we had graduated to a new level of flying. But every flight is a new learning opportunity.

With the increased availability of GPS approaches, all instrument students should gain regular exposure to actual IMC. In some desert locations, that's difficult, but the lack of real cloud experience during training remains a major shortcoming for many new instrument pilots.

Why Can't GA Pilots Fly as Safely as Airline Pilots?

There are some legitimate reasons. Airlines operate advanced aircraft with two professionally trained pilots flying in and out of roughly 500 well-equipped airports. GA pilots, on the other hand, fly less capable airplanes to and from nearly 5,000 public-use airports and thousands of private airstrips, each with varying conditions.

GA benefits from this flexibility and versatility, but it also carries higher risks. However, where GA operations mirror airline standards—using two pilots, flying turbine equipment, utilizing first-class flight simulators, and following Part 121 procedures with strong management support and intensive training—GA safety can match that of commercial air carriers.

Simulation and Training Devices

Training on the ground is far more efficient and safer than learning in the aircraft—though there are some caveats. Too many flight schools and instructors still operate under the old paradigm: minimal ground briefing, followed by, "Let's get the engine running, log flight time, and boost revenue."

Aircraft make terrible classrooms. They're noisy, hot, cold, or bumpy. They don't allow time to pause and process information, and they occasionally collide with other aircraft, objects, or the ground.

A quality simulator that accurately replicates handling characteristics and avionics configurations can save massive amounts of time. But note the qualifiers—if the simulator doesn't behave like the aircraft or if the avionics are significantly different, it's not helpful and may even lead to negative skill transfer.

Fortunately, as computers and visual systems improve, training devices continue to get better and more affordable. Instrument scanning techniques—whether for VFR or IFR procedures—can be greatly improved with simulation, making it an essential part of pilot training.

Little training Device, Redbird

Big Simulator, Airbus

In primary flight training, even a simple device can be immensely helpful at a much lower cost than actual flight time. Basic flight maneuvers are best taught through hands-on experience rather than just reading about them or watching videos.

After a familiarization flight, the first lesson—before moving to the aircraft—should focus on demonstrating how pitch and power are interrelated, how turns affect vertical lift, how to start the engine, and how to run checklists. Basic communications can also be introduced during these sessions. This process takes place over several lessons, allowing students to learn on the ground and develop the skills needed to apply in the air.

Newer avionics utilize ground-based simulation on PCs and tablets, reducing the need for costly flight time and fuel while learning the basics.

View-Limiting Device?

Traditional view-limiting devices (such as the hood) are marginal for teaching instrument skills. In the past, much of instrument training was done at night in VMC. There was less traffic, the trainee had extremely limited outside visibility, and ATC service was excellent since controllers often had more time.

A major flaw with view-limiting devices is that pilots develop a false sense of security by picking up subtle cues from sunlight movement and brief peeks beneath the device. This is a far cry from real IMC, where lurking vertigo becomes all too real.

At present, virtual reality headgear can accurately replicate IMC conditions in the aircraft. Using real avionics while experiencing G-forces, ATC complications, turbulence, and the potential for vertigo offers significant benefits. While virtual reality equipment won't replace simulation for tasks that are impossible or too dangerous to duplicate, nor will it match the efficiency of traditional simulators, it can realistically replicate the fully immersive experience of flying through clouds.

The All-Important CFI

Having seen the world from both sides of the cockpit, I believe that good flight instruction is one of the toughest flying jobs—and among the lowest paid.

Being an instructor is often an apprenticeship for flying bigger aircraft. Unfortunately, too many people with little experience, teaching skill, or genuine interest in instructing are forced into it as a stepping stone to a "real flying job." The current system offers few alternatives or incentives, but hopefully, that will continue to change.

View limiting device – not so effective, AOPA

A friend once noted that each generation of instructors learns only 90 percent of what the previous generation of CFIs taught. Over time, 90 percent of 90 percent of 90 percent results in 81 percent, then 73 percent, then 65 percent.

Despite that dystopian view, there are bright spots. Some new instructors turn out to be quite good, provided they receive significant supervision from experienced aviators and follow a solid curriculum. In other professions, new trial lawyers, doctors, electricians, or mechanics are not typically the teachers. However, in aviation, the economics of learning to fly are brutal for those not funded by the military or another outside entity. If the industry wants the best and brightest, it must address this issue. Our apprenticeship system could use improvement—just like it could for mechanics and avionics techs.

How Much Should I Fly to Stay Proficient?

How high is up? The answer depends on the type of aircraft, the weather conditions, the pilot's experience level, and whether the flying is for business or personal use. With so many variables, a one-size-fits-all answer is impossible.

That said, consider what the Air Force requires for desk-jockey pilots: a minimum of four hours per month, ideally broken into hourly segments each week. This serves as a guideline for experienced, well-trained aviators operating in a highly structured environment. Your situation may require more.

Even the pros fly at least 4 hours per month!

A long trip at the beginning of the month won't help much by the end of the month. I've found that about an hour a week is ideal for both the aircraft and me—it keeps us from rusting up.

Mix up your flights between VFR and IFR. While perfect days are gratifying, it's also valuable to fly in slightly grungy or windy conditions. If you're feeling a bit rusty, fly with another qualified pilot or an instructor.

To stay IFR current, take along a safety pilot at least every six months or complete an Instrument Proficiency Check. Remember, being *current* and being *proficient* are two different things.

Taking recurrent training at least once a year pays big dividends. Try something different to expand your skill set. If every pilot flew to at least FAA private pilot standards (and instrument rating standards, if applicable), the general aviation (GA) accident rate would drop significantly. Strive for higher proficiency, but that should be the minimum.

Pilots who follow these guidelines tend to have fewer accidents, though they may also be naturally more risk-averse. Until we find a more effective way to identify high-risk individuals, GA will continue to experience more human-factor-related crashes than commercial airlines. However, by adopting some airline and professional pilot limitations, procedures, and decision-making processes, you can significantly improve safety in GA operations.

Enough psychology—you've heard this before. Practice regularly and analyze every flight.

Key Points:

- Don't test common aviation knowledge the hard way—it's been done (think of the electric fence analogy).
- In light GA, pilots are the primary causal factor in 80 to 90 percent of crashes.
- Comparing airline safety records to light GA is misleading; the only real similarities are shared airports and some airspace. However, GA can still adopt airline-style risk mitigation strategies.
- Beware of Optimism Bias, Overconfidence Bias, Continuation Bias, and Fundamental Attribution Error.
- "See and avoid" has its limitations—stay vigilant. If your aircraft lacks ADS-B, consider getting it.
- Most of us are average pilots—sorry, but that's reality.
- Fully transition to new aircraft—small differences can have a big impact.
- Use high-quality simulation in all phases of flight instruction.
- Train regularly and fly often—rust builds quickly on both engines and pilots.

Colgan Calamity

A tragic series of failures changed the airline hiring world.

"Turn on, Tune In, and Drop out"

Timothy Leary/ Marshall McLuhan1960's (counter-culture figure and media studies professor)

While Leary and McLuhan were referring to mind-altering substances, too many pilots approach advanced avionics systems with a similar mindset. They turn them on, tune or program them to what seems correct, and mentally check out, assuming the system will handle everything. And most of the time, it does—until it doesn't. That's when complacency becomes a real hazard.

Q-400, Bombardier

As a landmark accident, this one ranks near the top. Human factors and aircraft automation were intertwined as causal factors. Airmanship played a key role, proving that no matter how sophisticated the aircraft or routine the flight might be, basic skills and adherence to procedure are essential. The failures here were both individual and systemic, which we'll discuss in the commentary.

For pilots considering an airline career, this crash had far-reaching consequences. Hiring requirements for first officers at regional airlines were significantly increased. While it was theoretically possible to be hired with just 250 hours, a commercial certificate, and a new multi-engine rating, that was far from the norm in the United States. Those minimums were often cited as justification for dramatically increasing the requirements.

Previously, regional airlines typically required between 700 and 1,200 total flight hours, a commercial certificate, and at least 50 hours of multi-engine time. FlightSafety ran an airline selection program years ago that included simulation, knowledge, and personality evaluations. This program provided an alternative way to identify high-potential pilots, and graduates never encountered issues transitioning to airline operations.

The Flight

On February 12, 2009, a Colgan Air Bombardier DHC-8-400 (Q400), operating as Continental Connection Flight 3407, was scheduled to depart Newark International Airport (EWR) at 7:10 p.m. Eastern Standard Time for a 53-minute flight to Buffalo-Niagara International Airport (BUF) in Buffalo, New York.

According to the NTSB report, as the crew prepared to taxi at 8:41 p.m., the first officer (FO) remarked, "I'm ready to be in the hotel room." She continued, "This is one of those times that if I felt like this when I was at home, there's no way I would have come all the way out here. If I call in sick now, I've got to put myself in a hotel until I feel better... we'll see how... it feels flying. If the pressure's just too much... I could always call in tomorrow. At least I'm in a hotel on the company's buck, but we'll see. I'm pretty tough." Her comments reflected the economic penalties some carriers imposed on pilots for calling in sick.

Flight 3407 was airborne at 9:18 p.m., with the captain flying and the FO acting as the pilot monitoring. The climb and cruise were uneventful. Propeller, airframe, and pitot-static deice equipment were operating normally, and the autopilot was engaged. The crew engaged in nearly continuous conversation throughout these phases of flight.

Ice Detection warning, NTSB

At about 9:53 p.m., the first officer (FO) briefed the airspeeds for landing: flaps set to 15°, 118 knots Vref, with 114 knots for a go-around speed.

At approximately 9:56 p.m., the FO remarked, "Might be easier on my ears if we start going down sooner." The Cleveland Center controller approved a descent from 16,000 to 11,000 feet.

At about 10:06 p.m., the flight was cleared to descend through 10,000 feet, where the sterile cockpit rule (FAR 121.542) came into effect, prohibiting conversation on non-operationally pertinent topics. The crew ignored it.

The flight was then cleared down to 4,000 feet. The captain asked the FO about her ears, and she acknowledged they were "popping." The conversation then shifted to icing.

At approximately 10:10 p.m., the FO noted ice on her windshield and asked if the captain was seeing any. The captain asked how her side of the windshield looked, to which she responded, "Lots of ice."

According to the NTSB, "The captain then stated, 'That's the most I've seen—most ice I've seen on the leading edges in a long time. In a while anyway, I should say.' About 10 seconds later, the captain and the first officer began a conversation that was unrelated to their flying duties. During that conversation, the first officer indicated that she had accumulated more actual flight time in icing conditions on her first day of initial operating experience (IOE) with Colgan than she had before her employment with the company."

The NTSB report continued: "She also stated that, when other company first officers were 'complaining' about not yet having upgraded to captain, she was thinking that she 'wouldn't mind going through a winter in the Northeast before [upgrading] to captain.' The first officer explained that, before IOE, she had 'never seen icing conditions… never deiced… never experienced any of that.'"

At about 10:12 p.m., the approach controller cleared the flight down to 2,300 feet. While performing flight-related duties, the non-pertinent conversation continued. At 10:13 p.m., the captain called for the descent and approach checklists, which the first officer performed.

The autopilot altitude hold mode was selected, and the autopilot leveled the aircraft at the preselected altitude of 2,300 feet at an airspeed of 180 knots. The captain called for flaps to be set at 5 degrees. The flight was subsequently cleared for the approach, and the captain began to slow the airplane.

According to the NTSB, "The engine power levers were reduced to about 42 degrees (flight idle was 35 degrees). At 10:16 p.m., both engines' torque values were at minimum thrust." The aircraft was slowing rapidly. (Degrees refer to the angle of the levers, not the power setting.)

The crew was directed to contact the Buffalo control tower. The first officer acknowledged, marking the last external communication. The landing gear was lowered, and the propeller condition levers were moved forward to the maximum RPM position. The autopilot applied nose-up trim to 14 degrees to maintain the preselected altitude. Airspeed was now about 145 knots. The autopilot continued to apply nose-up trim, and an "ice detected" message appeared on the engine display in the cockpit.

The captain called for the flaps to be set to 15 degrees and for the before-landing checklist. The Flight Data Recorder (FDR) showed that the flaps had been set to 10 degrees (there's a detent that must be passed to select 15 degrees) as the airspeed passed about 135 knots.

At 10:16:27 p.m., the Cockpit Voice Recorder (CVR) recorded the stick shaker, which warns a pilot of an impending wing stall through vibrations on the control column. The autopilot disconnect horn sounded at 131 knots as manual aircraft control was returned to the captain.

According to the NTSB (edited for length), "FDR data showed that the control columns moved aft at 10:16:28 p.m. and that the engine power levers were advanced to about 70 degrees… One second later… FDR data showed that engine power had increased to about 75 percent torque.... While engine power was increasing, the airplane pitched up, rolled to the left—reaching a roll angle of 45 degrees left wing down—and then rolled to the right. As the airplane rolled to the right through wings level, the stick pusher activated (about 10:16:34 p.m.), and flaps zero was selected. (The Q400 stick pusher applies an airplane-nose-down control column input to decrease the wing angle-of-attack [AOA] after an aerodynamic stall.)"

Final Dive, Q400 rolling to the right, NTSB animation

At 10:16:37 p.m., the first officer (FO) told the captain that she had put the flaps up. The flaps began retracting by 10:16:38 p.m., and the airspeed was approximately 100 knots. The roll angle reached 105 degrees, with the right wing down, before the airplane began rolling back to the left. The stick pusher activated a second time at 10:16:40 p.m. At that moment, the airplane's pitch angle was -1 degree.

Flight Data Recorder (FDR) data showed that the roll angle reached approximately 35 degrees, with the left wing down, before the airplane rolled again to the right. Shortly after, the FO asked whether she should retract the landing gear, to which the captain responded, "Gear up," followed by an expletive. The airplane's pitch and roll angles then reached approximately 25 degrees nose down and 100 degrees right wing down, respectively, before it entered a steep descent. The stick pusher activated a third time at 10:16:50 p.m. The flaps were fully retracted by 10:16:52 p.m. Around the same time, the captain stated, "We're down," followed by the sound of a thump.

The Cockpit Voice Recorder (CVR) recording ended at approximately 10:16:54 p.m., as Flight 3407 crashed into a single-family home, killing one person on the ground and all 49 people on board.

The Weather

According to the NTSB report, the 9:54 p.m. METAR for BUF indicated winds from 240 degrees at 15 knots, gusting to 22 knots, with visibility of 3 miles in light snow and mist. There were a few clouds at 1,100 feet AGL, a broken ceiling at 2,100 feet AGL, and overcast skies at 2,700 feet AGL. The temperature was 1°C, with a dew point of -1°C, and the altimeter reading was 29.79 inches.

Light to moderate icing had been reported earlier in the day, but nothing unusual for a winter evening when airline flights routinely operate. While there was potential for icing, the NTSB did not consider it a contributing factor.

The Captain

The 47-year-old captain held an airline transport pilot certificate and a first-class medical certificate. However, his flight record was concerning.

He had four certificate disapprovals, beginning in 1991 when he failed his instrument rating flight check. In 2002, he was disapproved for a single-engine commercial certificate in a Cessna Cardinal RG, failing takeoffs, landings, go-arounds, and performance maneuvers. In 2004, he failed his commercial multiengine flight certificate, and the entire exam was rated unsatisfactory. This disapproval was not reported to Colgan during his employment process, as required.

His airline training record also indicated ongoing weaknesses. He was hired by Colgan in September 2005 with 618 total flight hours, 38 hours of actual instrument time, and 71 hours of simulator instrument time. He passed his initial FO check in 2004, but as an FO, he required periodic additional training and failed at least one recurrency check in 2006.

His 2007 upgrade to captain also did not go well, as he failed his Airline Transport Pilot flight check. He struggled with single-engine approach and landing procedures in the Saab 340 but passed the re-check a few days later. It is unclear whether this re-check was conducted in an aircraft or a simulator.

He was type-rated as a captain on the Q400 in November 2008. At the time of the crash, he had accumulated 3,379 total flight hours, including 3,051 hours in turbine aircraft and 1,030 hours as pilot-in-command (PIC). However, he had only 111 hours in the Q400. In the 90, 30, and 7 days before the accident, he had flown 116, 56, and 16 hours, respectively.

According to the NTSB, the check airman who conducted the captain's Q400 simulator and line-oriented flight training described his decision-making abilities as "very good" but noted that he over-controlled the roll axis during unusual attitude training. However, he showed improvement with subsequent simulator experience. The IOE check airman described the captain's performance as "methodical and meticulous." It was noted that he had difficulties with the Q400's flight management system, but such challenges were typical for pilots transitioning to the aircraft.

FOs who flew with the captain earlier in February 2009 stated that he "handled the airplane well, used checklists, and did not miss callouts." They also described the cockpit atmosphere as "relaxed" but noted that he adhered to the sterile cockpit rule.

The First Officer

The 24-year-old FO held a commercial pilot certificate, a first-class medical certificate, and qualified as second-in-command on the DHC-8 in March 2008.

According to her Colgan employment application, from August to December 2006, she worked part-time as a flight instructor at an FBO in Arizona. From January 2007 to January 2008, she instructed at an airline training center in Phoenix. She was hired by Colgan in January 2008. Her résumé listed flight experience in the Piper PA-44, PA-34, PA-28, Cessna 152, 172, Beech BE-19, BE-23, and Diamond DA-40.

She was hired by Colgan with 1,470 total flight hours, including 6 hours of actual instrument time, 86 hours of simulated instrument time, and no turbine experience. At the time of the accident, she had accumulated 2,244 total flying hours, including 774 hours in the Q400. In the 90, 30, and 7 days before the accident, she had flown 163, 57, and 16 hours, respectively.

Her training record was significantly better than the captain's. She had only one failure—her initial CFI flight check in May 2006. Her initial proficiency check as an FO occurred on March 16, 2008, and her airline training record showed no failures.

According to the NTSB, the check airman who conducted her IOE stated that she had no issues handling the airplane and described her as a "good pilot who was sharp, assertive, and thorough." A captain who had flown with her multiple times rated her as "average to above average" for her experience level. He noted that she was always ahead of the airplane when monitoring and cross-checked her actions. Other captains believed she had the ability to upgrade to captain.

None of the captains interviewed reported any concerns about her adherence to sterile cockpit procedures, nor did they recall her making unprompted configuration changes while they were the flying pilot. Her recurrent training instructor stated that she had good knowledge of the airplane, and another FO in her class remarked that she had more technical knowledge than the average FO.

Fatigue

The air carrier lifestyle often involves constant travel and some degree of sleep disruption. Many Newark-based Colgan pilots commuted to work, and the report included considerable discussion about fatigue.

The captain was at the end of a multi-day trip and had been sleeping overnight between flights in the Newark crew room, which violated company policy. The NTSB noted, "In addition, the first officer's decision to begin a transcontinental commute about 15 hours before her scheduled report time without having an adequate rest facility affected her ability to begin the trip as rested as possible." She had traveled from Seattle (SEA) to Memphis (MEM) and then from MEM to Newark (EWR) on a red-eye flight.

She arrived in Newark seven hours before her scheduled report time, but crew-room rest is notoriously inadequate. Company policy did not discourage long commutes on the same day a trip was scheduled to begin.

The crew engaged in active conversation during the flight and just before the accident, indicating that they were mentally engaged. The NTSB concluded that the crew was fatigued but could not determine to what extent, if at all, it was a factor in the accident.

The Aircraft

The Bombardier Q400 was a state-of-the-art regional turboprop built in 2008, with only 1,819 total flight hours. It was designed to carry 74 passengers and four crew members. No pre-impact defects were noted.

The aircraft was delivered with full "glass" instrumentation. Its airspeed indicator tape was equipped with a trend vector arrow, predicting where airspeed would be in 10 seconds. Additionally, it featured a low-speed cue—a red bar rising from the bottom of the tape—while the airspeed tape numbers would change from white to red, signaling an approaching stall. The aircraft was equipped with both a stick shaker and a stick pusher. The stick shaker was designed to warn of an impending stall, while the stick pusher assisted in stall recovery. When the stick shaker activated, the autopilot automatically disengaged.

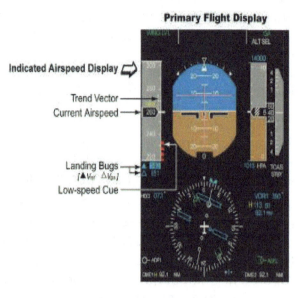

Reference, or "ref," speeds on landing are a fundamental aspect of airmanship. The Q400 was equipped with an ice detection system to alert pilots of possible icing and to remind them to adjust normal weight-dictated ref speeds

Q-400 Primary Flight Display, NTSB

upward due to the potentially higher stall speed. For the accident flight, the normal ref speed was 118 knots. With the ice warning, it should have been set to 138 knots. However, the crew apparently failed to notice the ice warning and did not adjust the ref speed accordingly. The NTSB found no fault with the aircraft.

Training

The NTSB found Colgan Air's training processes lacking. Specifically, the airline did not prepare the flight crew for an unexpected stall in the Q400 or the actions necessary to recover from a fully developed stall. It also failed to familiarize crews with the stick pusher and the aircraft's response during such events.

NTSB Probable Cause

The NTSB determined the probable cause as follows:

"The captain's inappropriate response to the activation of the stick shaker led to an aerodynamic stall from which the airplane did not recover. Contributing to the accident were: (1) the flight crew's failure to monitor airspeed in relation to the rising position of the low-speed cue, (2) the flight crew's failure to adhere to sterile cockpit procedures, (3) the captain's failure to effectively manage the flight, and (4) Colgan Air's inadequate

procedures for airspeed selection and management during approaches in icing conditions."

The NTSB issued 46 findings in addition to the probable cause. While too numerous to list in full, the full report is essential reading for anyone considering a career in aviation: NTSB Report.

Key Findings

- **Aircraft Performance:** *"The minimal aircraft performance degradation resulting from ice accumulation did not affect the flight crew's ability to fly and control the airplane."* This was noted because, shortly after the crash, speculation arose that icing may have been a significant factor.

- **Crew Awareness:** *"The reason the captain did not recognize the impending onset of the stick shaker could not be determined... but the first officer's tasks at the time the low-speed cue was visible likely reduced opportunities for her timely recognition of the impending event."*

- **Captain's Response:** *"The captain's response to the stick shaker activation should have been automatic, but his improper flight control inputs were inconsistent with his training."*

- **First Officer's Actions:** *"Although the reasons the first officer retracted the flaps and suggested raising the gear could not be determined from the available information, these actions were inconsistent with company stall recovery procedures and training."* (This was the only finding that reflected any shortcoming by the first officer).

- **Pilot Monitoring:** *"The monitoring errors made by the accident flight crew demonstrate the continuing need for specific pilot training on active monitoring skills."* The importance of maintaining a sterile cockpit during critical phases of flight is further discussed in Chapter 3.

- **Fatigue Factor:** *"The pilots' performance was likely impaired because of fatigue, but the extent of their impairment and the degree to which it contributed to the performance deficiencies that occurred during the flight could not be determined."*

- **Colgan Air's Responsibility:** *"Colgan Air did not proactively address the pilot fatigue hazards associated with operations at a predominantly commuter base."* (Many pilots could not afford to live in Newark, their home base, due to the low pay regional airlines offered at the time.)

- **Operator Responsibility: (**"*Operators have a responsibility to identify risks associated with commuting, implement strategies to mitigate these risks, and ensure that their commuting pilots are fit for duty.*")

Commentary

The NTSB held the captain accountable for a critical lapse in airmanship. Neither he nor the first officer was paying attention as the Q400 slowed to a stall while on autopilot, with the engines at low power. The autopilot disengaged, and the stick shaker and stick pusher activated exactly as designed.

Before the crash, Colgan Air had no requirement to teach stick-pusher or shaker procedures. Regardless of the aircraft, it is essential for pilots to familiarize themselves with stall characteristics and any safety systems designed to prevent loss of control. The FAA's oversight was lax in allowing such a deficiency in training programs, which is further discussed in the Automation chapter.

There was no indication that the Q400 was susceptible to tail stalls, but one theory suggests that the captain reacted as if it were a tail stall—pulling on the control column, which is the correct response in that rare scenario. Alternatively, he may have simply been startled by the sudden pitch-down movement and instinctively pulled back, an incorrect response.

Few training scenarios allow pilots to enter a stall without realizing they are near the edge. This challenge is also reflected in general aviation (GA) crashes. Replicating the element of surprise in training is difficult due to the *Hawthorne Effect,* a social science principle stating that individuals behave differently when they know they are being observed. In a simulator or under an instructor's watchful eye, pilots remain on guard.

Pilots training on aircraft equipped with autopilots featuring vertical speed hold, airspeed hold, or altitude hold must understand how the aircraft reacts to declining airspeed and power settings. A more effective design would involve an earlier warning system—announcing an approaching stall rather than waiting until it becomes imminent. Fortunately, advancements in flight envelope protection continue to improve safety.

The air carrier should not have allowed this captain to take command, in my opinion. His flight check record was dismal before he started with the airline and remained spotty throughout his career. His inexperience with the aircraft meant he had not yet developed solid habit patterns, and he wasn't paying attention during an instrument approach.

The fatigue debate continues. Both the captain and first officer had long commutes before their regular duty time. There are no easy answers to the commuting conundrum. The pressure on airlines to keep expenses and fares low works against solving this issue. Pilots are responsible for being ready to work, but occasionally, they will show up exhausted, regardless of company policies.

Aviation has an inherent Catch-22: everyone wants pilots to have experience, but many can't gain the necessary experience until they are hired. Substituting flight instruction hours in light, unprotected general aviation aircraft for the heavy-weather flying that is the staple of regional airline work is unrealistic. College, university, and flight school training environments have limitations, as they lack airline tools and must operate within students' financial realities. More simulator time in regional jets or turboprops will help, but what truly makes a difference is exposure to real weather in real-world operations—experience that is difficult and expensive to obtain outside the airlines.

Many airline-pilot hopefuls left college or flight training with $150,000 or more in debt, only to earn absurdly low pay as new first officers in a volatile industry. While pilot shortages periodically improve pay, the path remains long. Extended Initial Operating Experience (IOE) and rigorous selection for low-time new hires address much of what the regulation aims to fix, in my view. These measures require more training captains but tackle the root cause—flying airline aircraft in variable and challenging weather. Requiring first officer candidates to have that experience before being hired is unrealistic. Flying cargo with an experienced captain might be the best solution before transitioning into passenger service.

It was noted that the first officer retracted the flaps after the aircraft departed controlled flight, rolling past vertical. The NTSB did not consider this a critical factor for several reasons. My opinion: The captain was holding the aircraft in a stall, so flap position was irrelevant. The report did not indicate whether there was sufficient altitude to recover from what appeared to be an incipient spin, given the aircraft was only about 1,700 feet above the ground. The first officer retracting the flaps may have been a reflex from her experience in light aircraft, but departing controlled flight unexpectedly at such a low altitude in a heavy aircraft like the Q400 made recovery extremely unlikely. It had never been flight-tested under such conditions.

The victims' families were apparently led to believe that the first officer had made significant errors and was poorly qualified—contrary to the NTSB findings. Nevertheless, pressure was applied to Congress and the FAA to address this perceived deficiency by increasing hiring qualifications. This first officer had significantly more airline flight experience at the time of the crash than many new hires and was within a few hours of meeting the new regulatory requirements when she was initially hired. The captain, who was the pilot flying, had more than twice the required hours, much of it in airline equipment, but had a consistently poor training record.

As a result of pressure from the families, Congress passed the Airline Safety Act in 2013. The act introduced reasonable requirements for additional multiengine experience and at least 1,000 hours in air carrier operations before upgrading to captain—both good requirements. However, two rule-making advisory committees deadlocked over whether a fixed hourly requirement, regardless of pilot experience, was the best solution. They agreed on nearly everything else. The two largest aviation safety organizations—the Flight Safety Foundation and AOPA's Air Safety Institute—also believed that requiring 1,500 hours across the board was not the best approach. The air carrier pilot unions disagreed.

A statistical axiom states that correlation does not imply causation. Proponents of the 1,500-hour rule frequently cite the fact that no similar crashes have occurred since the Airline Safety Act was passed. However, they fail to acknowledge that there were also no crashes in the nearly four years between the Colgan Air crash and the passage of the act. Many first officers deemed "unqualified" under the new rules had already upgraded to captain before the act was passed, and there were no crashes involving those pilots. Hundreds of thousands of flights operated safely during that time. Proponents also overlook the fact that the first officer was not the causal factor in this tragedy—the captain was the pilot flying.

The Buffalo crash was a wake-up call for both airlines and the FAA, reinforcing that aviation is a performance-based profession. It cannot be reduced to rote procedures, and repeated substandard performance during evaluations cannot be ignored.

I believe performance and quality training are better safety measures than arbitrary hourly requirements. To my knowledge, there is no empirical evidence supporting a fixed-hour standard. Military pilots operate in far more rigorous environments, flying high-performance aircraft in much less time. The key differences are selection, training, and performance standards—not total flight hours. And they are supervised.

The families have every right to be upset—the system failed. We all share the same goal: ensuring qualified pilots are flying. The difference, as noted by the two leading safety organizations at the time, is that a fixed-hour minimum was not the best way to achieve that goal.

The Buffalo crash exposed multiple failures: the captain's poor flying skills, regional airline financial shortcuts, lax flight deck supervision, insufficient regulatory oversight, fatigue issues, significant training deficiencies, and limited background checks on new hires. A casual attitude and complacency on the flight deck, as well as in training and regulation, create a recipe for disaster. The final link in a long accident chain: no one was minding the store during a night instrument approach.

"Immediately" Means Right Now - (Midair Collision - C-150 & F16)"

See and avoid often doesn't always work

> ### *"Just one midair collision can ruin the rest of your day."*
>
> *A well-worn, but true aviation cliché*

This event was truly accidental, but human factors remain the primary contributor. It illustrates a human condition that we must work around. "See and avoid" has always been the primary means of in-flight separation in visual conditions, but honest pilots and controllers will tell you that it doesn't always work. There are hundreds of midair collisions to study and thousands of close calls. Midair collisions have steadily declined to about five per year as collision avoidance technology has proliferated.

Recounted here is a collision between a military fighter and a light GA aircraft.

The Flights and Collision

On July 7, 2015, at about 11:00 a.m. Eastern Daylight Time, a Cessna 150M (N3601V) and a U.S. Air Force F-16 collided near Moncks Corner, South Carolina. The weather was good VFR, with visibility better than 10 miles and scattered clouds.

The F-16 was conducting a post-maintenance flight check under IFR. The flight plan included practice instrument approaches at Myrtle Beach International Airport (MYR) and Charleston Air Force Base/International Airport (CHS), South Carolina, before returning to Shaw Air Force Base, SC.

The Cessna 150 departed Berkeley County Airport, SC (MKS), under VFR, headed for North Myrtle Beach, SC. MKS is located 17 nautical miles north of CHS, outside CHS Class C airspace. Radio communication was not required, and there was no contact with CHS Approach. The Cessna appeared on radar at 10:57:41, showing a Mode C altitude of 200 feet. It turned southeast and began climbing.

At 10:52, the F-16 pilot requested a tactical air navigation system (TACAN) approach to Runway 15. CHS Approach Control instructed the pilot to fly a heading of 260 degrees to intercept the final approach course. At about 10:55, when the fighter was 34 miles northeast of CHS, it was cleared from 6,000 feet down to 1,600 feet.

Radar detected a conflict between the two aircraft at 10:59:59. A conflict alert (CA) was displayed with an audible alarm at 11:00:13 and continued until after the collision. At that moment, the jet was 3.5 nautical miles from the Cessna, descending at about 250 knots, and only 400 feet above it.

Star marks approximate collision location

According to the NTSB:

"At 1100:16, the controller advised the F-16 pilot of 'Traffic 12 o'clock, 2 miles, opposite direction, 1,200 [feet altitude] indicated, type unknown.' At 1100:24, the F-16 pilot responded that he was 'looking' for the traffic. At 1100:26, the controller advised the F-16 pilot, 'Turn left heading 180 if you don't have that traffic in sight.' At 1100:30, the pilot asked, 'Confirm 2 miles?' At 1100:33, the controller stated, 'If you don't have that traffic in sight, turn left heading 180 immediately.'" (Emphasis added.)

As the controller was issuing the instruction and over the next 18 seconds, the F-16 began turning southerly toward the assigned heading. At 1100:49, the F-16 was one-half nautical mile northeast of the Cessna, at an altitude of 1,500 feet, and tracking about 215 degrees. The Cessna, at 1,400 feet, was tracking about 110 degrees.

At 1100:53, the controller advised the F-16 pilot, "Traffic passing below you, one thousand four hundred [feet]."

The last radar return from the Cessna was recorded at 1101:00, and it crashed almost immediately. Neither occupant survived. The F-16 pilot transmitted a distress call and successfully ejected shortly afterward.

Witness Accounts

Witnesses saw the Cessna flying eastbound, with the F-16 approaching from the Cessna's left rear position, roughly north to south. After the collision, debris began falling, and the F-16 attempted to "pull up." It then turned to the right and disappeared. The F-16 wreckage was located about six nautical miles south of the Cessna's impact point.

The Pilots and Controller

The F-16 pilot was current and qualified, with 2,383 total hours of military flight experience, including 624 hours in the F-16. His total flight time also included 1,055 hours operating Predator and Reaper drones. The remaining hours were in USAF training aircraft and flight simulators. In the 90 days before the accident, he had logged 35 hours, with 24 hours in the previous 30 days, all in the F-16.

The Cessna pilot held a private single-engine airplane certificate issued in December 2014. He had 244 total flight hours, with 239 hours in the Cessna 150. He had flown 58 hours in the 90 days before the accident and 18 hours in the 30 days before.

According to the NTSB:

"The Cessna pilot's primary flight instructor... [noted] the pilot was 'very careful' and 'responsive.' He stated that the pilot 'enjoyed' talking to ATC and was aware of the benefits. During his instruction, he would contact ATC for flight following without being prompted. A review of the pilot's logbook revealed that he communicated with... CHS ATC at least 21 times."

The CHS approach controller was hired by the FAA in August 2006 and initially worked at the Oakland Air Route Traffic Control Center. She resigned from the FAA in September 2007 and was rehired in February 2008 at CHS. She had also served as a USAF controller from 1998 to 2000.

At the time of the accident, the controller was working a regular shift from 07:00 to 15:00. The radar west and east positions were combined—a normal configuration at CHS during periods of low activity—with a radar assistant present. When the accident occurred, she had been working the position for about 1.5 hours.

No drugs, alcohol, or incapacitation were noted in the pilots or the controller.

The Aircraft

The Cessna 150M was equipped with a rotating beacon light, strobe lights, navigation lights, and a landing light, but it was impossible to determine if they were lit at the time of the accident. The aircraft did not have Automatic Dependent Surveillance-Broadcast (ADS-B) equipment installed, nor was it required. It had a single nav/com radio and a Mode C transponder. The aircraft's annual inspection had been completed in October 2014, and the airframe had accumulated 3,651 total flight hours.

The F-16 had 4,435 total flight hours and was not equipped with any collision avoidance equipment. A common misconception is that fighter aircraft can detect other aircraft, but slow-moving aircraft such as the Cessna 150 are typically not detectable. Not many adversaries use the 150.

According to the report:

"The F-16 was equipped with a radar unit… that could be used to locate and 'lock on' to other aircraft…. However, the USAF… did not believe the radar would locate a small general aviation aircraft at takeoff or climb speed. The radar… used aircraft closure rate rather than the airspeed of the other aircraft to filter out slow-moving targets."

The F-16's basic autopilot provided altitude hold, heading select, and steering select. In heading mode, the maximum bank angle did not exceed 30 degrees. At fighter speeds—250 knots in this case—that resulted in a wide turn.

Accident Reconstruction

The NTSB reviewed the ability of the pilots to see and avoid each other. They analyzed:

- Where the pilots would need to be looking
- How long the other aircraft would remain in view
- Whether aircraft structure would block the view at certain times
- How the aircraft would have appeared to each other

The F-16's flight recorder and ATC radar for the Cessna 150 were used to create a model that updated at one-second intervals.

At 1100:16, when ATC provided the initial traffic advisory to the F-16 pilot, the F-16 was in level flight at 1,570 feet, tracking 252 degrees, with a ground speed of 253 knots. The Cessna was 3.25 nautical miles from the F-16, at the 12 o'clock position, at 1,200 feet, tracking 109 degrees, and climbing at about 240 feet per minute.

According to the NTSB:

"The Cessna would have appeared to the F-16 pilot as a very small, stationary object just above the horizon and just above the heads-up display (HUD). The F-16 would have appeared to the Cessna pilot as a small, stationary object just above the horizon, outside the left cockpit door window, near the forward vertical post of the door frame."

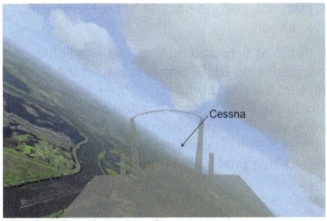

Simulated Cockpit view from F-16 at .6nm, NTSB

At 1100:26, when ATC advised, "Turn left heading 180 if you don't have that traffic in sight," the relative positions had changed only slightly.

At 1100:33, when the controller stated, "If you don't have that traffic in sight, turn left heading 180 immediately," and as the F-16 began banking to the left, the Cessna would have been obscured behind the left structural frame of the F-16's HUD. The F-16, however, would have remained visible through the left window of the C-150.

Over the next several seconds, just prior to the collision, the positions changed only slightly. On a collision course, the target's relative position remains constant, making it very difficult to spot—it just "blossoms" in the final seconds.

Just before the collision, as the F-16 turned, the Cessna would have been completely obscured by the lower cockpit structure. At impact, the F-16 was slightly above and approximately 80 to 90 degrees to the left of the Cessna. The closure rate was 246 knots—about 15 seconds before impact.

Post-Accident Interviews

The F-16 pilot had the onboard radar configured at a 20- and 40-mile range, manually alternating between them. He could not recall if it was set to receive civilian transponders. He reported scanning outside, checking flight instruments, and monitoring the radar display.

The F-16 had locked onto a radar target 20 miles away when ATC issued the alert: "12 o'clock, two miles, at 1,200 feet." The pilot noted that it was the "closest call I've ever received" and "a big alert for me."

Asking the controller to "confirm two miles," he began aggressively searching for the target. He recalled the controller instructing him to turn left "immediately." The pilot set the autopilot heading bug, and the F-16 began a 30-degree bank to 180. Seeing the Cessna directly in front of his aircraft, "within 500 feet," he applied full aft control input, but it was "too late." He estimated that less than one second passed between his initial sighting of the Cessna and the impact.

The controller described the traffic as light and routine. Several fighter aircraft were also making approaches to CHS. She had issued the F-16 a vector of 260 degrees to intercept the final approach course, keeping it south of MKS. She then directed the F-16 to descend and maintain 1,600 feet and was "pretty much done with him" while managing other traffic, including a flight of two additional F-16s. She stated that her usual technique was to clear aircraft to their final altitude quickly, as it "allowed more efficient use of her time."

Simulated cockpit view from C150 at .6 nm, NTSB

Initially, the controller believed the Cessna would remain in the MKS traffic pattern at around 1,000 feet. She sequenced the other two F-16s behind the accident F-16, asking them to expedite their descent to 3,000 feet to avoid other traffic. Then, she noticed the Cessna climbing above 1,000 feet. She advised the F-16 pilot of the traffic, but he did not see it. The F-16 was then instructed to turn left to heading 180 if the traffic was not in sight.

As the targets closed, she directed the F-16 pilot to turn to heading 180 immediately if the traffic was not visible. The controller assumed that "fighters could turn on a dime." However, this was a worst-case scenario—high speed with a shallow bank. She expected the fighter to execute a "max performance turn" when given an "immediate" clearance. She did not recall seeing or hearing the collision alert generated by the radar system and had elected not to climb the F-16, as the Cessna's altitude was unconfirmed.

Cockpit Display of Traffic Information

In April 2016, the FAA published an update to *Pilots' Role in Collision Avoidance* (AC 90-48D or its successor), which highlights aircraft systems and technologies available to improve safety and aid in collision avoidance. The following excerpt from the AC (edited for length) provides insight into these advancements:

"In 1977, the Massachusetts Institute of Technology published a report for the FAA…which describes how Pilot Warning Instruments (PWI) could be used…in the visual acquisition task. These instruments…present pilots with information about particular threats, focusing their attention where it was most needed. Target aircraft at a large crossing angle with apparent motion were far more likely to be seen. When the targets were head-on or tail-on, the subject pilots performed poorly."

In 1991, the Australian Transport Safety Bureau (ATSB) noted the limitations of the *See-and-Avoid* principle. Their findings reinforced previous observations: when pilots are busy, scan time is reduced. Even when actively looking, pilots frequently fail to see other aircraft. In many cockpits, structural components or high instrument panels severely limit visibility. Even aircraft with superb visibility, like an F-16, remain vulnerable to these limitations.

Visual scanning involves moving the eyes to bring successive areas of the visual field into the small area of sharp vision at the center of the eye. The process is often unsystematic and may leave large portions of the field of view unsearched.

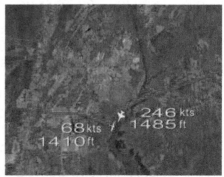

Midair between C150 & F-16 - NTSB

The physical limitations of the human eye mean that even the most careful search does not guarantee traffic will be sighted. An object smaller than the eye's acuity threshold is unlikely to be detected—and even less likely to be identified as an approaching aircraft.

Contrast between an aircraft and its background can be significantly reduced by atmospheric effects, even in good visibility. In many cases, an approaching aircraft presents a very small visual angle until just moments before impact. Additionally, complex backgrounds, such as ground features or clouds, further hinder the identification of aircraft.

Other studies have confirmed that alerted traffic is much more likely to be seen. Traffic alerts increased search effectiveness by a factor of eight (emphasis added). The ATSB report concludes, in part, that "the see-and-avoid principle in the absence of traffic alerts is subject to serious limitations...."

NTSB Probable Cause

"The approach controller's failure to provide an appropriate resolution to the conflict between the F-16 and the Cessna. Contributing to the accident were the inherent limitations of the see-and-avoid concept, resulting in both pilots' inability to take evasive action in time to avert the collision."

Commentary

As noted above, the radar detected a conflict alert (CA) for nearly a minute before the collision. The controller's manual, FAA Order 7110-65W, states:

"When a CA...alert is displayed, evaluate the reason for the alert without delay and take appropriate action.... Immediately issue/initiate an alert to an aircraft if you are aware of another aircraft at an altitude that you believe places them in unsafe proximity. If feasible, offer the pilot an alternate course of action. When an alternate course of action is given, end the transmission with the word 'immediately.'"

Some studies have examined false or "cry wolf" alarms from the CA algorithms, but the ones I reviewed concluded that controllers were diligent about taking appropriate action.

A contributing factor, in my view, was the F-16 pilot's apparent misunderstanding of the urgency of an "immediate" ATC instruction. The NTSB mentioned this in the report, but I believe it deserves more prominence in the probable cause.

FAA Advisory Circular 90-48C (or its successor) estimates the minimum time required to identify an approaching aircraft, recognize a collision course, decide on an action, execute the control movement, and allow the aircraft to respond to be around 12.5 seconds—*after* the target is spotted. At a closure rate of four miles per minute in this case, a two-mile buffer provides about 15 seconds to see the other aircraft and take evasive action.

The "big sky, little aircraft" theory works most of the time. But when it fails, it does so catastrophically.

After using ADS-B for more than two decades in various aircraft, I believe it's a major step forward in avoiding midair collisions. Most pilots who've flown with it never want to go back. These systems dramatically improve the ability to avoid collisions—including those pilots don't see.

The training and hundred-dollar-hamburger fleets have significant exposure and an urgent need for anti-collision equipment. They operate VFR, predominantly in non-towered airport environments, sometimes in very high-density areas. Near Class B or C airspace, the collision potential is high because VFR aircraft often skirt mandatory communication airspace. Despite that, they should be asking for flight following when it's available.

ADS-B equipment isn't cheap relative to the cost of an aircraft, but its value is immeasurable—even if it saves you only once. Looking strictly at the economics, the millions spent on litigation and lost aircraft could have equipped many, if not all, aircraft with collision-avoidance systems.

What Was He Thinking?

Was he thinking at all?

Aerobatics in a Baron 58

A low-time Baron pilot became enthralled with aerobatics and thought he could attempt it with a full load of people and luggage. The results proved otherwise.

This is a case of horrendous decision-making. Accident investigation seeks to answer *why* something happened and how similar catastrophes might be prevented. That's a tall order in this case.

> **"If you can't imitate him, don't copy him."**
>
> Yogi Berra, NY Yankees manager, ball player and dispenser of wisdom

The Flight

On April 22, 2007, at 2:51 p.m. Eastern Daylight Time, a Beechcraft Baron BE-58 broke up in flight near Hamilton, Georgia. The flight had originated from Gulf Shores, Alabama, around 1 p.m. Central Daylight Time, heading toward Georgia under good VFR conditions.

A witness heard an airplane approaching that sounded as if it was performing aerobatic maneuvers. While this is speculative—since the witness could not see the aircraft—he reported that as the engine noise increased, the aircraft appeared and was descending rapidly in a 45- to 60-degree nose-down attitude. A wing or part of the tail separated from the plane, which was shredded, scattering debris over a large area. The private pilot and four passengers were fatally injured.

The Pilot

The pilot's last flight review was on March 29, 2007, when he received his private pilot airplane multiengine rating. At the time, he had logged 1,113 hours of flight experience, with only 20 hours in twin-engine aircraft. He held a third-class medical certificate with no restrictions.

A friend noted that the crash pilot's flying skills were "below [the friend's] standards" because he was known for over-stressing the planes he flew. The friend even predicted that the pilot "would probably crash within the next year."

The Baron pilot was heard saying, *"I think I can roll this airplane."* According to the NTSB report:

"The pilot had been at Sun 'n Fun in Lakeland, Florida, a few days prior to the accident and had observed an [airshow] performer rolling a Beech[craft] 18, and the deceased pilot just kept the rolling issue in his head."

The accident pilot also bragged about flying with a retired airline pilot in a Beechcraft Baron 55 who had rolled his airplane.

A few days before the accident, on April 19, 2007, while returning from Sun 'n Fun with two other passengers on board, the accident pilot banked the airplane to the left, then back to the right, and stated, *"I believe it's possible to roll this airplane."*

As they began a descent, he pulled the aircraft up and rolled into a knife-edge attitude, according to

The image is inverted not the aircraft - Darwin Award! Beechcraft

the NTSB. Another pilot, seated in the right seat, grabbed the flight controls and leveled the airplane.

A short time later, the pilot deliberately shut down the right engine and feathered the propeller while in cruise flight. He later restarted the engine, and they descended for a normal landing.

The Aircraft

The Baron was deemed airworthy during its last annual inspection in June 2006. The airframe had over 9,200 hours, and maintenance logs indicated that wing spar cracks had been repaired about 10 years earlier. The aircraft had been used extensively as a trainer at an airline training school.

NTSB Probable Cause

"The pilot's exceeding the design stress limits of the airplane while performing aerobatics in a non-aerobatic airplane, which resulted in an in-flight overload failure of the airframe. A factor in the accident was the pilot's decision to perform aerobatics."

Commentary

Properly performed aerobatics can be conducted in normal category twins, but only with an aerobatic waiver, and it carries significantly higher risk than flying within the approved design envelope. Aerobatics in conventional multiengine aircraft were mastered by Bob Hoover, one of the greatest stick-and-rudder pilots of all time. However, since few pilots can match his skill, attempting to imitate him is reckless.

The report noted that the crash pilot had observed a retired airline pilot roll a Baron. This set a dangerous precedent for an impressionable individual, as it was both irresponsible and illegal.

Simulated engine-out procedures are essential for twin-engine flight training, but they should never be performed with passengers aboard—nor should aerobatics. The NTSB's investigation provided extensive detail (37 paragraphs) on the aircraft's condition post-crash. Once structural failure in normal flight is ruled out, additional investigation serves little purpose. However, proving improper aerobatic maneuvers was necessary. Examining the pilot's psychological factors leading up to this tragedy is beyond the current capabilities of accident investigation.

Collision Over Cerritos *(PA-28 Collides with DC-9)*

A Piper Archer brings down a DC-9 over a densely populated area, leading to a complete overhaul of the ATC system.

Why worry about midair collisions, especially when they are increasingly rare? Because they are among the few emergencies where control of the aircraft can be lost instantly, and a high percentage result in fatalities. Proper procedures, good visual scanning techniques, and modern technology significantly reduce the risk.

Cerritos, a suburb of Los Angeles, is remembered for an accident with far-reaching consequences. The midair collision between a Piper Archer and an AeroMexico DC-9 resulted in sweeping changes to airspace regulations, aircraft equipment requirements, and operational procedures—few accidents before or since have had such an impact.

What was left of Cerritos after the crash, NTSB

The Flights

At approximately 11:40 a.m. Pacific Daylight Time (PDT) on August 31, 1986, a Piper Archer carrying a pilot and two passengers departed Torrance, California, for a VFR flight to Big Bear, California. The proposed route was direct to Long Beach, then to Paradise VOR, and finally to Big Bear at 9,500 feet. Radar showed that after takeoff, the Piper turned east toward Paradise with its Mode A transponder (non-altitude reporting) set to the VFR code of 1200. The pilot did not request ATC assistance or clearance into the Terminal Control Area (TCA), now referred to as Class B airspace.

AeroMexico Flight 498, a DC-9 carrying 58 passengers and a crew of six, was inbound to Los Angeles (LAX) on an IFR flight plan. At about 11:47

56

a.m., the flight reported level at 7,000 feet to LAX Approach. At 11:50 a.m., the flight was instructed to reduce speed to 210 knots.

At 11:50:46 a.m., the controller advised Flight 498 of "Traffic, 10 o'clock, one mile, northbound, altitude unknown." AeroMexico acknowledged the call but did not report the traffic in sight. However, this was not the Archer target. At 11:51:04 a.m., Flight 498 was cleared to descend to 6,000 feet.

At 11:51:18 a.m., a Grumman Tiger pilot called approach control requesting VFR traffic advisories through the TCA (now Class B) at 4,500 feet. The controller assigned a squawk code at 11:52:04 a.m. and asked the pilot to verify his altitude shortly after. The Tiger pilot reported climbing through 3,400 feet, and at 11:52:36 a.m., the controller advised the Tiger that he was in the middle of the TCA, suggesting, "In the future, you should look at your TCA chart. You just had an aircraft pass right off your left above you at 5,000 feet, and we run a lot of jets through there at 3,500 feet."

The controller then noticed that radar was no longer tracking Flight 498. After several unsuccessful attempts at radio contact, the controller notified the arrival coordinator that both radar and radio contact had been lost.

At 11:52:09 a.m., Flight 498 and the Piper Archer collided over Cerritos at approximately 6,650 feet. The Archer had inadvertently penetrated the 6,000-foot floor of the TCA without clearance. The sky was clear, with a reported visibility of 14 miles.

There were no survivors on either aircraft, and 15 people on the ground were killed. Five houses were destroyed, and seven others sustained damage due to wreckage and post-impact fires. The DC-9's cockpit voice recorder captured only one comment by the captain relative to the collision:

"Oh, this can't be!"

Archer Pilot Information

The Archer pilot, age 53, held a private pilot certificate and had 231 total flight hours. In the 90 days preceding the accident, he had logged only two hours of flight time.

Interviews with other pilots and flight instructors described him as "conscientious and careful." His primary flight instructor stated that he was trained to "scan left, look at the instruments, scan right, look at the instruments, and repeat the procedure."

The pilot was familiar with the wing leveler in the Archer and reportedly used it while looking at maps, reviewing charts, or performing other in-cockpit activities and departure procedures. At the time of the crash, he was in possession of an L.A. Sectional Chart and a VFR Terminal Area Chart and was apparently referencing them.

DC-9 Crew

The AeroMexico captain had a total of 10,641 flight hours, with more than 4,600 hours in the DC-9. He had been employed by the airline since 1972.

The first officer had 1,453 flight hours and had been with the airline for two years at the time of the collision.

Wreckage Information

Based on paint smears and damage signatures, the NTSB determined that the Archer collided with the left horizontal stabilizer of the DC-9's tail.

The stabilizer separated from the DC-9 and fell intact about 1,700 feet from the main crash site.

The Piper Archer remained mostly intact, except for the upper portion of the fuselage, engine, and vertical stabilizer.

There was no fire damage and no evidence that either aircraft attempted evasive maneuvers before impact.

Visibility Studies

The NTSB examined whether either aircraft was visible to the other before the collision. Viewing angles for each airplane were plotted at five-second intervals.

The study showed that from 11:50:56 until 11:52:01, the captain of the DC-9 had a clear view of the Piper and was in the best position to observe the traffic. The Piper was located approximately 15 to 30 degrees to the left of the design eye reference point, or the center of the windshield.

The first officer could see the Piper about 50 percent of the time. The Piper pilot, however, could only spot the DC-9 by looking far to the right through the copilot's windshield. The Piper's front and rear seat passengers would have had a better view, but they were presumably not trained to spot other aircraft.

The Massachusetts Institute of Technology's Lincoln Laboratory studied pilots' abilities to detect other aircraft in flight. While this research was not conducted specifically for the accident, its findings were sobering. The study found that pilots identified an intruding aircraft only 56 percent of the time, with a median detection range of just under one mile. In contrast, an experimental group using a prototype traffic collision avoidance system (TCAS) detected targets 86 percent of the time at a median range of 1.4 miles.

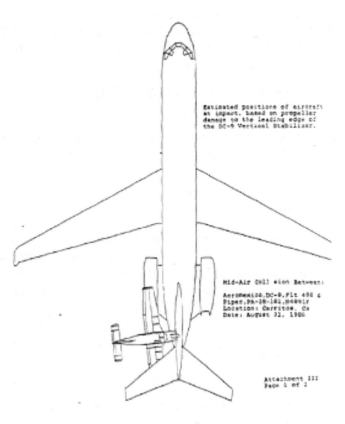

Collision Geometry, NTSB

Terminal Control Radar and ATC Procedures

Radar can only track a finite number of targets, but at the time of the collision, the system was not saturated and was operating properly. However, primary radar returns can be affected by temperature, pressure, and humidity—factors that influence the refractive index, which can bend or even split the radar beam.

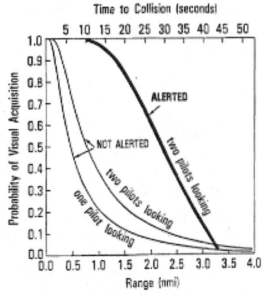

Benefits of TCAS, NTSB

Under certain conditions, this can create a hole in coverage, potentially causing a primary target to be missed. On the day of the accident, the refractive index was high at the altitudes where the collision occurred. The NTSB could not determine whether a primary target was present on the controller's scope.

NTSB Findings and Airspace Safety Improvements

The NTSB concluded that the controller, while not overly busy, had been distracted by coordinating a change for AeroMexico 498's arrival into LAX and by the sudden appearance of the Grumman Tiger, which posed a potential threat to other LAX traffic. Because of this, the controller apparently failed to see the Piper as a secondary target and did not advise Flight 498.

The controller had no reason to presume an altitude conflict, as most VFR traffic operated below the floor of the Terminal Control Area (TCA), and no other authorized traffic was in the vicinity. However, his failure to provide a traffic advisory to the DC-9 was considered a contributing factor. The Piper's transponder was set to 1200, and the pilot was not in contact with ATC.

NTSB Probable Cause

The NTSB determined the probable cause of the accident to be:

"...limitations of the air traffic control system to provide collision protection, including procedures and automated redundancy. Contributing factors were the inadvertent and unauthorized entry of the Piper into the Los Angeles Terminal Control Area and limitations of the see-and-avoid concept."

Commentary

The MIT study highlighted the inherent weakness of relying solely on visual detection. Pilots frequently miss targets due to aircraft structure, physiological limitations, distractions, procedural errors, fatigue, attention span, and lighting conditions. Despite good intentions, the PA-28 pilot violated restricted airspace, resulting in tragedy. There is no other way to put it. In the aftermath, the number of FAA airspace enforcement actions by the Los Angeles Flight Standards District Office increased tenfold.

Following the accident, the NTSB recommended, and the FAA ultimately mandated, the installation of a Traffic Alert and Collision Avoidance System (TCAS) on all air carrier and regional airline aircraft. While some called for banning light aircraft from high-density airspace, a more balanced approach emerged.

A major education campaign was launched in Southern California within two months of the accident to inform general aviation (GA) pilots of their responsibilities and procedural requirements when operating in or near TCA airspace. The FAA also intensified enforcement against pilots who improperly entered controlled airspace without clearance.

Airspace and Technology Advancements

The airspace around high-density airports was restructured, and Mode C transponders (which provide altitude reporting) became mandatory not only within a TCA but also within a 30-mile radius of primary airports—the so-called "30-mile veil." These measures allowed air carrier flights to receive warnings about unauthorized aircraft, which were previously assumed to be below the controlled airspace floor. Mode C also improved ATC's ability to track intruding aircraft and enforce entry requirements.

Today, collision avoidance technology has advanced significantly. Most light GA aircraft are now equipped with Automatic Dependent Surveillance–Broadcast (ADS-B) in addition to traditional transponders. ADS-B IN allows pilots to see most surrounding traffic on an installed or portable multi-function display, greatly enhancing situational awareness.

MIT's assessment remains valid—pilots still miss a significant amount of traffic. Having a real-time traffic display in the cockpit is a major safety advantage and is highly recommended. Since airlines began equipping aircraft with TCAS in the early 1990s, there has not been a single midair collision involving an air carrier in the United States until January 2025 when an Army helicopter collided with an airline on approach to Washington National (Reagan) Airport.

Buzz job goes awry (Glasair II does aerobatics over neighborhood)

Aerobatic flight over a populated area isn't just against the rules – It's a really bad idea. "It seemed like a good idea at the time!" A Poor Explanation after the Fact

The Flight

This report is brief, as was the flight.

On February 12, 2006, at approximately 11:30 a.m. Pacific Standard Time, an experimental Glasair II-S FT crashed into a house while performing low-level aerobatic maneuvers in Roseville, California.

Exemplar Glasair II, Glasair

Prior to departure from the nearby Lincoln Airport (KLHM), the pilot-rated passenger told bystanders, "We're going to show you guys what flying is about."

The **Glasair** departed KLHM at approximately 11:15 a.m. Shortly afterward, witnesses on a nearby golf course and in the residential area surrounding the crash site reported that the airplane was performing aerobatics at altitudes between 200 and 800 feet above the ground.

Witnesses stated that on the first pass, the airplane was traveling at high speed and entered a series of barrel rolls before pulling up into a climb and banking steeply to reverse course. On the final pass, the airplane completed the roll series, pulled up, and began climbing. It then stalled and entered a nosedive, "spinning or spiraling" until it crashed into the house. The pilot, the passenger, and a 19-year-old inside the house were fatally injured. There were no other ground injuries.

According to the NTSB, "Radar data showed...multiple passes in a northeast-to-southwest direction, with course reversals. Mode C altitude reports associated with the targets showed that the airplane approached the area between 1,900 and 2,000 feet mean sea level (MSL), then descended to below 1,000 feet for the northeast-to-southwest passes."

The Aircraft

The investigation found no defects related to the aircraft.

The Pilot

The 49-year-old owner held a private pilot certificate issued on November 26, 2000, along with a third-class medical. He had logged 287 flight hours, including 27 hours in the accident aircraft.

NTSB Probable Cause

"...*The pilot's failure to maintain an adequate airspeed while performing low-level aerobatics, which led to a stall/spin.*"

Commentary

Let's add to the probable cause: poor judgment and inadequate skill.

Members of the Aircraft Owners and Pilots Association (AOPA) called on the organization to "do something about the news coverage," which was, in fact, remarkably balanced. The reporters were merely covering a tragic and sensational story.

As then-AOPA President Phil Boyer put it, "You cannot defend the indefensible." How does one explain killing someone on the ground while performing aerobatics over a residential neighborhood?

The AOPA Air Safety Foundation (now Institute) conducted a seminar in the area a couple of weeks after the crash. The known facts, as identified in the NTSB's preliminary report, were presented, followed by a discussion on the responsibility of pilots to follow both common sense and regulations. The seminar was well received by the pilot community, and the media understood that, unfortunately, this was an outlier event. However, history shows there are still a few pilots who don't quite get it. It's not a good legacy to leave behind.

Collision at Quincy – Ground collision between a Beech 1900 and a Beech King Air

A perfect example of an accident chain where a senior pilot ignored the most basic non-towered airport procedures.

The ground collision between a Beech 1900 airliner and a Beech King Air A90 at Quincy, Illinois, in November 1996 illustrates the accident chain concept—change any one factor, and the crash is unlikely to occur.

Fortunately, at the time of this writing, this collision remains unique. In NTSB records dating back to 1982, there has not been another ground collision between an airliner and a corporate GA aircraft at a non-towered airport.

The Beech 1900 made a straight-in approach to Runway 13 and collided on the ground at the intersection of Runway 4 with the Beech King Air A90, which was on its takeoff roll.

The collision occurred at 5:01 p.m. Central Standard Time at the non-towered Quincy Airport, which is served by a common traffic advisory frequency (CTAF). A Piper Cherokee, flown by a newly licensed pilot, was also preparing for departure. His use of the CTAF became a contributing factor in the collision.

All 12 people aboard the Beech 1900—two pilots and ten passengers—along with the two pilots on the King Air, perished in the accident.

Beech 1900 Great Lakes Flight 251,Wkipedia   *Exemplar King Air A-90, Beechcraft*

Weather

The weather was good VMC. Ambient light conditions were nearly ideal for spotting traffic from the north. In the evening twilight, the 1900's landing and strobe lights were readily visible from several miles out on the approach to Runway 13, according to witnesses.

Pilots

There were five qualified pilots involved in this collision scenario:

The Beech 1900 captain, age 30, held an airline transport pilot (ATP) certificate and was type-rated in the aircraft. She had upgraded to captain nine months earlier, with about 4,000 hours of flight time, including 700 as PIC in the 1900. She was handling the radio during the flight as the pilot monitoring.

The first officer, 24, held a commercial certificate with instrument ratings for single- and multi-engine aircraft. He had accumulated about 1,950 hours of total time, with 800 as second-in-command of the Beech 1900. He was handling the controls at the time of the accident.

The King Air pilot, age 63, was a retired TWA captain and a pilot in the U.S. Air Force Reserve. He was type-rated in the Boeing 377, 707, 720, and 747; Douglas DC-9; Lockheed 382 and 1011; and North American B-25. He had retired from TWA in 1992, flew part-time as a flight instructor at an Air Force aero club, and did some charter work. He held a first-class medical certificate that required him to wear glasses and was seen wearing them when he boarded the aircraft just before the crash. The pilot had accumulated more than 25,000 hours of total flight time, with 22 hours in the accident airplane.

A pilot of a Cherokee, waiting to take off, had received his private pilot certificate in February 1996 and had 80 hours in a Piper Cherokee.

About six months earlier, the King Air pilot had been involved in a gear-up landing accident while giving dual instruction in a Cessna 172RG. The FAA offered him a remedial training option instead of enforcement action. The training had not yet taken place. The FAA inspector who investigated the accident stated that the pilot had "expressed an extremely negative attitude toward the FAA's questioning him about his landing. His statements were to the effect that he was a retired U.S. Air Force colonel with almost 30,000 hours of flying time and that a gear-up did not mean anything." Additionally, TWA had demoted him in 1991 from captain to flight engineer due to flying deficiencies.

The 34-year-old pilot-rated passenger in the King Air A90 was employed by FlightSafety International Airline Center in St. Louis. She was a ground instructor and a part-time CFI at the same Air Force aero club as the other King Air pilot. She had accumulated nearly 1,500 hours and was working to build multi-engine experience in preparation for an airline job. This was her first flight in a King Air, and she appeared to be receiving instruction, according to some prospective buyers of the King Air who had just gotten off the aircraft in Quincy. It could not be determined from the wreckage which seat she was sitting in or whether she was handling the controls.

The Crash

The only record of this accident came from the cockpit voice recorder on the Beech 1900. The NTSB transcript of the tape recounts the situation as it developed. Non-pertinent comments are omitted:

King Air A90 N1127D — 27D (identified as the female voice of the pilot)

Beech 1900 (Great Lakes Flight 251) external transmission — Lakes 251

Intra-cockpit— Capt (Captain), FO (First Officer)

Cherokee N7646J — 46J was on the taxiway and pulled up on the run-up pad behind the King Air.

1656:44 Capt: "You're planning on [Runway] 13 still, right?"

1656:46 FO: "Yeah, unless it doesn't look good. Then we'll just do a downwind for [Runway] 4, but right now plan for 13."

1656:56 Lakes 251: "Quincy area traffic, Lakes Air Two-Fifty-One is a Beech airliner currently 10 miles to the north of the field. We'll be inbound to enter on a left base for Runway 13 at Quincy. Any other traffic, please advise."

1657:45 Capt (running before-landing checklist): "...and the landing lights and logo lights?"

1657:47 FO: "They're on."

1659:04 27D: "Quincy traffic, King Air One-One-Two-Seven-Delta holding short of Runway 4. Be uh, takin' the runway for departure and heading, uh, southeast, Quincy."

(The King Air then moved onto the runway, pulled forward slightly, and stopped for about a minute, apparently to complete some last-minute items.)

1659:19 Capt: "She's takin' Runway 4 right now?"

1659:29 Lakes 251: "Quincy area traffic, Lakes Air Two-Fifty-One is a Beech airliner currently, uh, just about to turn onto about a six-mile final for Runway, uh, 13—more like a five-mile final for Runway 13 at Quincy."

1700:16 Lakes 251: "And Quincy traffic, Lakes Air Two-Fifty-One's on a short final for Runway 13. Um, is the aircraft gonna hold in position on Runway 4, or are you guys gonna take off?"

1700:28 46J: "Seven-Six-Four-Six-Juliet, uh, holding, uh, for departure on Runway 4."

1700:34 At this point, the ground proximity warning system in the Beech 1900, which provides warnings when the flight gets close to terrain, mechanically called out, "200 feet," interrupting the transmission from Cherokee 46J. That transmission concluded, "...on the, uh, King Air."

1700:37 Lakes 251: "OK, we'll—we'll get through your intersection in just a second, sir... We appreciate that."

1700:42 Capt: "Landing gear's down—three green; flaps are at landing. Your yaw damp is off. Final's complete."

1701:01 Capt: "Max reverse. Oh (expletive)."

1701:03 FO: "What? Oooh (expletive)."

1701:07 FO: "(Expletive)."

1701:08 End of recording.

Wreckage

Both airplanes came to rest along the east edge of Runway 13, with their wings interlocked approximately 110 feet east of where the skid marks converged near the intersection of Runways 4 and 13. According to witnesses, both airplanes remained on their landing gear after impact.

The Beech 1900 left continuous tire skid marks for 475 feet before converging with scuff marks left by the King Air's tires. The scuff marks veered to the right of Runway 4 for 260 feet before intersecting the 1900's skid marks.

Fire erupted, and while rescuers reached the aircraft almost immediately and were able to converse with the captain of the 1900, they were unable to open the cabin door.

Crash Diagram, Author.

NTSB Investigation

NTSB investigators performed tests at the accident site to determine visibility from a Beech 1900 on final approach and a King Air positioned on Runway 4. Additionally, they conducted computer-model simulations to show the views from various cockpit seats.

From the King Air, the pilot's view of the Beech 1900 could have been fully or partially obstructed by the side window post. Likewise, the captain of the 1900 did not have a clear view of the King Air. The relative position of an aircraft on a collision course does not change, so if it was hidden behind a window post, that relationship would have remained constant.

The 1900's copilot could have seen the King Air right at the point of touchdown, but as the A90 accelerated, it likely moved behind a windshield post. The first officer was very likely concentrating on the landing.

The only consistently clear view was from the right seat of the King Air. As mentioned earlier, it was impossible to determine who occupied which seat in the King Air, but a reasonable assumption is that the female instructor was sitting on the right side and operating the radio. Since this was her first flight in the A90, her situational awareness would have been relatively low.

NTSB Probable Cause

"...the failure of the pilots in the King Air A90 to effectively monitor the common traffic advisory frequency or to properly scan for traffic, resulting in their commencing a takeoff roll when the Beech 1900C (United Express Flight 5925) was landing on an intersecting runway.

Contributing to the cause of the accident was the Cherokee pilot's interrupted radio transmission, which led to the Beech 1900C pilots' misunderstanding of the transmission as an indication from the King Air that it would not take off until after Flight 5925 had cleared the runway.

Contributing to the severity of the accident and the loss of life were the lack of adequate aircraft rescue and firefighting services, and the failure of the air stair door on the Beech 1900C to open."

Commentary

The NTSB concluded that the 1900's flight crew made frequent radio broadcasts of their position during the approach and that the decision to land straight-in to Runway 13 was appropriate.

Both King Air pilots not only failed to listen on the CTAF but also failed to visually clear the intersecting runway and yield the right-of-way to landing traffic.

The Cherokee pilot's transmission in response to the 1900's request was unnecessary. Had he included the aircraft type—"Cherokee"—it might have alerted the airline crew that there was confusion about whom they were hearing. His response misled the crew into believing they were communicating with the King Air. The Cherokee pilot, not being in the number one slot on Runway 4, had no reason to respond to the 1900's query directed at the King Air. He was still learning. Communicating at the wrong time, even with good intentions, can create serious misunderstandings.

The primary human factor in this accident chain was a highly experienced King Air pilot with an "attitude."

Proper communication could have helped—announcing the departure was critical. Most importantly, failing to clear the departure path on either a single or intersecting runway is a sure recipe for disaster.

A Hasty Departure

A business overachiever ignores basic preflight preparation, goes off road, and loses control on a snowy evening.

Lousy Day to Fly - On a snowy February morning in 2008, the pilot called the Augusta, Maine, FBO and requested that her Cessna CitationJet (C525) be fueled and moved into the hangar—presumably to keep it out of the snow and ready for a late-afternoon departure.

"In Human affairs of danger or delicacy, successful conclusion is sharply limited by hurry."

John Steinbeck, Author, East of Eden

A local air carrier canceled its afternoon flight due to weather, and at 4 p.m., the CJ was moved out of the hangar to make room for the carrier's airplane.

Exemplar Citationjet, AOPA

Around 5 p.m., the pilot called Flight Service, received a standard briefing, and filed an IFR flight plan to Lincoln, Nebraska.

The briefer advised of icing, fog, mist, precipitation, and turbulence—from the surface to the entire requested route of flight. While not all of these conditions were present at every altitude, some could be expected throughout the route.

The pilot commented that the weather was "just cruddy."

The official weather observation taken just after the crash at Augusta, six miles northeast of the crash site, reported the following conditions:

Winds from 020 degrees at 3 knots, Visibility: 3 statute miles, Freezing rain and haze, Overcast clouds at 1,800 feet, Temperature: -6 degrees Celsius, Dew point: -6 degrees Celsius, Altimeter: 30.32

According to an NTSB weather study, the cloud bases were at 2,100 feet, with tops at 10,000 feet. Below the lowest cloud base, in-flight visibility was six miles in mist. Light freezing rain with moderate mixed/clear icing was present below 5,000 feet, and light to moderate turbulence was reported below 10,000 feet. A SIGMET was in effect for occasional severe mixed/clear icing in clouds and precipitation over the accident area.

The pilot arrived at the airport around 5:15 p.m. Despite the snow and ice covering the ground, she declined deicing when asked by an FBO representative. Witnesses reported that freezing rain had been falling for the past 90 minutes and that cars in the parking lot had accumulated a quarter-inch coating of ice.

Confusion on the Ground

At 5:31 p.m., the pilot contacted ATC for an IFR clearance. After reading back the clearance, she advised that she would be departing from Runway 17 in about five minutes. However, at 5:36 p.m., she mistakenly used the clearance delivery frequency to report that she was taxiing to Runway 17, then stated she was taxiing for Runway 35.

Although it was dark, the pilot did not use the common traffic advisory frequency (CTAF) to activate the pilot-controlled taxi and runway lights. With the taxiways barely visible, witnesses observed the jet veering off the taxiway and onto an adjacent grass area.

Off-Road taxi path, FAA

The situation escalated when the engines spooled up, and the CitationJet powered through a ditch on its way to the runway. A post-accident investigation revealed that the left main tire had broken through the ice and become stuck in the ditch. Attempting to power out of such a situation is not a recommended procedure.

The pilot announced on the CTAF that she was departing from Runway 35, prompting an FBO employee to turn on the runway and taxiway lights. Shortly after, she announced a change in plans, stating she would depart from Runway 17 instead, and proceeded to taxi to the opposite end of the airport.

At 5:41 p.m., the pilot advised the controller that she was ready for departure and received an IFR release with clearance to 10,000 feet. At 5:43 p.m., she announced her departure from Runway 17.

Short Flight

At 5:44 p.m., the pilot reported climbing out of 1,000 feet for 10,000 feet. One minute later, the departure controller instructed her to ident, and radar contact was established two miles southwest of Augusta. The flight was then cleared direct to Syracuse.

At 5:46 p.m., the pilot declared an emergency, reporting, "We've got an attitude indicator failure." The controller asked for her intentions. One minute later, at 5:47 p.m., she stated that she wasn't sure which way she was turning. Her transmission abruptly cut off mid-sentence, radar contact was lost, and an emergency locator transmitter (ELT) signal was detected.

Boston Center radar data showed the CJ departing from Runway 17 at Augusta and entering a climbing right turn, tracking about 260 degrees. It maintained that heading for 38 seconds while accelerating and climbing before the pilot declared an emergency. At that moment, the aircraft was at 3,500 feet, flying at 267 knots. Thirteen seconds later, when the pilot reported losing control of the aircraft's orientation, radar showed the plane entering a tight, rapidly descending left turn. The descent continued until radar contact was lost.

The aircraft impacted a wooded area during heavy snowfall. It struck the ground at approximately an 80-degree nose-down angle at high speed, making component identification and recovery difficult—especially after the post-crash fire. The pilot and her 10-year-old son did not survive. Investigators determined that the engines were producing power at impact, but little else could be analyzed due to the severity of the crash.

The Pilot

The pilot, age 45, held a private pilot certificate with airplane single-engine land, multiengine land, and instrument ratings, as well as Cessna 525S and Cessna 500 type ratings. Her most recent medical certificate was issued in December 2004. If obtained before turning 40, it would have been valid for five years. However, an aircraft insurance application indicated that a third-class medical certificate was issued in November 2007, though the FAA had no record of it.

At the time of the insurance application, she reported a total flight experience of 3,522 hours. Her last documented recurrent simulator training for the C525 was completed in December 2005. Since her logbook was not recovered, it is unclear what additional flight reviews or training she received. However, it appears that more than two years had elapsed since any documented training. She was scheduled for recurrent training in March 2008.

The Aircraft

The CJ, serial number 525-0433, was manufactured in 2001 and certificated for either single-pilot or two-pilot operation. It was equipped with a Rockwell Collins Pro Line 21 avionics system and was approved for flight into known icing conditions.

The aircraft had three pitot-static systems and three separate attitude indicators—one for the pilot, one for the co-pilot, and a standby indicator—each powered by separate and independent attitude and heading reference system (AHRS) units. Additionally, the aircraft was equipped with a standby airspeed indicator, altimeter, and compass.

The pitot-static and fuel vent system heat were controlled by a single switch, with the circuit for all six elements designed so that any failure would trigger an annunciator light. The pitot-static heating system was not automatic, and the checklist required it to be turned on just before takeoff.

NTSB Probable Cause

"...The pilot's spatial disorientation and subsequent failure to maintain airplane control."

Commentary

The NTSB's findings are straightforward, but several factors and hypotheses warrant consideration. It is possible that the aircraft's performance was degraded by ice-contaminated wing and tail surfaces. Deicing on the ground is essential when an aircraft has been sitting outside with snow, ice, or frost accumulation.

Operating at night in instrument meteorological conditions, combined with an accelerating climbing turn and the pilot's demonstrated impatience, created an environment conducive to spatial disorientation. Given the aircraft's altitude and speed, the pilot would have had only seconds to recognize, counteract, and recover from disorientation.

If the autopilot was engaged, it may have masked compromised handling and returned an out-of-trim, marginally controllable aircraft to the pilot— an extremely disorienting situation. For this reason, autopilot use is not recommended in icing conditions. There is no way to determine whether the anti-ice or deice systems were active. Additionally, as noted earlier, pitot heat in some aircraft is not automatic, and failure to turn it on could result in confusing instrument indications. Jets routinely operate in icing conditions and have the power to climb through them quickly. In this case, the clearance to 10,000 feet presented no air traffic control (ATC) restrictions.

A total flight instrument failure seems unlikely. However, a malfunction of the flight data computer or pitot heating system remains a possibility— especially given the aircraft's off-taxiway excursion before takeoff. Still, the failure of more than one system would be extraordinary.

Cessna's C525 emergency procedures address both dual and single attitude reference failures, but it is unknown when the pilot last practiced these rare yet critical actions. According to the FAA, "It may take as much as 35 seconds to establish full control after loss of visual reference by qualified pilots. The spatially disoriented pilot may place the aircraft in a dangerous attitude, which can lead to a rapid, uncontrollable, near-vertical descent."

Beyond technical factors, the pilot's mindset may have played a critical role—if we allow some armchair psychology. She was a highly successful entrepreneur. A business partner described her as "fearless in her pursuits, both professional and personal." The Boston Globe reported that she was named the wealthiest woman under 40 by Fortune in 2001, with a net worth that exceeded those of actor Tom Cruise and golfer Tiger Woods. While this is speculation, it is plausible that she was accustomed to operating on her own terms.

The NTSB noted that the pilot had been in Maine for a week while her son attended a skiing academy. A family member stated that she did not express concerns about time constraints but mentioned several times that the area was "boring." An FBO employee observed that she "looked like she wanted to get out of here" and was in a hurry. The failure to deice, the taxiway incident, and the last-minute runway change suggest distracted thinking at best. The paradox is that the mindset that contributes to economic success may, at times, interfere with good decision-making in aviation.

Selfie Silliness - (C150 pilot loses control in night IMC)

A Cessna 150 pilot took several friends on flights while dividing attention between flying the aircraft and using a camera.

Distraction is a factor in nearly every accident or mishap. When it stems from operational issues, it's understandable. But when a pilot deliberately engages in distracting behavior during a critical phase of flight, that's poking the bear.

Exemplar Cessna 150, AOPA

While distracted driving has reached epidemic proportions on the highways, there have been only a few widely publicized instances of distraction in aircraft. However, this crash makes a clear case for maintaining a sterile cockpit during takeoff and landing. Flying VFR in instrument conditions is inexcusable—doing so with passengers is even worse.

> ***"What could possibly go wrong?"***
>
> *A hopeful statement often made just prior to an ill-considered act.*

The Flight

A Cessna 150K departed Front Range Airport (KFRG), CO, just after midnight, with ASOS weather reporting 7 miles of visibility and overcast clouds at 300 feet above ground level (AGL). The aircraft was reported missing at about 3:30 a.m. and was located two miles west of the airport around 7:30 a.m. The pilot had been giving multiple rides to his friends.

A GoPro camera mounted in the aircraft recorded five successful and uneventful flights. On the final flight, according to the NTSB:

"Video recording began at the hangar area. A new passenger ... was in the cockpit with the pilot. The pilot's associates were seen standing near the hangar behind the airplane. The following ATIS recording was heard through the GoPro's internal microphone: Denver Front Range Airport automated weather observation zero-six-zero-five (0605) Zulu: weather – wind calm, visibility [seven], ceiling three hundred (300) overcast, temperature one-four (14) Celsius, dew point one-three (13), altimeter three-zero-two-zero (30.20)."

"The pilot taxied to a runway.... During the climb-out portion of the flight, the pilot used his cell phone to take a self-photograph. The camera's flash was activated, illuminating the cockpit area. The pilot's cell phone appeared to be on a user screen consistent with a camera application. The pilot landed and can be seen using his cell phone during the landing rollout."

The recording ended normally. However, the sixth flight did not.

No IFR flight plan was filed, and no contact was made with ATC. The radar track showed the flight became airborne at 12:04 a.m. According to the NTSB:

"At 12:18:56 a.m., the airplane again departed Runway 26 and began to drift to the left of the runway centerline. At 12:20:06, the airplane turned right to the northwest, ascended at 300 feet per minute, and reached an altitude of about 640 feet AGL. The airplane began a left turn and reached an altitude of about 740 feet AGL. At 12:21:24, the left turn tightened, and the airplane descended about 1,900 feet per minute. The last radar point was recorded at 12:21:43, about 140 feet AGL."

The pilot and passenger did not survive.

Aftermath

The NTSB reviewed the recorded images from the GoPro. While the actual crash was not recorded, the footage provided a rare look at predictive behavior leading up to the accident. The camera, mounted above the instrument panel facing aft, captured images of the pilot and passenger. This incident highlights the potential benefits of in-cockpit cameras—something worth discussing further.

The Pilot

The 29-year-old pilot held a commercial certificate with single- and multi-engine land and instrument ratings. His total flight time was 726 hours, with 38 hours logged in the last 30 days and 4.5 hours in the 24 hours before the accident. He had 27 hours of night flight and 15 hours in actual IMC. The autopsy revealed no drugs or alcohol in his system. However, he was not current to carry passengers at night or operate under IFR.

The Aircraft

The 1970 Cessna 150, with just over 7,000 flight hours, had recently undergone an annual inspection. No malfunctions were noted.

NTSB Probable Cause

"...was the pilot's loss of control and subsequent aerodynamic stall due to spatial disorientation in night instrument meteorological conditions.

Contributing to the accident was the pilot's distraction due to his cell phone use while maneuvering at low altitude."

Commentary

The pilot was proud of his skills and wanted to share his accomplishments with his friends—a natural inclination. Passengers are free to take pictures and selfies as long as it doesn't interfere with flight operations. However, using a camera flash at night is a bad idea.

Not shutting down an engine when boarding passengers has frequently led to fatal accidents. And, as noted in the introduction, this is a severe case of "rapture of the sky." Departing into IMC at night with a passenger, while not being proficient, current, or on an IFR flight plan, is beyond irresponsible.

The pain caused by this tragedy extends to everyone involved—the families, the friends, and all who knew these young people. As aviation professionals, we often discuss crashes clinically, using technical language, and may dismiss pilots as reckless or incompetent. However, no one wants to believe they could make such a mistake.

Close-knit groups of friends may struggle to recognize potentially disastrous decisions in the moment. Do fellow pilots have an obligation to privately and respectfully dissuade others from engaging in such self-destructive behavior? In my view, that's a rhetorical question.

Vertical Thinking - (C172 CFI fails to consider terrain at night)

What works in the flat lands doesn't work in mountainous terrain.

"No matter how sophisticated you may be, a large granite mountain cannot be denied – it speaks in silence to the very core of your being."

Ansel Adams, renowned mountain photographer

According to the FAA, Controlled Flight into Terrain (CFIT) is the third leading cause of fatal accidents in general aviation. It's important to remember the difference between mean sea level (MSL) and above ground level (AGL),

They disappear at night

especially when flying anywhere other than near sea level. In this case, the basics of a routine flight were overlooked, and an overreliance on unavailable equipment led to catastrophe.

Good old-fashioned planning and execution still work, just as they have for over a century. In this case, we got a rare inside look at how the accident developed, as one of the pilots survived. While the weather was VFR, the flight

conditions required instruments— something that often happens on dark nights over sparsely lit terrain.

The Flight

On a clear November evening in 2014, a flight instructor and student set out on a night cross-country flight in a Cessna 172. The student had planned a route from Frederick, MD (KFDK), around the DC restricted airspace, then over to Winchester, VA (KOKV) for some pattern work. From there, they would continue to Charlottesville, VA (KCHO) before returning over the Linden VOR back to KFDK.

Flight Path, NTSB

However, when the CFI and the student met at about 4 p.m. Eastern Standard Time, the CFI changed the destination to Hot Springs, VA (KHSP).

For private pilot training, FAR Part 61.109 requires a night cross-country flight with more than 100 nautical miles between airports. Belatedly, the CFI apparently realized that KCHO would not meet this requirement, but KHSP would suffice. Perhaps in the interest of time, the CFI did not have the student plan the new route. No flight plan was filed, nor was one required.

After several landings at KOKV, the flight turned toward KHSP on a heading of 240 degrees at 3,000 feet MSL. By the time they entered mountainous terrain, the sun had set.

According to the NTSB's post-accident interview:

"The student asked the instructor about terrain elevation in the area, and the instructor responded that he was not certain of the elevations because the airplane was not equipped with a [Garmin] G-1000 navigation system (emphasis added)."

The student pilot reported that there were no aeronautical charts aboard for the western portion of the trip, although a Washington sectional was tucked away in a map pocket. In the automation chapter, we'll discuss how even fully instrumented aircraft with moving maps get into trouble.

Everything proceeded routinely, and at a GPS-indicated distance of 68 nautical miles from KHSP, the CFI demonstrated autopilot heading and altitude modes at 3,000 feet. The instructor selected a 200-feet-per-minute (fpm) climb, then increased it to 500 fpm. The student noticed the drop in airspeed from 120 knots to 90 knots and asked about adding power, which the CFI then did.

The student recalled that it was "pitch black" outside as the aircraft climbed. Moments later, they crashed. The aircraft struck the ground at about 3,100 feet MSL, approximately 300 feet from the top of a ridge. The instructor died shortly after the crash, while the student survived but sustained serious injuries.

The NTSB noted that an Emergency Locator Transmitter (ELT) signal was received after the crash and passed along to an ATC facility by overflying aircraft. However, the supervisor did not notify search and rescue as required. Without a flight plan or VFR flight following, the rescue response was delayed, and the wreckage was not located until the following day.

Flight Instructor

The 49-year-old CFI held an ATP certificate, issued in August 2010, with ratings for airplane single-engine land, multiengine land, and helicopter. He also held a commercial certificate with ratings for airplane single-engine sea, airplane multiengine sea, and glider. His instructor's certificate was approved for airplane single-engine, multiengine, instrument, and glider operations.

His total flight time was 5,941 hours, with 1,182 hours as a flight instructor and 410 total hours at night. His most recent flight review had been conducted a month before the accident.

The crash site is about 3,700 feet while the MEF from the underlined previous sector is 3,600 feet. 5,100 is the MEF needed to the west.

Student Pilot

The student pilot, 51 years old, had started taking lessons in August 2014 and had accumulated nearly 24 hours of flight time, with 22 hours in the Cessna 172. He had 4.5 hours of night flying experience but had not yet soloed.

The Aircraft & Weather

The 2003 Cessna 172S (180 hp) had 4,263 hours and was equipped with conventional instruments, also known as "steam gauges." All inspections were current, and no pre-impact malfunctions were noted. Weather was not a factor, with clear skies and light winds.

NTSB Probable Cause

"The flight instructor's decision to conduct a night training flight in mountainous terrain without conducting or allowing the student to conduct appropriate preflight planning and his lack of situational awareness of the surrounding terrain altitude, which resulted in controlled flight into terrain."

Commentary

We got a valuable look into both pilots' thought processes thanks to the student's survival—a very unusual outcome in a CFIT crash. Had the student prepared a revised flight plan, perhaps terrain would have been considered, which is a critical idea. The sectional chart showed a maximum elevation figure (MEF) of 5,100 feet, with a mountain near the accident location topping out at 3,700 feet. (In the sectional chart excerpt, the MEF of 3,600 feet where the flight had come from was not enough to clear the ridge.)

This sets a terrible example when instructors do not model the procedures they're teaching. This includes not making a flight plan, not having appropriate charts—though anyone with an EFB should have everything needed—and not following the guidance that in mountainous terrain, at least 2,000 feet of vertical clearance should be allowed. Some might also question the wisdom of flying over such inhospitable terrain in a single-engine aircraft after dark.

With no charts for the area, night conditions, and no terrain awareness, the outcome was seriously in doubt. In the post-crash interview, the student noted that he felt very comfortable flying at night in a G-1000-equipped 172, which he had flown before, because of the additional situational awareness provided by the large moving maps and terrain warning system. That aircraft was unavailable, so a conventionally equipped C172 was used.

The student noted, "Everything got weird the night before the flight." He called the CFI at approximately 8:30 p.m. the night before to discuss the plan, but the instructor's wife said he had gone to bed. Around 7:30 a.m. the next morning, the CFI emailed the student, saying that the flight plan was good and indicated no need to change destinations. The report noted, "The student decided that all was okay because he trusted the CFI, so he continued on course and altitude."

While glass cockpits are a huge benefit, when we fly without all the latest equipment, more basic skills are needed. The electrons are seductive. While learning some of the old skills might seem old-fashioned, they've withstood the test of time.

Flat Land Mindset

A theory on how the instructor made such a fundamental mistake: Basic flight instruction takes place close to the home airport, and as creatures of habit, many of us get an "altitude mindset." Because terrain and obstructions are well known, we use a rote altitude for enroute VFR flight or maneuvers. In the relatively flat country around Frederick, MD, if one doesn't venture too far west, 3,000 feet will be sufficient for VFR operations. The highest terrain is about 1,700 feet. However, looking at instrument charts, if one goes just a little farther west, the minimum enroute or off-route altitudes jump to 5,000 to 6,000 feet. The VFR Maximum Elevation Figures (MEFs) rise significantly as well.

As a flatlander, I've had to consciously re-calibrate my altitude mindset. 3,000 feet MSL is no longer adequate. When flying to Denver, I've had to remember that 10,000 feet is only about half of what it means back east and not worth much once flying toward the Front Range foothills.

Night Flying Is Instrument Flying

Night flying is instrument flying in sparsely lit areas and over water. Use IFR procedures after dark, even when VFR. CFIs—teach your students well. Our Electronic Flight Bags (EFBs) and glass cockpits, while not officially enhanced ground proximity devices (and there are bold print disclaimers in all the manuals saying they are advisory only), will help a lot, if used intelligently. Regardless, you just have to know where the terrain and towers are. On the route from KOKV to KHSP, my EFB shows even 4,000 feet is too low. The proper VFR cruising (and safe) altitude was 6,500 feet.

NTSB Safety Alert

In January 2008, the NTSB issued a Safety Alert (SA) titled "Controlled Flight into Terrain in Visual Conditions," with the subheading "Nighttime Visual Flight Operations are Resulting in Avoidable Accidents." The SA stated that recent investigations identified several accidents that involved CFIT by pilots operating under visual flight conditions at night in remote areas, with pilots appearing unaware that the aircraft were in danger. It also noted that increased altitude awareness and better preflight planning likely would have prevented the accidents.

The SA suggested that pilots could avoid becoming involved in a similar accident by accomplishing several actions, including:

- Proper preflight planning
- Obtaining flight route terrain familiarization via sectional charts or other topographic references
- Maintaining awareness of visual limitations for operations in remote areas
- Following IFR practices until well above surrounding terrain
- Advising ATC about potential inability to avoid terrain
- Employing a GPS-based terrain awareness unit

Sure sounds familiar, but NTSB investigators continue to pick up the pieces on a regular basis. This flight instructor had excellent credentials, but when building hours and years of successful flights, complacency often sneaks in and waits quietly until all the links in the accident chain are in place.

Wrong Runway - (Comair 5191 departs on a short runway)

A momentary lapse in basic airmanship led to a tragic takeoff.

Lining up on the proper runway seems so easy. There are usually runway entrance signs with large numbers painted on the pavement. In this case, two professional pilots were ready to conduct a routine flight in an extremely familiar aircraft. However, subtle distractions, complacency, and a few systemic problems contributed to a disaster.

Exemplar CRJ on Takeoff, Delta

The Flight – On August 27, 2006, a Bombardier CL-600-2B19 (CRJ) was scheduled to fly from Lexington, Kentucky (LEX) to Atlanta Hartsfield International (ATL) in Georgia. Comair flight 5191 received an IFR clearance at 5:49 a.m. Eastern Daylight Time and was instructed to taxi to Runway 22. Sunrise wouldn't occur until about 7:05 a.m. This was the first officer's (FO) leg to fly.

During the pretaxi briefing, the FO asked which runway they would be using, and the captain replied, "Runway 22." The briefing continued, and three additional references to Runway 22 were made. The first officer said, "Let's take it out and … take … [taxiway] Alpha. Two two's a short taxi."

The engines were started, and the heading bugs were set to 227 degrees, the correct magnetic heading for Runway 22, according to the accident report. From 6:03:16 to 6:03:56, the crew engaged in non-pertinent conversation, contrary to the sterile cockpit rule.

At 06:04:33, the captain stopped the airplane at the hold-short line for Runway 26 and asked for takeoff clearance. It was the wrong runway! Neither the FO nor the controller stated the runway number during the request for takeoff or the subsequent clearance, and no one verified the runway. The captain crossed the hold-short line for Runway 26 and turned the aircraft onto the runway while running the lineup checklist.

NTSB: "At about 06:05:58, the captain told the first officer, 'All yours,' and the first officer acknowledged, 'My brakes, my controls.' FDR data showed that the magnetic heading of the airplane at that time was about 266 degrees, corresponding to the magnetic heading for Runway 26. About 06:06:05, the CVR recorded a sound similar to an increase in engine rpm. Afterward, the first officer stated, 'Set thrust, please,' to which the captain responded, 'Thrust set.' About 06:06:16, the first officer said, '[That] is weird with no lights,' and the captain responded, 'Yeah,' two seconds later. About 06:06:24, the captain called, 'One hundred knots,' to which the first officer replied, 'Checks.'

At 06:06:31.2, the captain called, 'V1, rotate,' and stated, 'Whoa,' at 06:06:31.8. FDR data showed that the callout for V1 occurred 6 knots early, and the callout for VR occurred 11 knots early. Both callouts took place when the airplane was at an airspeed of 131 knots. FDR data also showed that the control columns reached their full aft position at about 06:06:32, and the airplane rotated at a rate of about 10 degrees per second. The airplane impacted an earthen berm located about 265 feet from the end of Runway 26, and the CVR recorded the sound of impact at 06:06:33.0. FDR airspeed and altitude data showed that the airplane became temporarily airborne after impacting the berm but climbed less than 20 feet off the ground."

The CRJ impacted a tree about 900 feet from the end of the runway, broke up, and caught fire. The FO, the only survivor, suffered serious injuries. There were 49 fatalities.

The Crew – The 35-year-old captain held an ATP certificate and had accumulated 4,710 hours of total flight time. He had 3,082 hours in the CRJ and 1,567 hours as pilot-in-command. Comair records showed that he had flown into LEX six times before the accident, with two night departures. His training records showed no discrepancies, and he was considered a capable and disciplined pilot.

The 44-year-old FO held an ATP certificate with 6,564 hours of total time and 3,564 hours on the CRJ. His training record also showed no discrepancies. He had flown into LEX 12 times since 2004 and had one night departure. All pilots, instructors, and check airmen were complimentary of both crewmembers.

The Aircraft – The CL-65 CRJ was built in 2001 and had accumulated about 12,048 hours. It had been loaded properly, within weight and balance limits, and no mechanical discrepancies were noted.

The Airport – Weather was not a factor, but the NTSB noted that it was dark, with sunrise still about an hour away at the time of the accident.

Runway 22 was 7,003 feet long, 150 feet wide, and fully lit, except for centerline lights, which were notamed out of service. Runway 26, however, was 3,500 feet long and 150 feet wide, though it was marked as only 75 feet wide and unlit. It was only used for light general aviation operations in daylight VFR conditions.

The airport charting was incorrect due to construction. NTSB: "In addition, the August 3, 2006, NACO chart for LEX had been updated by the FAA to show the taxiway configuration at the completion of the construction project (future taxiway A7 was shown but not taxiway A5 and taxiway A north of runway 8/26). The program manager of the LEX construction project at the FAA's Memphis Airports District Office stated that he did not recommend publishing an interim chart because the chart would have been inaccurate during the time both before and after the construction project.

"Neither the Jeppesen nor the NACO chart reflected the taxiway identifiers that were present on the airport at the time of the accident. However, the CVR did not record any discussion between the flight crew members indicating confusion about the airport taxiways, and the discrepancies between the Jeppesen chart available to the flight crew and the taxiway

KLEX – The crew turned onto the first runway (wrong one) instead of going to Rwy 22, NTSB

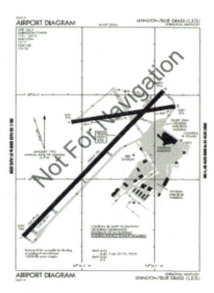

KLEX – Airport Diagram – NOT at time of crash, FAA

identifiers on the airport at the time would not have misled the pilots to Runway 26 instead of Runway 22." NTSB: "Nevertheless, the Safety Board recognizes that, under different circumstances, up-to-date charts might be necessary for a flight crew's successful navigation at an airport. For those circumstances, the charts must be complete, accurate, and timely. The Safety Board concludes that, due to an ongoing construction project at LEX, the taxiway identifiers on the airport chart available to the flight crew were inaccurate. Additionally, the information contained in a local NOTAM about the closure of taxiway A was not made available to the crew via the ATIS broadcast or the flight release paperwork."

Air Traffic Control – Staffing & Procedures

The tower was staffed by a single controller from midnight to 8 a.m. The controller on duty had pulled back-to-back shifts, having gone off duty at 2 p.m. and returned at midnight. While his performance showed no particular degradation, the NTSB noted that even two hours of sleep deprivation can result in a loss of attentiveness. At low-activity airports, staffing a tower with two controllers—or even keeping it open all night—is an economic challenge.

After clearing Comair 5191 for takeoff and transferring a previous flight to Indianapolis Center, the controller turned around in the cab to perform an administrative task instead of monitoring the departure of Comair flight 5191. At that time, the airplane had not yet deviated from the issued clearances, but had the controller been watching, he could have advised the crew that they were on the wrong runway.

That was the final opportunity to notice the flight crew's error and prevent the accident. The NTSB stated: "The controller most likely considered the importance of monitoring the Comair flight's takeoff as low because he assumed that there was little opportunity for a surface navigation error and no other traffic was on the airport surface to pose a conflict. The controller's recent duty times and sleep patterns indicated that he was most likely experiencing fatigue at the time of the accident. Task prioritization could be improved by modifying the then-existing FAA guidelines to explicitly state that active control tasks, such as monitoring departing and arriving aircraft, are a higher priority than administrative record-keeping tasks, such as a traffic count."

NTSB Probable Cause

The probable cause of the accident was the flight crew members' failure to use available cues and aids to identify the airplane's location on the airport surface during taxi and their failure to cross-check and verify that the airplane was on the correct runway before takeoff. Contributing to the accident were the flight crew's non-pertinent conversation during taxi, which resulted in a loss of positional awareness, and the Federal Aviation Administration's failure to require that all runway crossings be authorized only by specific air traffic control clearances. (This has since been changed to require positive crossing clearance for every runway.)

Commentary

Pilots have occasionally become misaligned since runways were numbered. Think of taxiing to the runway as the first leg to a waypoint; it's just as important as any flight segment. A contributing factor, which was listed in the Board's findings, was the controller's likely fatigue and failure to monitor the flight's taxi. My ATC friends might disagree.

Use an airport taxi diagram. These are published for all airports with control towers and many with IFR approaches. VFR-only pilots can easily download taxi diagrams from a variety of websites. All Electronic Flight Bag (EFB) programs have taxi diagrams, and most include a moving map.

Anytime there is more than one runway going more or less in the same direction—either parallel or within 45 degrees—the odds of lining up on the wrong one naturally increase. Equally troublesome are dyslexic runway numbers such as 13 and 31, 2 and 20, 23 and 32, which are prone to pilot confusion. Granted, they are not even close to being physically aligned, but just a short bout of confusion or inattention can be problematic. Any doubt about the path or limited visibility of intersections warrants a request for "progressive" taxi instructions. Verify again. Ditto on landing.

A technique suggestion: Always set the heading bug to the departure runway heading before taxi. Once lined up or after takeoff, reset the bug if needed. This method encourages one final check upon runway lineup to ensure the heading indicator, flight director, and bug are all in agreement. The crew did this but didn't verify. Don't have all that hardware? At least check the magnetic compass (in more sophisticated aircraft, they almost never get checked) and the directional gyro. Everything should point in the same direction as the runway numbers.

The NTSB also noted the complexity and confusion surrounding NOTAMs, which have been an ongoing challenge for the FAA. It's gradually improving, especially with the ability of EFBs to filter out irrelevant notices, such as laser light shows and low-level unlit towers miles from airports not even remotely connected to the flight. However, this is an area where additional improvement from the FAA was—and still is—long overdue.

On most large aircraft, ground steering is done only by the captain, with a wheel or tiller located on the far left side of the cockpit. This can create an out-of-the-loop situation for the first officer (FO). In Lexington, the captain taxied out, lined the aircraft up on the wrong runway, and handed control over to the FO. In this scenario, the FO moves from "pilot monitoring" to "pilot-flying" in a matter of seconds.

Depending on company procedure, the FO may have been "pilot-distracted"—heads down and dealing with paperwork, checklists, and other myriad details that accompany a jet launch. While automation and paperwork are valuable, the need to manage them often adds to a high-workload environment.

Primary Flight Display showing discrepancy between assigned runway heading (bug) and wrong runway heading, NTSB

For single-pilot operations, obtain the clearance and program the avionics before starting to taxi. If unable to talk to ATC on the ramp or if the clearance is "on request," wait until reaching the run-up pad. Taxiing while copying a clearance simultaneously is a bad idea. Complex airport designs, construction, and poor visibility conditions such as darkness or fog, all demand full attention.

Sterile cockpit—It's so simple and yet often ignored, both in general aviation (GA) and in commercial flight operations. While job-seeking resumes often tout the ability to multitask, humans are not good at it. The importance of concentrating on the task at hand cannot be overstated.

Does the tower have some responsibility to ensure that pilots take off on the proper runway? In my opinion, yes, after discussing it with several of my ATC friends. The tower's primary job is to sequence traffic for arrivals and departures, and while the controller did that, they should also monitor what's happening at the runway and assist as needed. There was an earlier event in Lexington involving runway misalignment. It's the backup that makes our system as safe as it is.

NTSB: "The controller was most likely fatigued at the time of the accident, but the extent to which fatigue affected his decision not to monitor the airplane's departure could not be determined, in part because his routine practices did not consistently include the monitoring of takeoffs."

On the midnight shift, if two controllers are required, might some communities with lower-activity airports decide to limit hours of operation due to the higher staffing cost? What effect does that have on safety? That must be answered on a case-by-case basis. However, pulling two shifts back-to-back, as the Lexington tower controller did by finishing at 2 p.m. and returning to duty at midnight, makes no sense from any human-factors perspective.

There are tens of millions of takeoffs every year, and the system works very well. A cursory review of FAA statistics shows that there have only been about half a dozen attempted takeoffs on the wrong runway or on a taxiway at towered fields in the past several years. In the final analysis, much of what we do in aviation is routine and repetitive. As pilots-in-command, we must also remember that certain activities must be done correctly every time or there will be a bad outcome. Runway identification falls into that category.

Chapter 3
TAKE OFFS ARE OPTIONAL,
LANDINGS ARE MANDATORY

"Takeoffs and landings must come out even, if you want to use the aircraft again."

Ancient aviation lore

Takeoffs and landings provide some of the greatest satisfactions and challenges of flying. The aircraft operates close to the edge of the flight envelope—and so, it would seem, do many pilots.

Boeing maintains a 10-year rolling average of worldwide commercial jet aircraft fatal accidents. A recent report notes that for an average 1.5-hour flight, only two percent of the time is spent in takeoff and initial climb—yet that phase accounts for 13 percent of crashes. Final approach (aptly named) and landings make up only four percent of flight time but a staggering 55 percent of fatal air carrier accidents. General aviation has similar numbers—close to the ground is where the problems occur. After all, no one ever collided with the sky.

Some well-known airline landing excursions include the Southwest Airlines 737 at Chicago Midway that slid off the end of a snowy runway (Overrun, this chapter) and an American Airlines MD-82 that hydroplaned while landing during a thunderstorm in Little Rock, AR (Bowling Alley Blues— this chapter). Runway contamination or adverse winds make landings more difficult—sometimes impossibly so.

Bad day in Toronto - runway excursion, Canadian TSB

Much of general aviation operates out of short or obstructed runways— places where optimistic thinking can take you where no pilot has successfully gone before. At the small airport where I learned to fly, a transient pilot would "tiptoe through the tulips" about every six months. The trees and a railroad embankment did a fine job of stopping airplanes but didn't do much for preserving them in flyable condition. There was great incentive to be on speed and altitude.

The Goldilocks parameters—not too much, not too little, just right—are known technically as a "stabilized approach." Airlines and corporate operators emphasize this concept, and it applies to all flying machines. The idea is simple: speed, alignment, and altitude should allow the aircraft to touch down within the first third of the runway.

The Citation Jet Pilots Association (CJP) implemented a program several years ago to improve the landing safety performance of its members. This largely owner-flown group had a somewhat checkered history of landing mishaps, and their insurance rates reflected that. As you'll read in the Automation chapter, the benefits of monitoring and analyzing performance can hardly be overstated.

CJP reported that in a recent three-year period, there had been no accidents or incidents among its members, who fly nearly 1,000 of Cessna's smaller jets. They partnered with ForeFlight to develop a data analysis program to track exceedances. Less than two percent of their approaches fell into that category—below the industry-wide standard of three percent. Exceedances were far more likely to occur on visual approaches than on instrument approaches. There's a message here for all pilots, regardless of what type of aircraft you fly: for light aircraft, maintain a stable approach at 500 feet AGL on visual approaches and at the final approach fix in instrument conditions.

You'll see some interesting interpretations of "stable" while watching arrivals at aviation gatherings or just observing at any airport. It's best to watch from the sidelines.

Simple is good, but one size rarely fits all situations. Can light aircraft adjust more quickly, even when below 500 feet AGL? Of course. Many CFIs won't call for a go-around at 400 feet unless the student is completely outside the parameters. With bigger machines, inertia exercises its prerogatives. On shorter runways, touching down on the 1,000-foot marker (the big white ones) leaves a lot of useful—and perhaps necessary—runway behind. However, in low visibility conditions, it's best to stay on the glideslope. Instrument runways usually have adequate length, except when contaminated or when there's a tailwind.

"Practice makes perfect."

A common misconception

It does not. Again, quoting football coach Vince Lombardi, "Perfect practice makes perfect." Doing it wrong just reinforces bad habits. Pros work at it, while amateurs accept mediocrity.

Speed can be your friend or your enemy. Approaching at 65 to 80 knots on final is not likely to endear you to ATC or other pilots at a major airport. It takes finesse to transition from fast to slow, but please stay away from Big Town Municipal until you've mastered the skill. Practice where there's plenty of pavement but not much traffic. Too much speed is just as bad for touchdown.

A well-rounded pilot progressively reduces power during approach, adding gear and flaps at the right points—VFR or IFR. Style points are deducted if more power is needed before touchdown. In windy conditions, it's tough—sometimes impossible—but when executed well, it's energy management at its best.

Auto-throttles on newer jets and turboprops make this process much easier, but pilots must understand the modes. This was something the Asiana pilots failed to grasp, as noted in the online supplement ("Too Low – This chapter").

"Performance figures from the POH are engineering fact and operational fiction."

Author

Adhering to the handbook or flight manual brings us closer to book performance, but most of us won't achieve that level. A recommended reference is the FAA's Advisory Circular AC 91-79A, or its successor, *Mitigating the Risks of a Runway Overrun Upon Landing.* While it primarily applies to larger aircraft, the formulas are both eye-opening and useful for all pilots. That excellent guidance will be interspersed throughout this discussion.

For performance calculations, forget interpolation. Always round up to the next higher number—then add some margin. While the FAA used to emphasize interpolation on knowledge tests, it's not worth the effort. The real world has too many variables to measure takeoff and landing distances down to the last ten or even 100 feet.

When I worked for Cessna, we used to tout that our aircraft could outperform the competition by 35 feet. In hindsight, I'd chalk that up to youthful enthusiasm and ignorance.

Multi-function graphs encourage approximation and are not entirely accurate. The same caution applies to performance tables, which imply precision down to the foot or the tenth of a gallon. When it comes to avoiding solid objects or ensuring enough fuel to reach the destination safely, adding a margin is just common sense.

If conditions are NOT identical to what the factory used (and they won't be), the performance will NOT be the same.

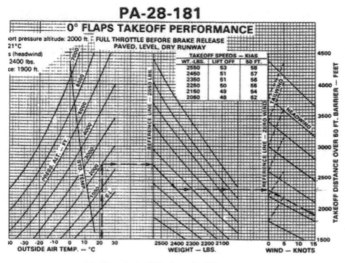

Multivariate TO Graph, Piper

- The runway is level, dry, and smooth—but many runways are not.

- Wind is not a factor beyond what's listed—but winds don't usually conform to test conditions.

- The engine(s) is/are producing rated power—but many times they don't.

- Full power is applied before brake release—how often is that actually done?

- The tires are properly inflated—when was the last time you checked?

- Airspeeds are executed exactly as listed, corrected for weight—sometimes.

- The 50-foot obstacle is much taller—have you seen the pine trees around here?

- On landing, maximum braking is used—are there square tires in your future?

- The aircraft weight and balance are accurate—don't bet on it as they age.

- The controls are perfectly rigged—let's hope they are.

- The airframe is clean, with no extra drag-producing hardware—probably not.

- The test pilot knows the aircraft intimately and has practiced the technique that yields the absolute best results—you may be that good, but most of us are not.

- The test pilot expects to be measured on all parameters—there are multiple tries.

- Distances are precisely measured—but most of us rely on "Kentucky windage."

"The Air Safety Institute's 50-50 rule - Add 50 percent to whatever distance the Pilot Operating Handbook computes to clear the 50' obstacle for both takeoff and landing."

Author

Begin at the beginning. Takeoffs sometimes end badly when pilots don't use the full runway length. As a taxpayer, you're entitled to all of it. It'd be a shame to crash off the end due to a midfield departure and an unanticipated hiccup. The pros won't do it, and ATC will always approve full length if you ask. The temptation to take the shortcut is strong, but resist it.

50 percent keeps the math simple. The obstacle may be more than 50 feet tall, and the amount by which the aircraft clears it is not added to the calculation—i.e., zero clearance. Flight tests are conducted with optical measuring devices instead of solid objects, so the consequences for test pilots are not real. In testing, all sorts of techniques can be tried with no damage or injuries. But when the obstacles are real, margins are a good thing.

Most of us can't estimate when we're 50 feet above the runway. Clearing the obstacle by 50 feet or more is reassuring, though it may add roughly 600 feet to the 50-foot obstacle clearance distance. The guidance on landing is to add 200 feet of rollout for every 10 feet above the obstacle or Threshold Crossing Height. It might be less for certain aircraft, but that's tough to estimate unless you have a radar altimeter. With Part 25 Transport Category aircraft, some factoring is done to address pilot variation. However, jets still periodically overrun the allotted pavement on landing, so speed control remains essential.

An example: A light aircraft POH predicts it will take 1,800 feet (under known test conditions) to just clear the obstacle on takeoff. Adding 50 percent to that yields a more realistic distance of about 2,700 feet—a margin to live by. Can it be done in less? Of course, under test conditions.

Landing performance charts are also optimistic. Each aircraft is different due to its drag characteristics, but as a rough estimate for light GA piston aircraft, adding 500 feet to compensate for an extra 5 knots of airspeed or groundspeed is reasonable. Very roughly, it's about 100 feet per knot. Is that relevant? It could be—ask anyone who's participated in a spot landing contest.

Another variable that isn't often mentioned is ground effect. Often described as a "cushion of air" when the aircraft gets within half a wingspan of the ground, it's technically the reduction of induced drag due to changes in the flow patterns of wingtip vortices. If the pilot doesn't bleed off excess airspeed before entering ground effect, the aircraft will happily whiz along just above the surface, burning up lots of runway. Low-wing aircraft are more susceptible because the wing is closer to the ground— ask any Mooney pilot. Forcing it down invariably leads to bad outcomes. When in doubt, it's time for a go-around.

On takeoff, we occasionally hear of an aircraft becoming airborne but failing to climb out of ground effect. Density altitude degrades performance and can quickly put us in a bad situation.

Landing distance is based on maximum braking, although the FAA includes a disclaimer about not causing undue tire wear. However, they don't explain how to determine when that line is crossed. Without anti-skid brakes, this technique often leads to square tires or flats.

Extreme flat spot (actually flat) tire, Goodyear aerospace

The notes on wind conditions or turf runways, usually in fine print in the performance section of the POH, will invariably affect the base calculation. Be sure to account for these factors before adding a safety margin.

Runways aren't always level—slope can either help or hinder performance. As a general rule, a one percent slope changes the effective runway length by 10 percent, either positively or negatively. The Chart

Supplement (formerly the Airport Facility Directory), instrument approach chart, or your Electronic Flight Bag (EFB) will display runway gradients. For those less mathematically inclined, a one percent slope equates to a 10-foot change in elevation per 1,000 feet of runway.

Let's take a normally aspirated Cessna 172 into the mountains for a day trip to Salida, Colorado (KANK). Runway 6-24 is 7,351 feet long, but the elevation at each end differs significantly—7,523 feet at one end and 7,384 feet at the other.

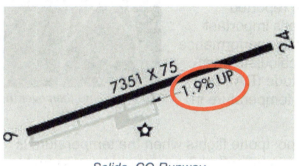

Salida, CO Runway

The Chart Supplement notes: "Runway 24 recommended for landing; Runway 06 recommended for departure—weather and traffic permitting." The gradient is 1.9 percent, but for easy figuring, we'll use two percent.

Landing in the morning, the temperature is 10°C with calm winds. The calculation from the POH is as follows: a 20 percent gradient = 1,232 feet + 50 percent safety factor = 1,848 feet. That's plenty of runway, and we'll make the first turnoff.

Later in the day, after touring the sights, the temperature has risen to 20°C. The airplane is heavier since we had a big lunch, added fuel, and bought some gifts, including a sculpture weighing an estimated 50 pounds. With some low clouds off to the northeast but VFR conditions everywhere else, we elect to depart on Runway 24—the same one used for landing. Takeoff safety distance is now just over 5,800 feet. (POH: 3,375 feet + 50 percent safety factor + 20 percent uphill = 6,075 feet). Now suppose there's a 7-knot tailwind, which adds about 700 feet. Obviously, we've stacked the deck, but the circumstances have changed dramatically.

Don't forget the rising terrain that may be off the end of the runway. Just getting airborne is only the beginning. Too many mountain flights take off only to realize they cannot out-climb the terrain. That's an "Aw shucks" moment, and it happens almost every year.

Density Altitude – The Meteorological Aerodrome Report (METAR) temperature at the ASOS or AWOS site is not necessarily what the wing, propeller, or engine experiences over hot pavement on the runway. Reported temperature is measured in the shade of an instrument shelter, some distance from heat-reflecting pavement.

To estimate real hot weather performance, add 5 to 10 degrees Celsius when the reported temperature is above 30°C. That's important when operating at the edges of the performance envelope. Many locations now calculate and automatically report density altitude (DA), but that calculation is based on the temperature in the instrument shelter.

Cooler in here then over the runway

In desert locations, airlines will postpone flights when the temperature is literally off the charts. A dangerous takeoff at 3 p.m. might be reasonable at 8 p.m., but flying over the mountains at night in light aircraft especially singles—increases the risk. Mornings in the mountains are usually cool, mostly smooth, and have much less convection.

Operating near gross weight at high density altitudes is hugely disappointing in many light aircraft, especially reciprocating-engine models. A somewhat anemic climb when lightly loaded will be nearly nonexistent when heavy. Many articles have been written on mountain operations, but a mountain flying course is essential and eye-opening when heading into high-altitude regions.

The normally aspirated Cessna 182 has been involved in several high-density altitude crashes, according to an Air Safety Institute study conducted years ago. The aircraft is a strong performer in the lowlands, and pilots often extrapolate that to higher altitudes. However, performance drops significantly when only 60 percent power is available. A normally aspirated Bonanza or Piper Saratoga will also underperform. High density altitude periodically claims a few aircraft, even in lower elevations, on hot days when heavily loaded. By the time a pilot realizes departure was a bad idea, a crash is often inevitable.

Leave your favorite anvil collection at home and manage luggage wisely. Give UPS or FedEx the business by shipping the extra 50 to 100 pounds of items you don't need but can't live without. Make an extra fuel stop—

this is NOT the time to be tankering fuel. Carry enough to reach your destination with a reasonable reserve, but don't plan a long leg when departing from hot or high-altitude locations unless there is an abundance of runway and no nearby terrain.

Technique

What's the best normal takeoff technique for retractable? Here's an opinion—yours may differ. Retract the gear once safely airborne. Get away from the ground as quickly and safely as possible because the first minute of flight is the riskiest. Altitude is more valuable than excess speed because it provides more options sooner. However, if the engine fails, the angle of attack will increase rapidly toward a stall if you hold the normal climb attitude. Pitch over immediately if the engine even burps. Add a quick memory reminder to the run-up checklist for use just after takeoff: **Pitch, boost pump on** (if required by your checklist).

Once certain the aircraft is not going to settle back on the runway (which is always annoying), select gear up. In crewed aircraft, the pilot monitoring will often call "positive rate," meaning the vertical speed indicator (VSI) shows the aircraft is climbing, and it's safe to retract the gear.

With short-field procedures, follow the POH recommendations until clear of obstacles if the aircraft has large, floppy gear doors. Some aircraft have significantly lower stall speeds and may clear an obstacle sooner with the gear down. However, if a safe takeoff depends on leaving the gear down because of slow retraction speed or increased drag during the retraction cycle, the runway is simply too short.

There is one exception—if the takeoff was from a snowy or slushy runway, leave the wheels down and let them spin to allow the airstream to dry them before retraction. Otherwise, they may freeze in place after a quick retraction, making the next landing "interesting."

Once high enough to have solid forced landing options, pitch over to cruise climb for better engine cooling and forward visibility. The improbable and often impossible turn is discussed in Emergencies, Chapter 6. Leaving the gear down to land on the remaining runway—anticipating that the engine will quit at just the right (wrong) time—is betting on long odds. My preference is to have altitude for landing options where they're needed most—off the end of the runway. You may decide differently.

Don't Be in a Hurry to Reduce Power

There's some debatable wisdom that engines are more likely to quit at the first power reduction after takeoff. However, there doesn't seem to be statistically valid data to support that claim—at least, I haven't found any. But there is validity in maintaining takeoff power until reaching a safe maneuvering altitude.

The aircraft spends less time in that vulnerable no-man's (or woman's) land of limited options. Most engines are designed to run for hours at high power, and while some have a five-minute limit at full power, an extra minute or two (within that limit) to reach a safe altitude won't wear it out any faster. Full power often provides fuel enrichment, helping to keep temperatures down. Yes, you'll be burning fuel faster than a three-year-old opening presents on Christmas morning, but that's a good trade-off.

> *"Wind does not discriminate it touches everyone, everything."*
>
> Lish McBride, Author, *Necromancing the Stone*

It gets windy up here! uwyo.edu

Wind Factors

If there's a tailwind or runway contamination, double the computed distances—at minimum. Tailwind and any form of contamination mean a different runway is needed.

Once wind, water, snow, or ice enter the equation, those neatly derived Pilot Operating Handbook (POH) numbers become rough estimates at best. To see the difference firsthand, try a takeoff and landing on a long, dry runway with just **5 to 7 knots** of tailwind. Your results may be quite different from expectations.

On landing, if the aircraft is equipped with GPS, compare the final approach indicated airspeed (IAS) with GPS groundspeed. If IAS is higher than groundspeed, great. But if groundspeed is higher, beware of the tailwind component.

Gusty crosswinds will extend landing distance. Too much speed is often the problem. The general advice is to add half the gust factor to the normal final approach speed. The float begins just before touchdown, and the wind pushes the aircraft sideways, which may lead to unintended runway light modifications.

In the previous example, we started with the weight-appropriate POH obstacle distance of 1,800 feet and added a 50 percent safety factor, bringing it to 2,700 feet. With a 10-knot gust factor, an additional 5 knots for controllability adds another 500 feet, making the new safety distance 3,200 feet. With this buffer, there's no need to smoke the tires, stress over clearing the obstacle, or expect perfect performance from the aircraft or the pilot.

If (when) a downdraft is encountered near the end of the runway, be grateful for that extra speed and the additional distance factored into the equation.

Many aircraft land better with less flap in strong winds due to reduced wind resistance. While this extends landing distance somewhat, the increase in approach speed is usually minor.

Some POH landing speed data points:

- Cessna 172N – Full flaps: 60 kts, Flaps up: 70 kts
- Beech Bonanza – Full flaps: 80 kts, Flaps up: 90 kts
- Piper PA-28-181 – Full flaps: 66 kts, Flaps up: 76 kts

Splitting the difference by using partial flaps often works well. Most high-performance aircraft pilots typically carry a little power over the fence, which further reduces stall speed. However, aircraft documentation doesn't account for all variables, so it's worth exploring different configurations to increase comfort and confidence.

Regardless of crosswind, once the wind exceeds 60 to 70 percent of the aircraft stall speed, short runways become a bad deal—especially if conditions are gusty. It may be a ride to remember. This is not the place for newbies or occasional fliers. Fortunately, strong winds rarely last more than 12 hours unless they are part of a major weather system.

Demonstrated Crosswind Component (DCC) is not a limitation, but there are limits. DCC represents the strongest crosswind tested during aircraft certification, not the absolute maximum the aircraft can handle. It's best to adhere to the demonstrated max until you and the aircraft are completely in sync.

There is a real physical limit—when there isn't enough rudder authority to keep the nose aligned with the runway. Lowering the upwind wing too far may also cause a wingtip or a wing-mounted jet engine pod to drag.

For light aircraft, the slip method is often the easiest way to handle crosswinds. Crabbing also works well, provided the pilot straightens out before touchdown. Jetliners typically use the crab method.

If you're cross-controlled and the aircraft still weathercocks into the wind despite full opposite rudder and wing-down input, find another runway.

There is an actual limit, Airbus

With crosswind components, consider gusts as a possible limiting factor—not just the steady-state wind. You and a gust may arrive at the critical point simultaneously. Most of the time, pilots get away with it, but armchair quarterbacks will claim the crash was predictable. That's debatable. The reality is that sometimes we just don't know in advance. If you decide to take a look, be prepared to bail out—whether on takeoff or landing.

Read the next two paragraphs carefully. They come from the NTSB's report on a Boeing 737 crosswind takeoff accident in Denver (Sideways Slide - this chapter):

"The NTSB notes that a manufacturer's demonstrated crosswind is based on the successful accomplishment of three takeoffs and landings by a highly skilled test pilot and reflects the wind conditions that were available to the manufacturer for testing during the certification process.

The NTSB also notes that an evaluation of an airplane's crosswind takeoff and landing performance (and perceived handling qualities) in very gusty wind conditions is not required by federal regulations, nor is the manufacturer required to publish information about the gust factor present during testing. Airplane manufacturers are not required to establish crosswind guidelines above the maximum demonstrated crosswind component, and there are no FAA standards for the establishment of such guidelines."

Another way to quickly estimate a crosswind component is by using degrees off the runway heading:

- 30 degrees = 50 percent of the reported wind
- 45 degrees = 70 percent
- 60 degrees = 100 percent

For example, if runway 18 is in use and the wind is blowing from 210 degrees (30 degrees off the runway heading) at 30 knots, the crosswind component is about 15 knots—50%. If the wind is from 225 degrees (45 degrees off the runway heading), the crosswind component is about 21 knots—70%.

An airport in Kansas, with little disrupting terrain, may have much easier landing conditions in high winds, while an airport in Colorado, Vermont, or the Sierra Nevada, surrounded by mountains, may be far more challenging with even lower total wind.(Nobody's Flying Today – This chapter) Introducing the term "orographic turbulence" will impress friends at adult beverage parties.

		My Crosswind Component is...					
		Aircraft Indicated Airspeed					
		50	60	70	80	90	100
Crab Angle Relative to Landing Strip Centerline	5	4	5	6	7	8	9
	10	9	10	12	14	16	17
	15	13	16	18	21	23	26
	20	17	21	24	27	31	34
	25	21	25	30	34	38	42
	30	25	30	35	40	45	50
	35	29	34	40	46	52	57

My maximum demonstrated crosswind component is: _____

This reference table will estimate your crosswind component if you know your indicated airspeed and your crab angle relative to the landing strip centerline.

A quick way to estimate crosswind component, FAA

But is the wind really that strong? Wind speed and direction are typically measured near midfield, which can be a long distance from the takeoff and touchdown zones. Even if the tower or automated system reports that it's blowing cats, dogs, and small boulders, trees or terrain may shelter the landing zone, reducing the wind's effects.

Conversely, a manageable wind at the reporting point may turn nasty and choppy due to obstructions near the runway, raising concerns about controllability. Be prepared for a diversion—got gas? Winds tend to be lightest at sunrise and sunset and strongest in the early afternoon, unless a front is arriving. Go earlier or later.

Don't Get Sheared – A rapid change in wind direction close to the ground can wreak havoc on controllability and alignment. When surface winds differ significantly from those just a few thousand feet up (or less), shear is always a possibility. Ten, or even twenty, knots can be scrubbed off or added in the blink of an eye—so much for those carefully computed POH numbers. This often leads to enthusiastic turbulence during descent, which is why staying out of the yellow arc is a great idea. You can anticipate this by comparing the winds aloft with surface winds.

In IMC, do not underestimate the potential impact of shear and moderate turbulence on light aircraft controllability. With light wing loading and less sophisticated autopilots, spatial disorientation may also become a factor. The ride will be uncomfortable, and autopilots often do exactly what we tell pilots NOT to do—fight the displacements instead of riding the waves. Some advanced autopilots have a soft ride mode, which helps.

On final approach, this can become extremely challenging, with little altitude for recovery. If there's a significant change in wind direction and/or speed, be prepared. Factor wind into your minimums for an IMC approach. If turbulence is expected—especially if LLWS (Low-Level Windshear) is predicted—you might want to raise your minimums by a couple of hundred feet. A light aircraft will be jumping around, and keeping the needles centered may become more than a full-time job. In crashes, wind is frequently found to be a contributing factor.

The accident cited in *Betting the Farm* (this chapter) might not have happened if not for the bad timing of a gust—and it wasn't even that strong. This isn't the kind of shear associated with thunderstorms; that's a different animal altogether.

Tailwinds on Base Leg Are a Hazard

Too many base-to-final stalls occur when pilots fly their normal distance from the runway on the downwind leg and end up overshooting the extended centerline. I've done it myself. A visual illusion can lead us into trouble—groundspeed picks up with the tailwind, and being close to the ground on base, the landscape appears to be whizzing by. The unwitting pilot, thinking airspeed has crept up, instinctively reduces power or adjusts pitch to slow down.

Airspeed and groundspeed can be radically different. Suddenly, those boring ground reference maneuvers studied so long ago become critical. We talk about speed, but it's always about angle of attack. High groundspeed affects the turn radius when trying to line up with the runway.

Next time you encounter a quartering tailwind on base, ask yourself a few questions:

- Did I anticipate this?
- Should I have widened the pattern?
- Did I over-bank a bit?
- Did I increase AOA to keep from losing altitude?
- Did I plan for a go-around?

Consider modifying the pattern with a Pilot-in-Command safety decision if there's a strong tailwind component on base. Fly a much wider pattern, enter on base, lead the final turn earlier, or fly a straight-in final. If you deviate from the normal pattern, announce it on the CTAF. When it's that windy, there likely won't be much other traffic flying. Be mindful of obstacles, courteous to other pilots, and, in the highly unlikely event the FAA has a question, be prepared to explain your logic.

Rounding the Bases

Here's another technique that may help with runway alignment: round off the base leg into a constant-radius turn. The U.S. Navy uses this pattern with their carrier pilots, as do many airlines. The FAA diagrams a square base leg in the *Airplane Flying Handbook*, mimicking the rectangular ground reference maneuver for new students.

Square base, FAA - but there's a better way!

I unconsciously adopted the Navy's technique years ago to maintain a tight downwind leg. It turned out to be less work than rolling into and out of two turns. It also provided a constant descent rate, allowing me to focus on aligning with the runway.

However, visually locking onto the runway in the pattern can sometimes lead to a collision in a low-wing aircraft, where the final approach view is blocked. It's far better to clear the final on downwind before turning onto a conflicting path. It's much easier to extend downwind behind a target on final than to maneuver out of the way once you're at 90 degrees and in conflict. We also don't want to cut anyone off.

Additionally, a turning aircraft presents a more visible target to others. You can shallow or steepen the bank as needed, but limit it to a maximum of

Water makes a difference, Boeing

30 degrees. If you're uncertain, try it with an instructor—you might like it.

Water is a fantastic lubricant when it comes to landing, and over the decades, many runway excursions have occurred on wet surfaces. Most small airport runways are not grooved and may not be crowned to drain water effectively. As a result, they can retain water for quite some time.

The FAA notes that as little as one-eighth of an inch of water can cause hydroplaning, meaning you'll be slip-sliding for a bit before the aircraft slows down enough for braking to be effective. By FAA definition, the runway is considered flooded at that point. In other words, even a merely wet runway should be regarded with caution.

Light aircraft typically have little or no guidance on landing distances under these conditions, making it even more critical to approach wet runways with suspicion and preparedness.

More Than You Wanted to Know

Many of us don't know the hydroplaning speed for our aircraft, so it's best to figure it out in advance. Hydroplaning—where braking is completely ineffective—occurs at a ground speed equal to 9 times the square root of the tire pressure. For my aircraft, that works out to about 57 knots, based on 40 psi main gear tire pressure in a no-wind condition.

To estimate groundspeed at touchdown, check your GPS on short final if time allows. However, it may be better to guestimate the headwind component while on downwind. The critical speed could be slightly higher depending on tire tread configuration, condition, or runway contamination (such as tire residue). This isn't simple, which is why plenty of pavement is your friend in wet conditions.

Bounces

They invariably happen:

a) to students and new pilots,

b) to experienced pilots, and

c) when there's an airport audience or someone you're trying to impress.

Typically, bounces result from too much speed, an unfortunate gust, and/or a too-rapid flare. Even airlines experience them occasionally, and the FAA wants them trained in simulators only. In light GA, we have no choice but to practice them in the aircraft. CFIs need to introduce them—even when students consistently make near-perfect landings—so they're prepared. It also means being ready to take control quickly when things go sideways.

Hangar Story

The student and I were sitting on the taxiway on the right side of the runway, waiting for takeoff. A Cessna Cardinal entered a high bounce a few hundred feet from our intersection, and the pilot rapidly applied full power. Unfortunately, there wasn't much airspeed, and the nose was held high. The Cardinal pulled hard left and collided with three parked aircraft off the left side.

The multi-thousand-hour CFI admitted he had his left hand resting on the student's seatback and his legs crossed instead of being in the "ready position." As a new CFI (me), this was a lesson I never forgot. One can be casually alert without hovering or riding the controls. Glad we weren't on the left side of the runway.

Botched Go-Around

Go-arounds get fouled up regularly. Many aircraft do not climb well with full flaps—some don't climb at all.

Our flight school only allowed new students to use 20° of flaps on Cessnas—no retraction required for a go-around. Later, they were taught full flaps procedures. The sequence that works for most aircraft:

1. Pitch (forward—just enough to keep it level) and power together (smoothly applied, with right rudder).

2. Pitch to a shallow climb and set flaps as directed by the POH/flight manual.

3. Trim as needed.

Early attitude adjustment in a go-around is critical because most aircraft will pitch up vigorously with power when trimmed for landing, and stall attitude may be close to level. Re-trim, transition to normal climb, and—when well clear of the ground—select landing gear up.

With regular practice, this can be done smoothly and calmly. Power, right rudder, and pitch together—then flaps, then gear. After a go-around and raising the gear, triple-check gear down for the subsequent landing.

Go Left

Accident investigators routinely find wreckage from go-arounds off the left side of the runway—almost never to the right. When power is applied and the aircraft pitches up (because it's trimmed for landing), all those left-turning tendencies come into play.

The bigger the engine, the slower the aircraft is moving, and the more abruptly power is applied, the more pronounced the effects. Short-coupled aircraft with big engines—those with a short fuselage—will have an even stronger left-turning tendency. Think Cirrus.

Ramming the throttle forward can also cause the engine to stumble, further complicating the aerodynamics and control forces. You've heard all this before, yet it keeps NTSB investigators busy almost every week.

Flaps

Takeoff flap configuration is critical. Set it as if your life depended on it.

On takeoff, full flaps can turn a great day into a disaster. In large aircraft, improper flap settings can lead to serious trouble at the end of the runway.

A Northwest Airlines DC-9 and a Delta Boeing 727 both crashed due to improper flap settings—none at all. In airliners and jets, the issue is usually not having enough flaps, while in light aircraft, it's sometimes having too much.

At the time of this writing, there have been at least four fatal accidents involving a Cessna 172, two Cessna 150s, and a Lockheed Electra (which was still in preliminary status with the NTSB). In each case, the pilots attempted to take off with full flaps, failing to retract them after the preflight check.

Full flap, Great for landing, bad for takeoff, AOPA

I'm not aware of any recent fatal crashes caused by flap malfunctions. In a well-maintained aircraft, such failures are extremely unlikely. However, it's common for pilots to need a friendly reminder while taxiing for takeoff that their full flaps are still deployed. This should have been caught on the checklist, but ultimately, it comes down to ensuring everything is ready. In hindsight, these mistakes seem obvious, but most of us have missed something on the *Before Takeoff* checklist at some point.

I'm not a big fan of checking flaps on every preflight. Some aircraft manufacturers have mandated it—likely due to legal departments losing lawsuits they shouldn't have lost. If flaps are needed for takeoff, that will be determined at the run-up pad. In light aircraft, flaps are only necessary for short runways when landing. If that's your destination, fine—check them. But excessive checking wears out components faster, which increases the chance of failure when you actually need them. Legality versus common sense? You decide. On check rides, always adhere to the POH. (*Check ride procedures don't always have to make sense.*)

Sterile Cockpit & Preflights

When in the airport vicinity—during checklists, taxiing, takeoff, landing, or the preflight inspection—avoid unrelated discussions about politics, lunch, or the size of the houses below. Brief passengers in advance and stay focused on traffic, aircraft configuration, and maintaining a stabilized approach when landing. If necessary, use isolation mode on the audio panel to silence onboard chatter. One hundred percent concentration is required during these phases.

When taxiing, avoid multitasking, such as programming avionics or running checklists. Airlines have learned the hard way, with resulting runway incursions and occasional minor collisions. In general aviation, more than a few pilots have wandered off the taxiway and into a ditch or a runway light.

Taking a bite out of a barrier, Author

The taxiway closure marker shown here has some conspicuous chew marks from when a pilot turned right instead of left. He stopped when the barrier prevented him from taxiing any further. According to manufacturer guidance, this incident would have required an engine teardown. Instead, he departed into IMC. Let's hope a sudden engine failure doesn't occur years down the line—possibly with another owner.

The same caution applies to the traffic cones that FBOs often place to mark parked aircraft. Move them during the preflight inspection.

Checklists – "Read and Do" vs. Flows?

Pick one and use it. Some pilots follow checklists religiously as a "Do List." When new to an aircraft or flying infrequently, this may be the best option. Another approach is to use a checklist as a follow-up after completing all or certain items as a group. Establish a logical flow, moving across the cockpit and accomplishing items in sequence. Air carrier and corporate jet crews often operate this way, especially with shorter lists—one crewmember completes tasks, and the other verifies. If flying with another pilot, this is a good way to manage tasks efficiently.

Single pilots should "do" first and then follow up with the checklist when time allows. If interrupted before completing the list—using either method—start over from the beginning. Gear-up landings happen because of interrupted checklists and distractions. Leaving your hand on the gear switch until the green lights are smiling back at you on landing, or have extinguished after takeoff, ensures the hardware complied.

The manufacturer's checklist is law, but legal departments often complicate things. One light aircraft requires pilots to check the fuel pump four times before takeoff—twice would be sufficient. When checklists become cumbersome, they don't get used as often.

Consider making your own version that "flows" well. A few additional items might be included when avionics power is turned on, such as setting the fuel totalizer, verifying waypoints, confirming the nav source mode, and checking the VFR transponder code.

A note on nav source for the FMS: If the previous approach was via VLOC and the next flight plan is via GPS, some avionics must be manually switched back to GPS mode. Otherwise, you and ATC will have "words." Worse, navigating into unfriendly terrain could mean an uncomfortably short flight. It's better to switch the mode back to GPS after landing but always verify before takeoff.

There are some killer items that must be verified. These mantras don't replace using a checklist—they're just a quick verification:

- Before takeoff: Immediate troublemakers are controls (free), fuel (proper tank and mixture, if leaned before takeoff), flaps, and trim. Mess up any of these, and you may end up with a dirt sandwich.

- Brief that if there's an engine failure right after takeoff: pitch (attitude), switch (tanks), and pump (fuel pump) to prepare for engine nonperformance. The exact requirements will vary by aircraft.

One final point: Always take off using the same tank from which you did the run-up. Switching tanks just before takeoff almost guarantees that a non-feeding tank will provide just enough fuel to get airborne—before quitting.

Mind the Humble Gust or Control Lock

G-IV controls locked, NTSB

Gust locks are highly effective at keeping controls immobile, but too many crashes have occurred because pilots forgot this most basic item. Even highly experienced pilots have made this mistake.

The remains of a Gulfstream IV serve as a stark reminder. The aircraft had an interlock designed to prevent this issue by limiting thrust lever control, but it failed. Long-term flight data recordings revealed that the crew failed to check controls as "free and correct" over 90 percent of the time. It had always worked before—until it didn't.

About once a year, the NTSB picks somebody out of the weeds who used a rusty nail or bolt in place of the factory-approved (and sometimes absurdly priced) gust lock. No matter—if it's lost or damaged, get a suitable replacement that blocks the start switches.

Another little ditty that's helpful: Lights, Camera, Action. "Lights" means turning on the landing light and strobes (it's better not to use strobes when taxiing, except when crossing a runway, to avoid blinding other pilots). "Camera" refers to setting the transponder with the proper code. "Action" reminds us to look both ways for traffic, even after the tower has given takeoff clearance, just to verify that the way is, in fact, clear. This is especially important at non-towered airports. If the previous approach was via VLOC and the next flight plan is via GPS, some avionics must be manually switched back to GPS mode prior to the next takeoff.

The GUMPPPS (I've added a few P's) before the bump has served pilots for decades as a quick check that the aircraft is ready to land. Piper places the before-landing checklist on the panel for some of their aircraft, making it more likely to be used. Arrival into the pattern is one of the busiest times, so anything that ensures everything gets done with minimal distraction is helpful.

- **Gas** – Set to the proper tank. This should be done when beginning the descent to avoid surprises in the pattern.

- **Undercarriage** – A British term that reminds us the wheels should be down. It takes much less power to taxi. Missing this one is both noisy and expensive to fix. Military towers always add: "Check wheels down – Cleared to land." When flying VFR, put the gear down on downwind abeam the touchdown point, although it may serve well as a speed brake on the 45-degree entry to downwind, or even earlier if airspeed/altitude is high. On instrument flights, lower the gear at the Final Approach Fix (FAF) or glide slope intercept unless ATC has asked to keep the speed up, but never later than the FAF. Check again on short final. Some EFBs and newer avionics will announce "500 feet" on final—a great reminder to recheck gear position. At a towered airport, also verify to yourself that you have clearance to land. If uncertain ask the tower.

- **Mixture** – Set appropriately, especially if a go-around may be necessary. However, ramming it to full rich may cause the engine some indigestion on a routine landing or at high-density altitudes.

- **Prop** – In piston aircraft, move it to full forward on short final and reduce power. Avoid jamming the prop control home on downwind; it just creates lots of noise as the blades flatten out and irritates the neighbors. (Yes, I know the airport was there first, but be considerate and follow what the checklist says.) Wait until the prop is out of the governing range before doing this—usually on final at reduced power.

- **Pump** – Fuel pump ON (if appropriate for your aircraft). Some CFIs will say to do this earlier.

- **Pressurization** – Cabin depressurized before landing (for pressurized aircraft).

- **Path** – Let's add one more critical item. Is the approach and landing path clear? With all that GUMPing going on, it's easy to stay inside and develop tunnel vision toward the approach end of the runway. Keep your eyes outside and really look around—side to side and down the runway. Maybe someone's taking off in the opposite direction, especially on light-wind days. An aircraft on a collision course, or one you're overtaking, will have no apparent movement and will be very difficult or impossible to spot until a paint swap is almost inevitable.

- **Safety belts and shoulder harnesses** – Ensure everyone is properly belted in.

"Haste Makes Mess"

Timing is everything, and one thing that invariably leads to trouble is rushing. NTSB accident reports are full of haste-making errors. Hurrying is almost always self-inflicted—during preflight, loading, obtaining clearance, programming the FMS, and communicating.

Too many controllers and pilots attempt to communicate way too fast. A misheard or misunderstood clearance creates difficulty or worse. It saves no time to have to repeat it. New York Approach, for example, has a few controllers who have pacing down to an art—fast enough to keep traffic moving but slow enough that "Say again" is seldom heard. Ditto for pilots.

Both sides need to enunciate. Some controllers are careless in pronunciation. You'll also hear too many corporate pilots mumble responses, sounding totally bored with the whole business. Airlines seem to do a better job of teaching their pilots that lives, jobs, and efficiency depend on saying it right. Another key point is to chunk information so the other side can absorb the details. Two to three items—such as location, altitude, and maybe aircraft type—make for a perfect transmission. More than that is too many eggs in one communication basket.

Get Ready

You've heard this many times. On landing, start preparations before leaving cruise flight, especially for instrument approaches. Set up the FMS, program the frequencies, and study the charts. With Flight Information System Broadcast (FIS-B) weather, it's possible to get an advanced look at the runway in use. It's all obvious, simple, and frequently ignored. When flying with great pilots, observe the orderly progression.

This leaves mental margin for when ATC, the weather, or other traffic creates a problem. Feeling overloaded? Ask for a hold, a vector, or execute a 360 or two (with ATC approval, if you're in contact) to stuff the hurry-up genie back into the bottle. Slow down, but advise ATC if it's more than a 5-knot reduction. Just because the aircraft can go fast doesn't mean you should. If not proficient at higher-speed approaches, get some practice. Need time to program the avionics that you should have done sooner? Ask ATC for a delaying vector before getting totally balled up—or if they change something late in the process. If on a proficiency flight with multiple approaches to nearby airports, advise ATC that you'll need some programming (delay) time to get set up. Especially for single pilots, they'll understand. Practicing FMS programming on a simulator or computer helps.

If the airport does not have a run-up pad, perform critical pre-takeoff checks in a leisurely, out-of-the-way place. You shouldn't be rushed, but don't hold up other flights either. CFIs and flight schools, take note: checklists are often poorly written. Too many pilots, both new and experienced, will sit on the taxiway in the number-one slot, blocking everyone and wasting fuel, while doing tasks that could have been completed earlier—either before or right after engine start.

Getting the IFR clearance at a non-towered airport before leaving the ramp or on the run-up pad takes the stress off and is considerate to others. As mentioned, do not run checklists or copy clearances while taxiing as a single pilot.

The Post and Preflight –
While on the subject of haste, I consider a post-flight inspection essential. After landing, especially back at home base, much of the time pressure is relieved. On the outbound leg, we're usually meeting someone or heading somewhere. But once back at the hangar or tie-down, it's the perfect time to clean off the bugs and check the gear and

Better to find this a few days before flying

tires—especially after a landing that was less than silky smooth.

On engine shutdown at low power, just before pulling the mixture, a quick magneto grounding check is a great idea to ensure all the wiring is still doing its job. It does not need to be done after every flight but just periodically, in my opinion. Otherwise, you are just wearing out the ignition switch.

Oil and hydraulic leaks tend to show up after flying. If time and driving distance permit, a pre-preflight visit to the aircraft a day or two before a flight is a great opportunity to check in with the aircraft. Look again for leaks, check the oil and fuel levels, inspect under the hood, and assess tire condition and pressure—all the usual stuff. Airlines and corporate flight departments conduct their serious inspections the night before a flight rather than 10 minutes before departure.

> ***"If it's gonna happen, it's gonna happen out there."***
>
> *Capt. Ron (1992), starring Kurt Russell*

For rental or flying club pilots, the only good approach is to be meticulous and thorough during preflight. The previous pilot may have left you a mess. Takeoff or the first 10 minutes of flight is not when you want to discover that the intercom is misconfobulated (technical term) or that a switch is in an unexpected position.

Pay close attention to the maintenance of the club or rental fleet. You may not own the aircraft, but when you fly it, treat it as if you did. Report all squawks and observe whether they are fixed promptly. Most of the time, it won't matter—but in the event of a crash investigation, both the FAA and NTSB will closely examine the aircraft's condition, what the pilot and owner knew, and when they knew it.

Pattern Police – There are plenty of reckless pilot antics at non-towered airports. The "Me first, I'm important" types have infested highways, restaurants, and the skies. The FAA provides guidelines—not rigid laws— for pilots to follow. With so many variables at play, common sense and a little tolerance will enhance traffic flow and safety more effectively than strict, prescriptive rules.

FAR 91.126(b) specifies left turns in the pattern unless designated otherwise. As PIC, you have considerable latitude, but safety—not convenience—is the defining criterion. A straight-in approach or crosswind entry might make sense if it doesn't interfere with traffic already established in the pattern.

Pattern saturation often occurs on VFR weekends—the time when most midair collisions take place. Lecturing over CTAF only adds to frequency congestion and doesn't solve the problem after the fact. However, if you see a collision is imminent, broadcast a message that cannot be misinterpreted. See "Collision at Quincy"—Previous chapter—where a well-intentioned but improper radio transmission was a contributing factor.)That one transmission could make all the difference.

Taking position on the runway and holding at a non-towered airport is never a recommended practice. At some busy non-towered airports, a pilot in a hurry to stake a claim to the pavement—just as a previous aircraft is rolling out—might transmit, "King Fisher 85X taking the active runway 27, depart when clear." This is supposed to prevent an overtaking collision or redefine the right-of-way rules for landing aircraft. It doesn't.

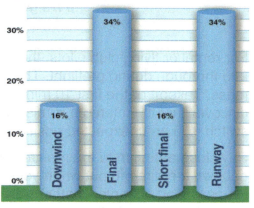

Midair collision locations, FAA

If pattern traffic is so tightly spaced that departures cannot leave, ask someone to extend downwind slightly to allow a departure or two.

Another misguided call is from an inbound pilot asking, "Any traffic in the pattern, please advise." If another aircraft is Nordo (no radio) or on a different frequency, they won't respond.

Let faster aircraft play through if they're close—extend the downwind, loiter outside the pattern, or delay takeoff until they land. Announce your intentions.

The absolute worst place to be is in front of a faster aircraft, as any fighter pilot will affirm. Do you really want to depend on them to see you? I'll cheerfully extend my downwind to avoid that highly vulnerable situation (see *Check Six, Bogey on Your Tail*, This chapter). Announce what you're doing, and odds are the other pilot will thank you for letting them play through.

Classic? Get a handheld radio and a headset, AOPA

In the midair collision histogram above, note that fully 50 percent of traffic pattern collisions occur on final approach or on the runway. That's tremendous exposure for a very short time—stay alert.

Classic aircraft with no electrical system? Please—invest in a handheld radio and a headset because the 1960s aren't coming back. Eyes outside, of course, but using two senses will greatly improve your ability to see and avoid. It will also help other pilots find you, and chances are, you're going to be a lot slower than what's behind you.

The rules allow no-radio operation at busy non-towered airports, but just because you can doesn't mean you should. ADS-B is also a great addition. It can provide an early indication of what runway is in use at non-towered airports.

A fatal midair collision occurred in Rock Springs, Wyoming—hardly a hotspot of activity—and neither pilot likely expected anyone to be near the airport. I like to remember that it's the ones I don't see or hear that will get me, and all it takes is once.

Hangar Story

Our airport manager had a strict rule: no touch-and-goes with more than four or five aircraft in the pattern or on weekends. This allowed transients to get in and out while still enabling student landing practice. A pilot who broke the rules could expect an immediate tirade on the CTAF, warning that they'd be "sent on down the road" if they didn't abide. It was a privately owned, public-use facility with "traffic control" at its most basic—but we never had a collision, even with very heavy traffic.

Sadly, after a collision, the airport sages will opine that it was "only a matter of time." Why not do something about it before lives are lost or aircraft are damaged? The CTAF is not the place for etiquette lectures. When someone gives you a break—or if you're the transgressor—pay it forward when there's an opportunity.

Persistent problem? Organize a local collision avoidance/pattern courtesy meeting with the flight school, airport manager, soaring club, helicopter pilots, and local flyers to discuss the subject respectfully. The FAA and ATC may also be happy to participate.

> **"Everything I really needed to know about traffic patterns, I learned as a student pilot.***"*

> *Corrupted from Robert Fulghum, "Everything I Really Needed to Know, I Learned in Kindergarten."*

Just as we forget the profound life lessons of childhood, we also forget the lessons of courtesy and basic airmanship. They don't change.

Key Points:

- Stabilized approaches, as appropriate for the situation, every time = "Goldilocks."

- Performance numbers – 50/50 – give yourself some margin, especially when conditions are wet, icy, windy, heavy, or affected by density altitude.

- Strong crosswind? More runway will be needed for landing. Beware the tailwind on the base leg.

- Practice go-arounds periodically.

- Don't foul up with flaps—either on takeoff or landing.

- Think like a controller and let faster traffic play through.

- "Line up and wait" does not apply at non-towered airports.

- Sterile cockpit during preflight, takeoff, landing, and taxiing—do not multitask.

- Checklists and flows—use them.

- Take your time—mistakes happen when rushed. Consider a thorough post-flight inspection or visit the aircraft a day early to review at leisure. Insert For renters and flying club pilots –you can't so just be extra fussy on the preflight.

- Be courteous and smart in the pattern. Leave policing to the police. Use ADS-B and CTAF before arrival to determine the active runway, but be prepared to adjust upon arrival.

"It's hard to stop a plane."

(Southwest 1248 slides off the end—computer issues) A Boeing 737 slides off the end of the runway after the computer fails to calculate properly.

A Southwest Airlines (SWA) Boeing 737-700 slid off the end of Runway 31C after landing at Chicago Midway International Airport (MDW) on a snowy evening. Tragically, a child's life was lost when his parents' car happened to be in the right place at the wrong time. The Boeing rolled through the blast fence and airport perimeter fences before stopping on the roadway—with the car underneath.

How could a diligent, professional crew—flying for one of the world's most successful airlines in one of the most reliable aircraft—find themselves sliding off the end of a runway they had landed on many times before, at an airport that has seen hundreds of thousands of landings without mishap?

Sometimes, situations that seem simple aren't. Arcane information involving braking action reports, confusing flight operation policies, inconsistent computer programming, and airport environment all played a part in this event. There's plenty to ponder, regardless of what size aircraft you fly.

Boeing 737 off the end, NTSB

Weather

Southwest Flight 1248 departed Baltimore/Washington International Thurgood Marshall Airport (BWI) on December 8, 2005, at around 5:58 p.m. Eastern Standard Time, about two hours late due to deteriorating weather in Chicago. The six to nine inches of snow predicted for Chicago began falling at Midway around 2 p.m. The forecasted surface winds of "calm" were revised to "090 at 11 knots," and runway braking action changed from "wet-good" to "wet-fair," which was relayed to the flight en route. While the winds were accurate, the braking action report was optimistic.

At 6:53 p.m., about 20 minutes before the accident, the weather was reported as wind 100 degrees at 11 knots, visibility one-half statute mile in moderate snow and freezing fog. The ceiling was broken at 400 feet and overcast at 1,400 feet.

Runway Conditions

The 6,522-foot-long runway was plowed 30 minutes before the accident, with an average runway friction reading of 0.67 (out of a possible 1.00, which indicates perfect conditions). After the accident, a second test revealed a friction reading of 0.40. The Aeronautical Information Manual stated, at the time, that when values drop below 0.40" aircraft braking action begins to deteriorate, and directional control becomes less responsive." Braking action advisories were in effect, requiring pilots and controllers to provide each other with the latest updates on aircraft stopping performance.

Why did Midway continue to operate on Runway 31 despite a significant tailwind? Changing runways might have interfered with operations at Chicago O'Hare International, located 13 nautical miles to the northwest, so air traffic considerations apparently factored into the decision to continue using Runway 31.

The Crew

The captain, age 59, was a retired Air Force pilot with 15,000 total flight hours, including 4,500 hours in the Boeing 737. He had flown for Southwest Airlines since August 1995 and had an exemplary flight record. He told investigators he had slept well and was not fatigued. He also noted in post-accident interviews that the weather was "the worst" he had experienced, but he had "encountered similar conditions about a dozen times" in his tenure at Southwest and "expected to be able to land safely."

The first officer, age 34, had flown as a regional airline pilot before joining Southwest in 2003. He had about 8,500 total flight hours, including 2,000 in the 737. He claimed extensive winter weather experience as a regional airline captain.

The Flight

It was the captain's leg to fly, and the crew discussed braking action reports, required landing distances, the use of autobrakes, and the possibility of diverting to an alternate airport. A confounding factor was the mixed braking action reports—braking was generally reported as "good to fair" on the first half of the runway and "poor" on the second half. The tailwind component was eight knots, and the crew discussed possible diversions.

Everything was routine, despite some holding delays for all flights entering Midway as the runway was plowed. About 30 minutes before the accident, the latest report indicated: "Runway 31—trace to 1/16th of an inch of wet snow over 90 percent of the surface, 10 percent clear and wet."

The landing itself was unremarkable. The flight data recorder (FDR) verified that the main gear touched down 1,250 feet from the runway threshold (well within the recommended touchdown zone) at 124 knots airspeed—right on target—with a ground speed of 131 knots. Ground spoilers and autobrakes deployed immediately as designed.

The captain had difficulty applying reverse thrust, which was essential for this challenging landing. As soon as the first officer noticed that the reversers were not engaged, he deployed them. According to the NTSB report, "The first indication of thrust reverser deployment was recorded at 15 seconds after touchdown, with full deployment at 18 seconds." The four airliners that landed in the 20 minutes preceding the accident deployed reverse thrust within four to six seconds of touchdown, according to their FDRs.

The autobrakes on Southwest Flight 1248 deactivated 12 seconds after touchdown as the pilot manually applied maximum braking and maximum reverse thrust. However, the aircraft departed the runway overrun at 53 knots and came to a stop 500 feet beyond the runway. It slid through two fences and ended up on a public road. The passengers and crew evacuated with only minor injuries, but tragically, a child in a car beyond the airport boundary was killed.

The Computer

Southwest equipped all of its aircraft with an onboard performance laptop computer (OPC) to assist crews with a variety of calculations, including landing performance and stopping margins. Pilots entered pertinent data such as the landing runway, prevailing wind conditions, aircraft weight, ambient temperature, and reported braking conditions. Up to 10 knots of tailwind was allowed with good or fair braking action, but only five knots was permissible under poor conditions.

The OPC computed a stopping margin of 560 feet based on fair braking action and just 40 feet based on poor braking action. A 40-foot margin is effectively no margin at all, but company guidance noted that any positive number was considered acceptable. Additionally, the computer did not clearly indicate that five knots was the maximum allowable tailwind—it accepted an eight-knot entry but did not compute the distances based on the actual eight-knot tailwind. Had the OPC used the actual tailwind entry, the calculation would have shown a 260-foot overrun.

There was another flaw in the computer programming. For the 737-700, stopping distances were based on the use of reverse thrust, but for the 737-300 and -500, they were not. As a result, crews might mistakenly assume that reverse thrust would shorten stopping distances by several hundred feet. This important difference was not clearly explained in training. The accident crew assumed, as did many other Southwest crews interviewed after the accident, that the OPC computed stopping distances the same way for all models. Pilots were expected to remember which formula applied to which aircraft, even though they might fly all three models in a single day.

Procedural Confusion

Southwest required crews to use the most adverse braking action report for planning purposes and not rely on a "blended" report. Thus, any part of the runway reported as poor should have overridden a good or fair report

119

from other sections. Post-accident interviews revealed that many Southwest crews did not understand this policy. With an eight-knot tailwind and any portion of the runway reported as poor, Southwest Flight 1248—and the flights before it—should have diverted. A five-knot tailwind was the maximum allowed with poor braking conditions.

The tower controller did not provide all braking action reports, as required by FAA procedure. A Gulfstream III that landed shortly before reported "poor" conditions, but the tower failed to pass that information along. While Southwest pilots were instructed to only accept braking action reports from other commercial aircraft, this omission may have deprived the crew of crucial information about deteriorating conditions.

Southwest did not allow the use of autobrakes at the time but was in the process of changing procedures to standardize their use across the fleet. Crews interviewed after the accident expressed confusion and uncertainty about the policy. The change was not scheduled to take effect until four days after the accident, but the crew had read about the procedural changes before the flight and was unsure whether to use autobrakes.

Midway Airport – At the time of the accident, Midway was not optimally configured for large airline jet operations. The ILS Runway 31C was the only approach available due to visibility requirements. The runway had a displaced threshold and a usable landing distance of 5,826 feet.

As early as September 2000, the FAA had determined that the Runway Safety Areas (RSA) were not in compliance with an airport design advisory circular, which considers a 1,000-foot overrun as standard. Runway 31C's RSA was only 81 feet.

In 2003, the FAA again requested a reassessment, but the city of Chicago stated that there were "no alternatives." Their contentions were that the runway could not be shortened while still allowing air carrier operations, and that extending the RSA beyond the existing airport perimeter would have a major impact on the surrounding community.

The use of an engineered materials arresting system (EMAS)—a soft surface designed to stop an overrunning aircraft by allowing it to sink in with minimal damage—was not feasible, as it would either encroach on the runway or extend beyond the airport perimeter.

After-accident simulations showed that a modified EMAS would have stopped SWA Flight 1248 on airport property. It was installed on several runways shortly afterward.

Braking Action Reports

The NTSB had "long been concerned about runway surface condition assessments." There is considerable variability between pilot braking-action reports, airport contaminant type and depth observations, and ground surface vehicle friction measurements. There may be little or no correlation between the three, and there is no standard for how a particular aircraft will perform under given conditions.

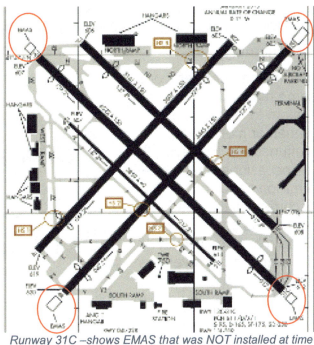

Runway 31C —shows EMAS that was NOT installed at time of crash, FAA

Air carriers are required to perform landing distance calculations before departure to determine if the aircraft can land safely. However, the NTSB noted, "The assessments do not attempt to comprehensively account for actual conditions." There are many variables to consider, and much subjectivity is involved. While the FAA requires operators to take the aircraft manufacturers' maximum performance data and add margins of 67 percent for dry runways and 92 percent for contaminated runways, these margins may not be sufficient for all conditions.

NTSB Probable Cause

"…was the pilots' failure to use available reverse thrust in a timely manner to safely slow or stop the airplane after landing. This failure occurred because the pilots' first experience and lack of familiarity with the airplane's autobrake system distracted them from thrust reverser usage during the challenging landing."

Numerous contributing factors included SWA's failure to provide clear guidance on landing distance calculations, programming of the OPC, implementation of the new autobrake procedure without appropriate familiarization, failure to include appropriate margins of safety to account for operational uncertainties, the pilots' failure to divert to an alternate, and the lack of an EMAS system given the restricted runway environment.

Commentary

In all air carrier operations, some margin is built into the system to allow for inevitable human or mechanical malfunctions. However, there will occasionally be a one-in-a-million series of occurrences that lead to an event. The odds of the captain having difficulty deploying thrust reversers are small, but it was a critical—if understandable—lapse.

EMAS at work, NTSB

Multiple crews misunderstood company procedures regarding mandatory diversion to alternates. The nonstandard programming of the OPC highlights the importance of standardization.

Everyone wants to cooperate. It is admittedly very difficult to go against the prevailing flow, and we've all gone along with the crowd. How do you explain to passengers that you diverted when all the other flights were getting in? It's so easy to second-guess.

For light aircraft, takeoff and landing performance estimates on contaminated runways are largely untested and usually unpublished. This seemingly simple accident demonstrates the many dynamics of approach, touchdown, and rollout, proving that physics never takes a day off.

Sideways Slide – Continental 1404 Gets Blown Off the Runway in Denver

Big wind and big aircraft at the limits.

> **"One should expect that the expected can be prevented, but the unexpected should have been prevented."**
>
> *Norman Ralph Augustine, Chair of the National Academy of Engineering, CEO of Lockheed, Author*

Boeing 737 exits to the left, NTSB

Crosswinds affect all aircraft, big and small. This well-documented event illustrates that previous experience and training don't always serve us well—nature is both capricious and powerful. It's also a reminder to those of us who comment on accidents, which includes most pilots, that hindsight is perfect.

The Takeoff

On December 20, 2008, Continental Airlines Flight 1404, a Boeing 737-500, was scheduled to fly from Denver to Houston around 6 p.m. Mountain Standard Time. Pushback and taxi were routine. The weather was VFR with strong westerly winds, although the current ATIS was only reporting 11 knots of crosswind.

At 6:14:27 p.m., the tower controller cleared the Boeing to line up and wait on Runway 34R following a prior departure. At 6:16:16 p.m., one of the pilots asked about the windsAt 6:17:26 p.m., the tower controller advised that the wind was from 270° at 27 knots, assigned a departure heading of 020°, and cleared the flight for takeoff.

The captain remarked, "Looks like… some wind out there." The first officer (FO) replied, "Yeah," and the captain added, "Oh yeah, look at those clouds moving."

As the takeoff roll began, the captain said, "Alright… left crosswind, twenty ah seven knots… alright, look for ninety point nine." This referred to the engine power setting. The captain focused outside to maintain the runway centerline. At 6:18:04 p.m., the FO advised that the power was set and shifted attention to airspeed to make the standard callouts.

Walking the Rudder

As the airplane accelerated, the flight data recorder (FDR) indicated increasing right rudder while the elevator and ailerons remained in their neutral position.

At 6:18:07 p.m., as the airplane accelerated through 55 knots, the nose began to move left, and the FDR recorded a large right rudder input that peaked at 88 percent of available forward travel.

This was relaxed to about 15 percent by 6:18:09 p.m. as the left aileron deflected. The nose moved to the right; however, at 6:18:10 p.m., accelerating through 85 knots, it began moving left again at about 1° per second. This continued for two seconds as the captain countered with strong right rudder input of 72 percent displacement at 6:18:11 p.m. It then

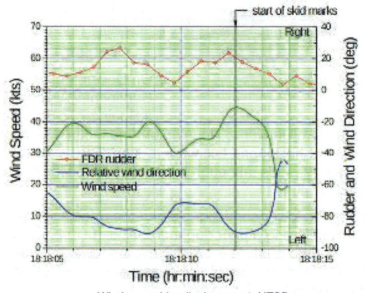

Wind vs. rudder displacement, NTSB

decreased to 33 percent at 6:18:13 p.m.

Airspeed was now above 90 knots.

NTSB: "During this second large right rudder pedal movement (at 6:18:12), the airplane's left-turning motion slowed for about one second, and then the nose began moving rapidly to the left again. A fraction of a second later (at 6:18:13.25), the right rudder pedal was abruptly relaxed, reaching its neutral position about one second later."

At 6:18:13.5, the CVR recorded one of the pilots exclaiming, "Jesus." According to the captain, after the airplane left the runway, he subsequently rejected the takeoff, stating they were "along for the ride."

Continental 1404 departed the left side of Runway 34R, reaching a maximum speed of about 120 knots. It crossed a taxiway, a service road, and some uneven terrain before coming to a stop. A post-crash fire ensued. The captain and five of the 110 passengers were seriously injured. The first officer, two cabin crewmembers, and 38 passengers sustained minor injuries, while one cabin crewmember and 67 passengers (including three lap-held children) were uninjured.

The Crew

The 50-year-old captain was hired by Continental in 1997 and previously served as a first officer on the DC-9 and Boeing 737/757/767 aircraft before becoming a 737 captain about 14 months before the accident. He completed a line check in April 2008 and recurrent training/proficiency checks in October 2008. Additionally, he completed the Continuing Qualification Syllabus, which included at least one takeoff and one landing in a 35-knot direct crosswind.

The captain had 4,500 hours of Navy flight experience before joining the airline. At the time of the accident, he had accumulated 13,100 total flight hours, including about 6,300 hours in the 737. He had flown approximately 915 hours in the past 12 months, 81 hours in the past month, and four hours in the past 24 hours. His flying skills, training records, and peer reviews were all rated as excellent.

The 34-year-old first officer was hired by Continental in March 2007 and was type-rated in the de Havilland DHC-8 and Boeing 737. He had previously flown for Horizon Airlines and had approximately 8,000 total flight hours, including about 1,500 hours in the Boeing 737. His training records and peer reviews were rated "above average."

The Aircraft

The Boeing 737 was delivered in 1994 and had accumulated just over 40,000 flight hours and about 21,000 cycles. No discrepancies were noted regarding the aircraft or its loading. Boeing guidance listed a maximum crosswind limit of 40 knots. Winglets installed a month before the accident were tested to a crosswind maximum of 33 knots—above the tower-reported wind of 27 knots but within the airline's published crosswind guideline of 33 knots for a clear, dry runway.

The Weather

The National Weather Service (NWS) surface analysis charts showed a low-pressure system near the Colorado–New Mexico border, with a stationary front extending north to south through those states. Westerly winds across the Rocky Mountains suggested downslope winds with moderately strong wind gusts and mountain wave activity.

To assess whether mountain waves played a role in the gusty surface wind conditions, the National Center for Atmospheric Research (NCAR) simulated conditions around the time of the accident using a high-resolution numerical model. The model indicated a well-defined wave that increased significantly shortly before the accident, with winds of 40 to 68 knots at the airport between 6:08 and 6:18 p.m.

This resulted in strong, localized, intermittent gusts at the airport's surface. NCAR's model depicted generally stronger westerly flow to the north of the airport, with large regions of relatively lighter winds over the central and southern portions. Embedded within the overall flow structure were multiple gusts moving west to east across the airport. A gust estimated at 45 knots crossed the accident site between 6:14 p.m. and 6:16 p.m. This gust was apparently captured by one of the airport's wind sensors but was not reported to the crew.

NTSB Findings

The NTSB reported:

"At the time of the accident, of the nine possible runway configurations available to ATC, the 'Landing North/West' configuration was in use. In this configuration, traffic was landing on Runways 35L, 35R, 34R, and 26, while departures were occurring from Runways 34L, 34R, and 25. The local controller, responsible for departing traffic on all three runways, cleared seven airplanes for takeoff in about nine minutes before the accident. (Two additional airplanes were holding in position awaiting takeoff clearance when the accident occurred.) The pilots of the other departing airplanes did not report any crosswind-related issues or difficulties."

Using data from the FDR, the NTSB estimated that winds during the accident sequence:

"...varied between 30 and 45 knots from the west, resulting in an almost direct crosswind for Runway 34R and a crosswind component that varied from 29 to 45 knots. A peak gust of 45 knots occurred at 6:18:12 p.m., about the same time the FDR recorded the right rudder pedal beginning to move aft from a position about 72 percent of its available forward travel, reaching a near-neutral position at 6:18:13.75 p.m. Although the pilot briefly made a small right rudder pedal input at 6:18:14.25 p.m., the FDR did not record any more substantial right rudder pedal inputs as the airplane continued to veer left."

The NTSB determined that the rudder was capable of producing enough aerodynamic force to offset the weathervaning tendency created by the wind gusts. This suggests that applying full rudder for as long as necessary could have counteracted the gust.

Denver International Airport was well-equipped with low-level wind shear monitoring equipment. However, at the time, FAA guidance did not require controllers to report the most adverse wind conditions—only the wind at the departure end of the runway. One of the NTSB's findings stated:

"If the accident pilots had received the most adverse available wind information (displayed on the Denver International Airport control tower's ribbon display terminal, indicating a 35-knot crosswind with 40-knot gusts), the captain would likely have delayed departure or requested a different runway, as the resultant crosswind component exceeded Continental's 33-knot guideline."

Following the accident, Continental analyzed its operational database for crosswind components encountered during takeoff. Reviewing data from 940,000 departures over eight years, the NTSB found that only 250 departures faced crosswinds of 25 knots or greater, and only 62 faced crosswinds of 30 knots or greater. That's roughly one event in every 15,000 takeoffs. Such strong crosswinds are truly outlier events.

Post-accident testing also revealed that Continental's B737 simulators were not accurately replicating gusty conditions. They were only programmed to produce steady-state winds—below 50 feet AGL—rather than realistic gust scenarios. Aircraft behave quite differently in gusty conditions compared to steady winds.

NTSB Probable Cause

The NTSB determined that the probable cause of the accident was:

"...the captain's cessation of right rudder input, which was needed to maintain directional control of the airplane, about four seconds before the excursion, when the airplane encountered a strong and gusty crosswind that exceeded the captain's training and experience."

Contributing factors included:

1. *An air traffic control system that did not require or facilitate the dissemination of key, available wind information to air traffic controllers and pilots.*

2. *Inadequate crosswind training in the airline industry due to deficient simulator wind gust modeling.*

Commentary

I empathize with the captain. Critical wind information was not provided, and prior training could not replicate the conditions encountered. Similar challenges exist in general aviation—actual winds in the runway zone may be unknown. Effective crosswind training in GA is often lacking, as wind conditions rarely cooperate during training flights.

Most GA crashes occur well within the aircraft's capability, making improvements in training a great opportunity. However, weather can sometimes deliver surprises, making both luck and skill critical. Anticipation is essential, as Mr. Augustine's quote suggests.

In light aircraft, the Pilot Operating Handbook (POH) provides demonstrated crosswind capability—essentially the maximum encountered during testing. Actual aerodynamic limits are higher but unknown. Testing those limits should be done cautiously, as mistakes can be costly.

On gusty days, extra airspeed can provide additional control—within reason. Pilots should remain flexible with runway selection and be prepared to divert, delay, or cancel when necessary. In hindsight, it's easy to say, "I would never have done that."

Nobody's Flying Today! - (Mooney M20E sheared on Takeoff)

A low time Mooney pilot tangles with big winds in New Mexico on takeoff with a nearly full load and out of balance.

> **"Experience is a hard teacher because she gives the test first and the lesson afterwards."**
>
> *Vernon Law, major league baseball player*

Exemplar Mooney M20E

A few years back, a Mooney pilot decided to challenge the strong winds in New Mexico. The results were catastrophic and serve as a reminder that local knowledge and published warnings about airport conditions should always be heeded. It's also a case of overconfidence and a lack of understanding of the realities of weather, despite limited experience.

The Environment – On March 3, 2013, in the early afternoon, the pilot and three passengers prepared to depart Angel Fire, New Mexico, airport (KAXX) for their return to the San Antonio, Texas, area, where they lived. The weather that day was not conducive to flying light aircraft, and the airport attendant questioned the pilot about his intention to fly in such conditions. According to the NTSB, "The pilot indicated that he planned to fly and that the winds would not be a problem."

The automated weather system (AWOS) reported winds out of the west from 250 degrees at 33 knots, gusting to 47 knots, with visibility of 10 miles, a clear sky, a temperature of 47 degrees Fahrenheit (F), a dew point of 17 degrees F, and a barometric pressure of 29.93 inches of mercury. The density altitude was calculated at 9,549 feet. The north-south runway is 8,900 feet long, and the airport elevation is 8,380 feet. The terrain just west of the airport rises 2,000 feet above the airport elevation.

The Flight – The short-body Mooney M20E, loaded with four people, taxied out to Runway 17. According to the NTSB, "The current wind and altimeter were relayed to the pilot by the FBO employee, which were repeated by the pilot. Due to snow piles on the airfield, the FBO employee could not see the takeoff and next saw the airplane airborne with a significant crab angle into the wind, about 40 degrees right of the runway heading. The airplane rose and fell repeatedly as its wings rocked. Then, the airport employee saw the airplane's right wing rise rapidly. The airplane rolled left and descended inverted, with the airplane's nose pointed straight down."

A witness driving by the airport reported that the Mooney was struggling, reaching an altitude of only about 100 feet, where it hovered momentarily before the left wing dropped and the aircraft descended nose-first to the ground. There were no survivors.

The Pilot – The 33-year-old pilot held a private certificate for airplane single-engine land and a current medical certificate. His insurance application, filed about six months before the accident, reported 459 total flight hours, with 384 in type. His occupation was engineering, and he had worked at Mooney Aircraft and Boeing and was currently employed by SyberJet.

The pilot told the airport manager that he had flown the accident airplane for five years and that KAXX was the highest-elevation airport he had visited. His experience flying to other high-altitude airports was unknown. The evening before the accident, the pilot's cousin, who lived in the area, talked about airplane crashes in and around Angel Fire. The pilot responded that flying in wind did not bother him.

The Aircraft – The logbooks of the 1966 Mooney burned in the post-impact fire, but receipts showed that the last annual inspection had occurred three months earlier, in December 2012, with a total airframe time of 4,752 hours. The engine had 1,736.7 hours since its major overhaul. At the time of the accident, the tachometer read 4,785.8 hours.

An old copy of the airplane's weight and balance found in the wreckage was used to estimate the weight and center of gravity. The luggage was destroyed but estimated to weigh about 60 pounds. With approximately half a tank of fuel, the takeoff weight was calculated at 2,519 pounds, about 56 pounds below maximum. However, the center of gravity was computed slightly aft of the envelope.

It's best not to keep pilot or aircraft logbooks on board. This makes crash investigations easier, and in the event of a ramp check, the FAA will have fewer items to immediately "contemplate."

Former Mooney engineering test pilot Bob Kromer provided guidance regarding crosswind landings: "I think you will find a common consensus of test pilot opinion that most Mooneys can be operated in 90-degree crosswinds up to 15 knots with an acceptable level of pilot workload. Crosswinds of 15–20 knots can be handled but require a much higher level of pilot proficiency and skill in crosswind landing techniques. At 20 knots or above, you should consider finding another airport to land."

After impact, the airspeed indicator read 81 mph. The attitude indicator showed a left-wing-low, inverted attitude. The tachometer read 2,000 rpm. (It was not noted in the report what the manifold pressure reading was.) There were no signs of mechanical failure.

The Weather – KAXX is located in a basin surrounded by mountainous terrain. According to the NTSB, "Mountains to the west and northwest of the airport have peaks between 10,470 and 13,160 feet. A weather study was compiled for the accident site. An upper air sound for 1400 Mountain Standard Time (MST) depicted an unstable vertical environment, which would allow mixing of the wind on the lee side of the terrain. Winds as high as 55 knots could occasionally reach the surface.

Satellite imagery between 1300 and 1400 MST recorded a large amount of standing lenticular cloud near all of the mountainous terrain around the accident site. These clouds indicated the presence of a mountain wave environment. At 3:22 a.m. and 11:34 a.m. MST, the National Weather Service issued wind advisories for the accident area, warning of a west or southwest wind between 25 and 35 miles per hour (mph), with gusts up to 50 mph."

A weather model simulating conditions at the time of the accident predicted a "turbulent mountain wave environment, with low-level wind shear, updrafts and downdrafts, downslope winds, and an environment conducive for rotors." There was no record of the pilot receiving an official weather briefing, nor is it known what other sources he may have consulted. The airport manager noted that no other flights had arrived or departed that day.

NTSB Probable Cause

"...The pilot's loss of control while flying in a turbulent mountain-wave environment. Contributing to the accident was the pilot's overconfidence in his ability to safely pilot the airplane in gusting wind conditions and his lack of experience operating in mountainous areas."

Commentary

The accident occurred on a Sunday, so perhaps there was some compelling job-related pressure to return before the beginning of the workweek. When flying light aircraft, if there is an urgent need to be somewhere on a schedule, flexibility is essential, or alternative transportation options should be considered. It's wise to allow a few extra days to depart earlier or later. If that luxury isn't available, having a Plan B is critical.

Note terrain west of the airport, FAA

Having flown an M20E around much of the country, I have great appreciation for the aircraft's cruise speed and fuel efficiency. However, even at low-density altitudes, it is not a strong climber when fully loaded. Although the official service ceiling is 19,500 feet, it would take a long time to reach that altitude.

The accident report noted the tachometer reading at 2,000 rpm, but since the aircraft had a constant-speed propeller, the tach should have read about 2,700 rpm. Manifold pressure, however, would have been significantly lower than sea-level numbers due to the high-density altitude.

To see how your favorite aircraft will perform in thin air, try a takeoff (on a long runway) using only 2,200 rpm on a fixed-pitch prop or 20–22 inches of manifold pressure (MP). The amount of runway required and the lackluster climb performance will be eye-opening. Want even more excitement? Try it at close to gross weight.

A mountain flying course somewhere along the Front Range is an excellent investment before attempting a solo flight in that region.

Exploring the flight envelope beyond the allowable center of gravity (CG) is an extremely bad idea. The manufacturer has already tested those limits and determined that there be dragons that exist beyond them. It's a safe bet that the aircraft would be uncontrollable under such wind conditions. It might also be uncontrollable in far more benign conditions. Stalls would be ugly.

High winds in the mountains require a different perspective from pilots accustomed to flat terrain. Many may not fully appreciate the power of updrafts, downdrafts, wave action, shear, and rotors. The winds at Angel Fire were well beyond anything this low-time pilot had ever experienced, and they quickly overpowered the Mooney.

Twenty-five knots in the mountains is far more treacherous than the same wind speed in the midlands or coastal plains. Flying early or late in the day is often a much better strategy. Looking at historical METARs for KAXX on that March day, a 7 a.m. departure recorded winds from 120 degrees at 8 knots. By 6 p.m., winds were 200 degrees at 8 knots with a lower temperature, lower density altitude, and a far safer scenario.

Weather warnings had been posted, and the airport manager clearly communicated that it was a bad day to fly. The crosswind was nearly double what an experienced test pilot would tackle. At 500 hours of total flight time, a pilot has only begun to encounter the challenges aviation can present. Sadly, some wings get broken in the seasoning process.

Betting the Farm – (The Bonanza A36 Did Not Meet the Numbers Here)

A Bonanza pilot took it past the limit.

"Betting the farm" – To be so sure of an outcome that you're willing to risk everything.

(A traditional American proverb about putting the family farm at risk on a bet.)

If there were ever a crash that proves the importance of not taking performance charts too literally, this is the one. The pilot was experienced, the aircraft was in good condition, and according to the documentation, this takeoff should have worked—even though the aircraft was slightly overloaded. It didn't. The result was two fatalities.

This is extreme Monday morning quarterbacking, but it highlights that performance charts do not account for every variable. They simply can't. Early speculation on several websites was completely inaccurate, as the subsequent investigation would determine. If it's on the internet, it just might not be true.

The crash was exceptionally well documented by:

1. A TV film crew that happened to be on the scene, ironically, to cover an earlier minor crash.

2. The pilot, who survived.

3. Witnesses.

4. Onboard engine instrumentation.

This gave the NTSB a detailed look at what happened.

The Flight

On August 30, 2007, at 12:35 p.m. Pacific Daylight Time, a Beech A36 Bonanza pilot attempted a takeoff from Cameron Air Park, California (O61), with three passengers on board. The pilot stated that he had topped off the aircraft with fuel and mentally performed a weight and balance calculation. He had successfully flown from the airport several times in similar conditions and saw no particular problem.

The engine run-up was normal, although the takeoff roll was slightly longer than usual. Video footage shows the aircraft lifting off about two-thirds of the way down the runway. According to the NTSB:

"Once airborne, the airplane climbed to approximately 40 feet, and the wings began to wobble before settling back down toward the ground. The airplane then descended into the rising terrain off the end of the runway, slid a short distance, and abruptly flipped over."

The NTSB further reported:

"During initial climb, the airplane was accelerating. At some point, the airplane stopped climbing, the airspeed indicated 84 knots, and the rate of climb had dropped off. The pilot lowered the nose and felt a gust of wind from the left side, at which point the wings started to wobble. He cut the power just as the airplane was settling into the rising terrain and runway overrun."

The pilot and front-seat passenger were seriously injured. The two passengers in the aft-facing seats did not survive.

The aircraft impacted about 500 feet from the end of the runway at an elevation of 1,314 feet MSL. The total distance from the initial point of impact to the main wreckage was 365 feet. At the 246-foot mark, a large boulder was dislodged and moved 60 feet before coming to rest next to the wreckage.

It was likely an impossible place to make a successful forced landing. The video clearly shows that the aircraft impacted with significant speed, and the collision with an immovable object resulted in a lethal, sudden stop.

(More in Chapter 6, *Emergencies*.)

Cameron Airpark, approx crash location at red circle, Google Earth

The landing gear handle was down, the flap handle was up, and the throttle, mixture, and propeller levers were in the full-forward position. This conflicts with the pilot's statement that he cut the power, although, given the crash trauma, memory can easily be affected with no intent to deceive. It's also possible that the throttle moved during the crash sequence. The fuel selector was set to the left tank position.

The baggage and cargo recovered from the wreckage consisted of a portable ice maker, four pieces of individual luggage, a case of soda, and a medium-sized cooler. The items were collected and found to weigh 271 pounds. They became deadly projectiles inside the cabin.

The Airport and Weather

The airport sits in a geographical bowl with slightly rising terrain at both ends. The field elevation is 1,293 feet MSL, and the single runway length is 4,051 feet. The aircraft engine data monitoring system accurately recorded an outside air temperature of 95 to 97 degrees Fahrenheit.

The windsock appeared limp in the video, but trees were swaying in the background at the northwest end of the runway, and the aircraft was crabbed to the northwest after liftoff. The density altitude was computed at 4,125 feet, and Runway 31 has a 0.05 percent upslope. It is important to note that windsocks only show wind conditions at their immediate location.

The accident video showed the airplane entering a left crab immediately after takeoff, while both the airplane and its shadow appeared to remain centered over the runway. Comparing the runway direction and the airplane's nose alignment, the crab angle was approximately seven degrees. Using a ground speed calculated in the video of 80 knots, a seven-degree crab corresponds to a 10-knot crosswind component.

The Pilot

The 64-year-old pilot held a commercial pilot certificate issued in 1989, with ratings for single-engine land, single-engine sea, multiengine land, multiengine sea, helicopter, gyroplane, and airplane instruments. He also held a flight instructor certificate with ratings for single- and multiengine airplanes, instrument airplanes, rotorcraft helicopters, and gyroplanes.

The pilot reported a total flight time of 2,000 hours, with 1,700 hours as pilot-in-command and 1,000 hours as an instructor. He had flown 16 hours in the accident aircraft make and model within the last 30 days. His total time in the Bonanza was not reported, and his third-class medical certificate was current.

The Aircraft

The 1996 Bonanza A36 had an airframe time of 1,031.5 hours and was retrofitted with a turbo-normalized 300 HP engine in 2000. The engine had 679.1 hours since overhaul, and the annual inspection was current. The maximum approved takeoff weight had been increased to 4,000 pounds due to an approved modification.

A standard late-model A36 is typically rated for a maximum takeoff weight of 3,650 pounds. According to the Supplemental Type Certificate (STC) documents, the accident flight had the following performance adjustments to accommodate the additional weight:

- Increased Takeoff Distance: Up to 30%
- Decreased Rate of Climb: Up to 13%
- Increased Stall Speed: Up to 7%
- Increased Landing Distance: Up to 15%

- Increased Takeoff & Approach Speeds: +2 knots
- Increased Vx and Vy Speeds: +2 knots

NTSB Findings

The NTSB reported:

"The empty weight of the airplane was stated to be 2,630 pounds. Using the following weights—removal of rear seats (32 lbs), full fuel load (431 lbs), pilot (162 lbs), right front seat occupant (204 lbs), third-seat occupant (195 lbs), fourth-seat occupant (234 lbs), and baggage/cargo (271 lbs)—the total takeoff weight was approximately 4,095 pounds at a CG of +86.15 in."

This was within balance limits but exceeded the revised maximum takeoff weight by 95 pounds.

"Applying a pressure altitude of 1,293 feet and an OAT of 35°C, and using factory takeoff performance charts, the ground roll plus 30% was 2,210 feet. The takeoff distance to clear a 50-foot obstacle plus 30% was 4,030 feet, and the takeoff speed plus 2 knots was 86 knots."

The horizontal distance between the liftoff point and the highest terrain directly ahead was approximately 2,128 feet, with an elevation gain of 80 feet. The total distance from the start of the takeoff roll to the highest terrain directly ahead was about 4,860 feet.

The airplane had an engine analyzer, and the data from engine start to impact was downloaded. The NTSB reported:

"The data revealed consistent engine operation throughout the takeoff power application. RPM was maintained between 2,669 - 2,672 rpm; manifold pressure (MAP) was maintained between 29.5 - 30.1 in Hg; fuel flow was between 33.1 - 33.8 gallons per hour. The recorded OAT was 93°F at engine start and 95-97°F during takeoff. Horsepower produced was between 91% and 94%."

The NTSB observed the following from the video:

- The ground speed was computed to be approximately 84 knots (±4 knots) at rotation and 80 knots (±4 knots) over the end of the runway.
- The takeoff distance, from the start of the ground roll to liftoff, was 2,732 feet—about 500 feet longer than the expected performance chart value.

The NTSB calculated the probable performance of a Bonanza weighing 4,100 pounds at a density altitude of 4,100 feet. With the engine producing an average power of 277.5 HP (92.5% of the maximum rated power of 300 HP), the theoretical rate of climb was estimated at 911 feet per minute (FPM) at 80 knots calibrated airspeed (CAS).

"The elevation difference between the liftoff point (1,270 feet MSL) and the highest terrain directly ahead (1,350 feet MSL) was approximately 80 feet. The distance between these points was approximately 2,128 feet. At 80 knots, the airplane would take approximately 15 seconds to travel this distance. At a climb rate of 911 FPM, the airplane would have climbed approximately 228 feet in that time, clearing the terrain high point by 148 feet."

The standard A36 POH notes that using approach flaps for takeoff reduces the distance required to clear an obstacle by approximately 20%. While this may have been beneficial, the wind effects remain unknown.

No mechanical anomalies were identified.

NTSB Probable Cause

"The airplane's sudden encounter with a wind shift during the initial takeoff climb resulted in degraded climb performance and a stall mush condition. Contributing to the accident was the airplane's over-gross weight condition, high density altitude, the pilot's inability to compensate for the sudden wind shift, and the rising terrain in the departure path."

Commentary:

As noted in the introduction, strictly by the numbers, this takeoff should have worked, even though the aircraft was slightly overweight. However, there was little margin—it all had to work perfectly. And it might have, had it not been for a gust of wind or a downdraft that appears to have sheared off some airspeed and climb ability at a critical point in the takeoff.

But wind is a "known unknown," to quote a former Secretary of Defense, so we must anticipate it. However, the engine was computed to be producing between six and nine percent less than full-rated power. That is also unknown to us on every takeoff, hence the need for generous margins.

The book distance for ground roll and the computed ground roll yield a significant difference—the book says to expect 2,210 feet, whereas the actual ground roll was 2,732 feet, a difference of 522 feet. Not developing full power, possibly less-than-perfect pilot technique, and some runway upslope all tend to make book figures optimistic.

Moving almost the length of a football field every two seconds at 84 knots, the pilot would have had very little time to recognize that the aircraft was not going to climb and find a suitable landing spot. In this case, there weren't many options.

Reducing the fuel/baggage load by 300 pounds might well have made the difference, as would delaying takeoff until the ambient temperature was 20 degrees cooler. Applying the Air Safety Institute's 50-50 solution would have yielded a 6,000-foot distance (4,000 feet plus 2,000 feet) from the start of the takeoff roll to the obstacle, which was 60 percent taller than standard. Would that have been sufficient to overcome the minor wind shear or downdraft that put the aircraft into the terrain off the end of the runway? I'm guessing it would have.

Still willing to bet the farm, your life, and those of your passengers on everything working perfectly and on book performance numbers? The book numbers are not necessarily wrong, but they were derived under conditions that we will not likely achieve.

Control Confusion (SR22 CFI Fails to Intervene)

An unstable approach to a too-short runway and some confusion as to who was on the controls led to a bad outcome.

> *"As soon as you see a mistake and don't fix it, it becomes your mistake."*
>
> *Anonymous*

Unstabilized approaches, regardless of aircraft size, are problematic. Positive transfer of control is also essential if that becomes necessary.

A short Labor Day vacation trip for a student pilot, his CFI, and the student pilot's wife turned tragic with a botched approach, but the causal factors may go well beyond those stated in the accident report. Probable cause in landing accidents is usually easy to ascertain, but other factors are sometimes in play.

The Flight

On September 1, 2012, a Cirrus SR22 departed Tweed-New Haven Airport in Connecticut on a VFR flight to non-towered Falmouth Airpark in Massachusetts. Just before 11 a.m. Eastern Daylight Time, the Cirrus entered the pattern at Falmouth.

According to the student, they overflew the airport at 3,000 feet and descended to join the right downwind for Runway 7. The weather— reported from four miles away—included a few clouds at 1,600 feet,

138

visibility of 10 statute miles, and wind from 066 degrees at 15 knots, gusting to 18 knots.

Witnesses described the final approach over trees as "unstable, with rocking wings," and one witness speculated the landing would be aborted.

Another witness wrote:

"Subject aircraft was on a short final when he came in over the trees. He was low and slow. He got into a high sink rate, and he went to full power and pulled the nose up abruptly—about 30 to 40 degrees nose up—and the airplane veered to the left, went into the trees, and exploded on impact."

Witnesses recalled that as the Cirrus neared the runway, its descent rate increased with some additions and reductions in power. The airplane veered left, power was added, and the left wing almost hit the ground. Touchdown occurred in the grass left of the runway. The aircraft went through a section of wooden fence, entered some woods, and burst into flames.

Flaps were fully down. The wife, seated in the back, pulled her husband from the wreckage while a fire ensued. The CFI was killed. The student pilot and his wife sustained serious injuries.

Recollections

The student thought he was flying the airplane but recalled the CFI's hands and feet on the controls, which he said happened often. He remembered clearing the trees at the approach end of Runway 7 when the CFI said they were "low and slow."

The NTSB report noted:

"The student pilot did not remember much thereafter, other than being 'jounced around a bit.' He did not remember 'seeking' the runway or touching down on or near the runway. He did not know if the CFI took control of the airplane or if he continued to fly it, nor did he recall the CFI saying anything else to him other than they were low and slow. The next thing the student pilot remembered was the airplane hitting trees, breaking up, and coming to rest."

Five of the last seven of the CFI's students had not passed their practical tests on the first attempt, but not because of pattern work or landings. Former students were complimentary of the instructor, describing him as "thorough, meticulous, and encouraging."

The NTSB report stated:

"Only one noted that he rode the controls occasionally. Because the student pilot indicated that the CFI would be on the controls with him at times, the question of riding the controls was asked of the other five student pilots. Three said he did not ride the controls, one said that he would be on the rudders, and one, who was only with the CFI before her solo, said he did. All but one of the student pilots flew with the CFI in a conventional, yoke-configured airplane."

On the morning of the crash, another flight instructor spoke with the CFI while walking out to the accident airplane. The report noted:

"The CFI seemed upset and, for the first time ever, made disparaging remarks about the president of the flight school. The other CFI did not ask what brought about the remarks."

The student pilot recalled they were delayed about an hour waiting for the CFI. He appeared "normal but slightly distracted" but said something like, "Ready to have some fun?" During the flight, the CFI "seemed to be his normal self but somewhat casual."

The CFI

The 24-year-old CFI held a commercial pilot certificate with single-engine, multiengine, and airplane instrument ratings. Cirrus training was completed on September 29, 2011. His logbook listed 1,519 total flight hours, with 1,407 hours of single-engine flight time and 1,002 hours of instruction given. The accident report did not include Cirrus SR22 flight time.

Student Pilot

The student pilot, age 55, reported 117 hours of flight time at the time of the accident, about 100 of which were in the SR22. His logbook was destroyed in the accident. He began flight training and bought a used SR22 shortly afterward.

The NTSB was curious why the pilot, with so much flight time, had not taken his private pilot flight test. His attorney responded:

"He was not in a rush to obtain his private pilot certificate and believed that the additional time and instruction would only make him a better, safer pilot."

The pilot had not yet completed night flight or solo cross-country tasks. He and his wife used the aircraft for transportation and hired the CFI to fly with them while providing instruction.

The Aircraft

The pilot purchased the 2008 Cirrus SR22 in April 2012. Total flight time was approximately 965 hours. No pre-impact malfunctions were noted, and no recorded data could be retrieved due to fire damage.

Falmouth Airport, Google Earth

The Airport

Falmouth Airpark has a single runway, 7/25, which is 2,298 feet long and 40 feet wide. The AOPA Air Safety Institute's accident database recorded several landing accidents involving wind or short-field mishaps. AOPA's Airport Directory reported obstacles for Runway 7 as: "Trees, 60 feet left of center, 33 feet high, 300 feet from end, 3.1 clearance slope."

NTSB Probable Cause

"The flight instructor's inadequate remedial action. Contributing to the accident was the student pilot's poor control of the airplane during the approach."

Commentary

A significant contributing factor was that this airport was too short for routine operations in an SR22. Falmouth Airpark can be challenging in gusty conditions. A short runway surrounded by trees with shifting winds makes approaches complex. There's little margin for a heavily loaded SR22.

The recommended short-field approach speed is 77 knots, slightly higher if turbulent. The student apparently was concerned about landing long, as indicated by his comment that the airspeed was 69 knots on final. The additional speed needed for controllability is problematic on a short runway, especially with dynamic winds.

The pilot's operating handbook estimated a no-wind landing distance of about 2,375 feet, which is longer than the Falmouth runway. Some reduction could have been factored in due to the headwind, but there would have been no margin. With some clear space at the end of the runway, it would be possible to land in a shorter distance, but many factors would have needed to align perfectly.

FAR Part 23, the certification rule governing Cirrus aircraft, requires that the landing distance account for crossing the end of the runway at 50 feet, at the correct speed, and with braking that does not cause undue wear on the brakes or tires. The 50/50 solution works out to about 3,600 feet. While it can be done in less, both pilot and aircraft performance must be exceptional.

Possibly, the argument with the flight school president distracted the CFI, but that is speculative. Much of the CFI's experience was in lower-powered, more lightly wing-loaded aircraft, which may have been a factor. However, instructors must clearly take command and declare their authority before control is in jeopardy. The key question is how far to let things go. Based on the SR22 landing distance numbers, this airport was unsuitable, in my view, even with perfect technique.

Owatonna Overshoot – (Hawker 125-800 Runs Out of Runway)

A downwind landing on a wet runway and overly optimistic Flight Manual numbers led to a fatal overrun.

While many runway excursions result only in damaged egos and hardware, they can also be deadly. Such was the case for a Hawker 125-800 business jet on a Part 135 charter to Owatonna, Minnesota.

The Flight

The flight departed Atlantic City, New Jersey, at 8:13 a.m. Eastern time on July 31, 2008, with six passengers and two crew members. About two and a half hours later, the flight was maneuvering around a line of severe thunderstorms east of Owatonna (OWA).

Exemplar HS-125-800, AOPA

At 9:37 a.m. Central time, the controller provided the Owatonna weather report, which was about 20 minutes old. The conditions were: winds from 320 degrees at 8 knots, visibility of 10 miles or more, scattered thunderstorms, clouds at 3,700 feet, an overcast layer at 5,000 feet, and lightning visible in all quadrants.

At 9:38 a.m., the controller advised that light precipitation was present along most of the remaining route, with a couple of heavy storm cells located about five miles north and northeast of Owatonna.

The crew conducted an abbreviated approach briefing, and the first officer (FO) unsuccessfully attempted three times to contact the FBO regarding fuel and passenger arrangements. The flight intercepted the Runway 30 localizer eight miles out, and at 9:42 a.m., the captain reported the runway in sight and canceled the IFR flight plan. He then asked the FO to try the FBO again—this time, the call was successful.

At 9:42:37 a.m., the before-landing checklist was completed, and at 9:43:05 a.m., the captain asked the FO to verify the landing configuration. The Hawker touched down on the wet runway at 9:45:04 a.m. at a speed of 122 knots. Two and a half seconds later, the cockpit voice recorder (CVR) captured a sound similar to the air brakes moving to the Open position.

According to the NTSB report: "At 9:45:08 a.m., the first officer stated, '(We're) dumped,' followed immediately by, 'We're not dumped.' [This referred to the deployment of the lift-dump feature of the air brake and flap systems, which is used to help decelerate the airplane upon landing.] About 1.5 seconds later, the captain replied, 'No, we're not,' and at the same time, the CVR recorded a sound similar to the air brake handle moving to the Dump position.

Ten seconds later, the CVR recorded a sound similar to the air brakes moving to the Shut position. The captain then stated, 'Flaps,' and around the same time, the CVR recorded a sound consistent with increasing engine noise. At 9:45:27 a.m., the captain stated, 'Here we go...not flyin'...not flyin'.' At 9:45:36 a.m., the CVR recorded an aural warning stating, 'Bank angle, bank angle.' The CVR stopped recording at 9:45:45."

Witnesses observed that the approach appeared normal but reported hearing a power increase near the end of the runway. According to the NTSB report: "The airplane ran off the runway end at 9:45:29 a.m. and lifted off the ground at 9:45:34 a.m., about 978 feet from the runway end. Subsequently, the airplane collided with the Runway 30 localizer antenna support structure, which was about 1,000 feet from the runway end, and eventually came to rest in a cornfield beyond a dirt access road that borders the airport, approximately 2,136 feet from the runway end."

The flaps were set to zero instead of the normal takeoff configuration of 15 degrees. Flight data recorder analysis showed the aircraft pitching up to about 20 degrees and rolling 50 degrees to the right just before impact. There were no survivors.

The Weather

At approximately 9:45 a.m., near the time of the accident, the OWA AWOS reported winds from 170° at 6 knots, visibility of 10 miles in moderate rain, scattered clouds at 1,800 feet and 2,900 feet, a broken ceiling at 3,700 feet, a temperature of 19°C, a dew point of 17°C, and an altimeter setting of 29.83. The remarks indicated that 0.09 inches of precipitation had fallen in the preceding 20 minutes, and lightning was detected in the distance from the east through the south.

The Crew

The 40-year-old captain held a multiengine airline transport pilot certificate with type ratings in the HS-125 and Learjet series. He had accumulated 3,600 total flight hours, including approximately 1,188 hours in the HS-125 and 874 hours in Learjets. In the 90, 30, and 24 days before the accident, he had flown 110, 24, and 0.3 hours, respectively. He had completed recurrent training in the Hawker a few months prior to the accident.

The captain had been off duty for 72 hours before the accident. According to his girlfriend, he went to bed around midnight the night before (after a poker game) and awoke between 4:45 and 5 a.m. the next morning.

The 27-year-old first officer held a commercial pilot certificate with a type rating in the HS-125. He had accumulated 1,454 total flight hours, with 297 hours as second in command, most of them in the Hawker. In the 90, 30, and 24 days before the accident, he had flown 86, 27, and 0.3 hours, respectively. His most recent Hawker proficiency check had occurred eight months earlier.

The first officer had also been off duty for 72 hours before the accident. His fiancée reported that he went to bed around 11 p.m. the night before and woke up around 5 a.m.

The Aircraft

The HS 125-800 was built in Great Britain in 1991 and had accumulated approximately 6,570 total flight hours. The NTSB found no mechanical discrepancies during the accident investigation.

The aircraft is typically landed with flaps set to 45 degrees. A separate handle deploys the air brakes (a British term; the U.S. equivalent is "spoiler") in flight. After touchdown, this same control is used to "dump" lift by deploying the air brakes to the ground-only or dump position. The lever is moved to the "Open" position and then up and over a detent, which increases the flaps to 71 degrees and further extends the air brakes. This system is used in lieu of reverse thrust. The aircraft is also equipped with anti-skid brakes, allowing the pilot to apply maximum braking without wheel lockup.

HS-125 Flap Indicator, NTSB

The Runway – Owatonna's Runway 30 is 5,500 feet long and 100 feet wide, with a 0.7-percent downslope. The concrete surface was not grooved but was crowned to enhance drainage, with 1,000 feet of grass overrun at each end. The NTSB determined that the runway had excellent drainage and that standing water from recent heavy rains did not cause hydroplaning.

NTSB Analysis – Based on witness statements, ASOS observations, and data extracted from the aircraft's Flight Management System (FMS), the tailwind component was estimated at eight knots. The Hawker crossed the threshold at a reference speed of 122 knots and touched down 1,128 feet from the threshold at a groundspeed of approximately 130 knots. At those speeds, the available friction on the wet runway was only about 20 to 30 percent of that on a dry runway.

HS-125 pedestal showing controls, NTSB

On the accident flight, the air brakes were moved to the Open position approximately 4.1 seconds after touchdown, and the lift-dump system engaged about 8.9 seconds after touchdown.

The NTSB provided a detailed analysis of how the landing distances were calculated and noted that the British formula is more optimistic than the FAR Part 25 formula used for certifying U.S. aircraft. The actual braking coefficients on wet, ungrooved runways may have been lower than those predicted by British certification tests. It is unlikely that any calculations would have precisely replicated the actual conditions.

Simulation Scenario	Wind Condition	Calculated Landing Distance (in feet)			
		Braking Coefficient Source AMJ 25X1591 or BCAR RWHS		Braking Coefficient Source 14 CFR 25.109	
		Total	+ 15 Percent Margin	Total	+ 15 Percent Margin
AFM air distance; AFM deceleration device deployment times	No wind	3,338	3,840	4,225	4,860
	8-knot tailwind	3,792	4,361	4,928	5,667

Rules and FAA certification regulations show 1,300' discrepancy with tailwind including margin, NTSB

According to the AFM, with a landing speed of 122 knots, the no-wind landing distance on a dry runway would have been approximately 4,216 feet—3,966 feet with a 10-knot headwind and 5,059 feet with a 10-knot tailwind component. The manual does not provide a correction for a wet runway.

The NTSB report states: "However, if the destination runway was expected to be wet or slippery, pilots were trained to add a 15-percent safety margin to the required factored dry-runway landing distance." Based on this, the Hawker would have gone off the end of the runway between 27 and 37 knots, coming to a stop in the grass 100 to 300 feet beyond the runway.

The onboard handheld computers calculated the landing distance at 3,940 feet with a tailwind and wet runway, which was optimistic.

Factors:

- Wet runway and tailwind
- Optimistic landing distances in the flight manual
- Cockpit distraction
- Possible crew fatigue

NTSB Probable Cause:

"...the captain's decision to attempt a go-around late in the landing roll with insufficient runway remaining. Contributing to the accident were (1) the pilots' poor crew coordination and lack of cockpit discipline; (2) fatigue, which likely impaired both pilots' performance; and (3) the failure of the Federal Aviation Administration to require crew resource management training and standard operating procedures for Part 135 operators."

Commentary:

I would add that the downwind landing and the optimistic flight manual numbers were contributing factors. The CVR recorded a less-than-complete before-landing check with no discussion of wind or runway conditions, which should always be considered. Had the crew landed into the wind, they would have had a much better chance of stopping on the runway. However, the winds were light (170 at 6 knots on ASOS), so it's understandable why the crew chose not to perform a circling approach.

When the crew selected the landing runway, the winds were reported from the northwest, and they may not have received an updated report. The weather was good VFR, and a circle-to-land approach would have only added a couple of minutes to the flight. Hindsight makes such judgment easy.

There was no landing distance assessment prior to landing, nor did the FAA require it at the time of the accident. However, FAR 91.103 (Preflight Action) requires pilots to review runway lengths along with takeoff and landing distance information.

Non-pertinent discussion with the FBO distracted the crew at a critical time when the cockpit should have been sterile.

Convenience—whether arranging ground transportation or getting the fuel truck out a little faster—pales in comparison to landing safely. Aviate first, communicate last, even when you really want the rental car ready and a quick fuel turn. Convenience for Part 135 passengers is nice, but safety is essential.

The NTSB cited fatigue as a factor, noting that the captain had only about five hours of sleep and the first officer about six. Both pilots had been off for several days and knew they had an early start the next day. While fatigue is a regulatory issue in professional aviation, that wasn't the case here—the responsibility for proper rest rested solely with the crew. Pilots should plan for at least seven to eight hours of solid sleep before an early morning flight and adjust their schedules accordingly. Save poker nights for after the flight!

As stated in the introduction to this chapter, it's wishful thinking to believe that most pilots will consistently perform as well as test pilots. In real-world flight operations, we don't. To achieve the numbers in the Airplane Flight Manual (AFM/POH), all conditions must be met. The landing distances were based

Owatonna Airport, Google Earth

on the air brakes moving to the "Open" position 0.56 seconds after touchdown, with the lift-dump system engaging at the same time. It takes about two seconds for the lift-dump system to fully deploy, and requiring manual deployment just half a second after touchdown stretches human performance limits.

The NTSB noted that at the time, regulations did not require flight crews to conduct landing distance assessments once airborne. The conditions at arrival—not during preflight—are what matter. Landing data for anything other than dry runways is currently calculated rather than flight-tested. While jet flight manuals include some "factoring," day-to-day operations often require a more cautious approach. A contaminated runway and a tailwind demand a different runway.

This charter company had a perfect safety record over a 10-year period. Everything works—until the final link in the accident chain clicks into place. The warnings can be subtle.

Wild Wings – High-Performance CEO Loses Control of Lancair IV-PT

A CEO loses control in a high-performance aircraft.

A rejected takeoff is a high-risk maneuver but sometimes necessary. Turning back to the departure runway after an engine problem is even more dangerous. Until maneuvering altitude is reached—whatever that might be—it's often safer to find something soft and cheap to hit off-airport.

Stall recognition and recovery proficiency are fundamental skills for all pilots, but executing them immediately after takeoff in a steep turn is especially challenging. This scenario, in some form, could happen to any of us.

Low-speed aircraft handling characteristics play a critical role in forced landings immediately after takeoff. Experimental aircraft can vary widely in these parameters. Amateur builders are rightly given significant latitude in design, performance, and construction. However, some designs prioritize high-speed cruise performance, sometimes at the expense of other handling qualities. A highly modified Lancair IV-PT fits that description.

Lancair IV-PT, Lancair

The Pilot

Steve Appleton was a celebrity CEO, leading the Fortune 500 company Micron Corporation, a major semiconductor manufacturer. His notoriety led to media discussions and considerable backlash regarding the safety of general aviation and whether CEOs should be allowed to participate.

The Flight

February 3, 2012, dawned bright and clear with light winds in Boise, Idaho. At 8:46 a.m., the experimental Lancair IV-PT began its takeoff roll on the 9,763-foot runway. It climbed to about 60 feet before touching back down. The pilot stated he was going to "land here and stop. We got a problem." The controller asked if he needed assistance. The pilot responded, "Negative, I'm going to taxi back and see if I can figure it out."

The pilot returned to his hangar area and performed some run-up activities. About nine minutes later, he asked to remain in the traffic pattern for a "couple of laps." The aircraft departed Runway 10R again, climbing to about 300 feet. Witnesses saw the aircraft roll left while rapidly losing altitude. It completed about one full revolution before impacting terrain in a nose-low attitude.

The NTSB report stated: "At 08:55:44, the pilot made his last intelligible transmission when he requested that he would 'like to turn back in and, uh, land, coming back in, uh, three.'" A subsequent fire occurred, and the pilot did not survive.

Based on simulations of the accident flight and witness statements, it is likely the pilot was attempting to return to the runway but did not lower the nose enough to maintain flying speed. The crash signature indicated the airplane was in a spin at impact.

The Pilot

Steve Appleton could be described as a risk taker. He participated in motocross, skydiving, race car driving, and flying high-performance airplanes—including the Extra 300, Aero Vodochody L-29 (a Soviet-era jet trainer), Hawker Hunter (a British 1950s single-seat fighter), and the Cessna Citation.

He had previously been involved in an aerobatic accident in July 2004, which, according to the NTSB, "did not allow adequate clearance from the ground." His total flight time was estimated at about 3,500 hours, but he had fewer than 14 hours in the IV-PT. There was no evidence Appleton had received formal instruction in the Lancair, although his insurance company had requested it.

The Aircraft

This Lancair IV-PT, a composite, pressurized turboprop, received its experimental airworthiness certificate in 2007. (The -PT stands for pressurized turbine. Lancair marketed the airplane as a Jetprop, and it has erroneously been called a IV-TP, including in the NTSB report.) The aircraft had been significantly modified from its original design, including a higher-horsepower engine and a larger fuel system.

Lancair's recommended maximum gross takeoff weight was 3,550 pounds, but accident documentation listed a maximum takeoff weight of 4,300 pounds. The NTSB estimated the aircraft's takeoff weight at 3,837 pounds, within center-of-gravity limits. The FAA allows amateur builders to set weight limits at any value they desire, but they must determine and document stall speeds. No such documentation was found for this aircraft.

The pilot had purchased the aircraft less than two months before the accident. He had mentioned to a lineman that he liked the speed but described the handling as "squirrelly." It's a common scenario for a subsequent buyer of an amateur-built aircraft to crash shortly after taking ownership. The builder typically knows the quirks and has spent time learning the aircraft, but new buyers lack that familiarity.

The Lancair was equipped with electronic flight instrumentation, which survived the post-crash fire and provided extensive data on engine and flight performance.

A former engineer and general manager of the kit manufacturer noted that if the engine failed during takeoff, airspeed would decay rapidly, and the nose would have to be lowered quickly to maintain flying speed. (*This is true of all aircraft.*) If no action was taken, the aircraft would stall within about five seconds, resulting in an immediate wing drop. Stall recovery was considered unlikely below 1,500 feet AGL. There was no data on the minimum altitude required for a turnback maneuver.

According to the NTSB: "A former (Lancair) employee further stated that a pilot could not use full engine power during takeoff because the IV-PT was not designed for such a high-horsepower engine and did not have enough rudder authority to compensate for the P-factor at full power. He noted that the aft fuel tank in the baggage area greatly affected the airplane's center of gravity, making it extremely sensitive in pitch."

The NTSB found no evidence of pre-impact mechanical malfunction or failure that would have prevented normal operation—though "normal" for this aircraft may have been significantly different from what most pilots are accustomed to.

NTSB Probable Cause

"…A loss or commanded reduction of engine power during the initial climb for reasons that could not be determined due to post-accident impact damage and fire destruction of engine systems and components. Also causal were the pilot's failure to maintain adequate airspeed and airplane control while attempting to return to the runway, despite unpopulated, flat terrain immediately ahead that was suitable for an emergency landing; his decision to take off again with a known problem; and his lack of training in the make and model airplane."

Commentary

Engines are extremely reliable, but when they show any signs of trouble, it's critical not to let wishful thinking interfere with decision-making. Flight data monitoring showed fuel flow and power fluctuations on both attempted takeoffs. The cause was never determined. Appleton correctly rejected the first takeoff but apparently believed that, since he had resolved a similar issue days before, it was a transient problem that could be managed.

The engine may have been misbehaving, and a power/controllability factor could have compounded the situation. Perhaps the pilot reduced power in response to a perceived engine issue, causing the aircraft to stall. In certificated aircraft, these factors are considered and tested to ensure adequate aerodynamic control and time to troubleshoot. In a highly modified kit aircraft, almost anything is possible.

Rejecting a takeoff shortly after becoming airborne is one of aviation's biggest challenges—there's little altitude, airspeed, or time. Part 25 multi-engine aircraft are designed to mitigate this risk. Once past takeoff safety speed (V1), there is enough energy with the remaining engine to continue. In general aviation aircraft, however, there is a strong natural tendency to turn back to the airport for safety, yet this maneuver can be deadly. It's called "The Impossible Turn," but because it occasionally works, pilots keep trying. (*See Chapter 6 - Emergencies.*)

Every single-engine aircraft has a minimum maneuvering altitude below which a turnback attempt carries much higher risk than landing off-airport. Unfortunately, manufacturers are not required to test or publish data on accelerate-stop or turnback altitudes and distances.

Aircraft design also played a role here. The Lancair IV-PT is one of the highest-performance amateur-built aircraft. As noted, stall recovery at low altitude requires at least 1,500 feet. Experimental aircraft wings are not required to meet FAA standards for certificated wing performance. By comparison, the TBM 900—a certificated single-engine turboprop with similar cruise performance—recovers from stalls in several hundred feet.

Wings are designed with different objectives. Some are forgiving; others can be unforgiving if mishandled. The Lancair Owners and Builders Organization offers active training programs, as do many type clubs. Regardless of a pilot's experience in other models, it makes sense to invest time and money in thoroughly learning any new aircraft—especially one with unique flight characteristics.

Steve Appleton was, by any measure, a highly successful risk taker and accomplished pilot. Society needs people like him to push boundaries and innovate. However, high-profile aviation accidents tarnish the reputation of general aviation and discourage others from participating. Overachievers must recognize that gravity and physics are absolute and unforgiving—just like some aircraft.

Head-to-Head - (Airbus 320 and King Air 350 almost meet head on)

An airliner hurries to get airborne at a non-towered airport. Communication confusion in the airliner cockpit and on the CTAF results in a close call and damage to an Airbus 320. The ever-present potential failure to communicate conveys lessons for all pilots.

In the earlier case study, Collision at Quincy, we examined a corporate captain's bad attitude and miscommunication or lack of communication on the CTAF. In this case, the loss of situational awareness—revealed by the cockpit voice recorder (CVR)—demonstrates how routine situations can lead pilots to perceive things as they think they are, rather than as they actually are. Left unsaid in the accident report are other human factors that may have played a role, which are addressed in the commentary.

Airbus 320   vs.   King Air 350

The Flight

On January 22, 2022, JetBlue Airways Flight 1748 was scheduled to fly from Yampa Valley Hayden Airport (HDN) to Fort Lauderdale, Florida. HDN is a non-towered airport and uses a Common Traffic Advisory Frequency (CTAF). While airlines primarily operate from towered airports, they also serve smaller airports.

The ASOS reported a 500-foot overcast, but witnesses noted that a very thin layer of clouds was confusing the system. The weather was good VFR.

At approximately 11:48 a.m. Mountain Standard Time, JetBlue 1748 announced they were leaving the ramp and taxiing to Runway 10 for departure. Shortly afterward, a Beechcraft B300 King Air, N350J, on an instrument flight rules (IFR) flight plan, reported in on the CTAF.

The NTSB's CVR transcript is lengthy but provides great insight into how the conflict developed.

Transcript Key:

- PF – JetBlue pilot flying (captain)
- PM – JetBlue pilot monitoring (first officer)
- KA-1 – King Air pilot
- KA-2 – King Air copilot
- CTR – Denver Center air traffic controller
- UNI – HDN UNICOM
- ** – Unintelligible
- # – Expletive

Transcript

11:48:41

PM: "And Hayden Airport, JetBlue 1748, we're pushing off the ramp for ten."

11:48:46

KA-1: "Hayden Yampa Valley, King Air 350 Juliet from the east, descending out of seventeen thousand, 350 Juliet."

11:49:04

UNI: "350J, Hayden UNICOM, we have multiple aircraft inbound, winds are calm, altimeter 30.36."

KA-1: "30.36, 350 Juliet."

At 11:49:11, the PM discussed with the PF that multiple airplanes, including the King Air, were inbound and all were using Runway 10. The PM then radioed Denver Center to report they were on the ground at HDN, preparing for engine start, and would be ready for departure shortly.

PM: "Denver, JetBlue 1748 on the ground, Hayden. Starting engines. We'll be ready for our departure in about, uh, six or seven minutes."

CTR: "JetBlue 1748, and uh, are you planning one zero?"

PM: "Affirmative."

CTR: "All right, just call me number one ready to go."

PM: "Wilco."

At 11:52:12, the King Air radioed Denver Center to cancel their IFR flight plan as they had visual contact with HDN. The King Air then stated they planned to land on Runway 28 instead of Runway 10, but this information was not relayed to JetBlue when they received their IFR clearance (**emphasis added**).

KA-1: "Denver, 350 Juliet."

CTR: "350 Juliet, go ahead."

KA-1: "Uhhh, would it inconvenience you if we canceled and landed two eight? We can see the runway ** almost five hundred overcast."

CTR: "It's up to you. If you want to cancel, that's fine."

KA-1: "All right, we're going to cancel right now and land two eight, Hayden Yampa Valley."

CTR: "350 Juliet, cancellation received. Squawk VFR, frequency change approved. Have a good day."

KA-2: "Squawk VFR, over to ** as well, 350 Juliet. Good day."

At 11:53:07, the King Air radioed the CTAF and stated they would be landing on Runway 28 (**emphasis added**).

KA-1: "350 Juliet's going to go ahead and land two eight, Hayden Yampa Valley. We're straight in, 28 right now."

At 11:53:18, the JetBlue PM radioed the CTAF and stated they were leaving the ramp area and taxiing to Runway 10 for departure.

PM: "Hayden Yampa Valley, JetBlue 1748, Airbus coming off the ramp for Runway 10, Hayden Yampa Valley."

At 11:53:26, HDN UNICOM reported that multiple airplanes were inbound, and the wind was calm.

UNI: "JetBlue 1748, Hayden UNICOM, we have multiple aircraft inbound. Winds are currently calm. Altimeter is 30.36."

PM: "Roger, 1748."

At 11:53:42, the JetBlue flight crew performed an after-start checklist.

At 11:53:45, the King Air radioed the CTAF and reported they were on a 12-mile straight-in final for Runway 28 (**emphasis added**).

KA-1: "350 Juliet's uh...twelve-mile final two eight, straight in. Hayden Yampa Valley."

At 11:54:00, the JetBlue flight crew discussed pre-departure activities.

PM: "We're number one [for departure]. Do you want to do anything else first?"

PF: "I think we're going to wait for these planes to come in."

The JetBlue flight crew performed a flight control free and clear check.

At 11:54:25, the King Air radioed the CTAF and asked if anyone was about to depart from Runway 10.

KA-1: "Anybody about to depart uh, ten at Hayden?"

PF: "JetBlue 1748 is uh...holding short of the runway, Runway 10 at the end of the runway, waiting for our clearance."

KA-1: "All right, we're on a uh, ten-mile final, two eight, straight in." **(emphasis added)**

PF: "All right, copy that. We'll keep an eye out for you."

KA-1: "Hayden Yampa Valley."

PM: "All right, so he's coming in right now."

At 11:54:48, the JetBlue flight crew began the before-takeoff checklist. During the checklist, they verbalized speeds of 142 knots, 142 knots, and 143 knots.

At 11:55:20, the JetBlue PM radioed Denver Center and reported they were ready for departure on Runway 10 at HDN.

PM: "Denver, JetBlue 1748, number one at Runway 10, ready for departure."

CTR: "JetBlue, ready at Hayden, understood?"

PM: "Affirmative, 1748."

CTR: "JetBlue 1748, cleared from Hayden to uh, Fort Lauderdale as filed. Climb and maintain one three thousand, squawk zero six two two, released. Clearance void if not off in two minutes."

PM: "All right, roger that. Fort Lauderdale as filed, one three thousand, squawk zero six two two, uh, off in two minutes, JetBlue 1748."

At 11:56:06, the JetBlue flight crew performed a takeoff briefing and finished the before-takeoff checklist.

At 11:56:25, the PM radioed the CTAF and stated they had clearance and would be taking off on Runway 10.

PM: "Hayden traffic, JetBlue 1748, uh, released and we are taking off Runway 10."

KA-1: "Uh, Hayden Yampa Valley, uh, you got uh, a King Air on final two eight. Hayden Yampa Valley. We've been calling."

When the Airbus taxied onto Runway 10, ADS-B data showed the King Air less than five miles out for Runway 28.

At 11:56:37, the flight crew discussed the King Air's position.

PF: "Yeah, but where is he at?"

PM: "Yeah, I thought you guys were still like eight...nine miles out."

KA-1: "Four miles."

KA-2: "Less than that."

At 11:56:48, the PF stated: "Well, we're taking off now."

PM: "And JetBlue's on the roll. JetBlue, uh, 1748 on the roll, Runway One Zero, Hayden."

KA-1: "All right, we're on short final. I hope you don't hit us."

At 11:57:02, a sound similar to an increase in engine thrust was noted.

At 11:57:06, the PM verified the airspeed indicator was functional:

PM: "Airspeed's alive."

At 11:57:13, the flight crew discussed the position of the King Air:

PM: "He's [the King Air] on Two Eight?"

PF: "Is he? Oh #."

PM: "Eighty knots, power set."

At 11:57:19, the flight crew continued to discuss the King Air as the Airbus accelerated down the runway:

PM: "Yes, he [the King Air] is on Two Eight. Do you see him?"

PF: "No."

At 11:57:23, the King Air radioed the CTAF and asked if JetBlue was going to make a quick turn after departure to avoid a collision.

KA-1: "You guys do a quick, uh, turn out 'cause we're…." [cut off by another radio transmission].

At 11:57:25, on the CVR:

PF: "Oh Jesus Christ."

PM: "You're slow. Airspeed, airspeed, airspeed, airspeed."

During this conversation, the captain pitched the airplane up steeply—about 24 knots before rotation speed—resulting in a tail strike. The Airbus began a climbing right turn away from the traffic indicated on the TCAS, which showed a 2.3-mile separation.

Both JetBlue crew members stated they never saw the King Air and believed it was landing on Runway 10. They called maintenance to report the strike, and maintenance recommended landing in Denver to assess the damage.

Flight Crew

The JetBlue captain, 45 years old, had over 11,000 hours of total flight time, with more than half in the Airbus. The 40-year-old first officer had 3,300 total flight hours, with 2,981 hours in type. Both crew members had no prior infractions or failed flight checks.

The Airport

NTSB Statement: "JetBlue Airways listed HDN as a special airport per 14 CFR 121.445 and provided their pilots with an Airport Briefing Guide (ABG) for the airport. General cautions included high terrain, high-density altitude, high grid minimum

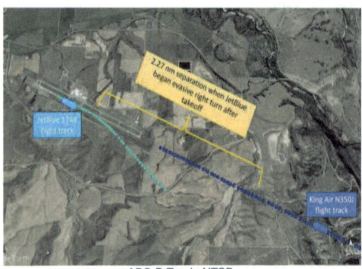

ADS-B Track, NTSB

off-route altitude (MORA), frequent use of opposing runways by arriving and departing aircraft, and high traffic (both commercial and general aviation). Pilots should make traffic calls and monitor radio traffic carefully."

JetBlue Actions Following the Tail Strike

JetBlue University instructors developed training curriculum for flight crews on non-towered operations.

NTSB Probable Cause

"The captain's premature rotation of the airplane before reaching takeoff speed, due to his surprise at encountering head-on landing traffic, resulted in an exceedance of the airplane's pitch limit and a subsequent tail strike.

Contributing to the accident was the flight crew's expectation bias, which led them to believe the incoming aircraft was landing on the same runway they were departing from. Additionally, the conflicting traffic's nonstandard phraseology when making position calls on the common traffic advisory frequency contributed to the misunderstanding."

Commentary

How could such an experienced crew find themselves in this situation? According to the NTSB, psychologists explain expectation bias as a cognitive tendency "...which causes an incorrect belief to persist despite available contradictory evidence." In this case, the crew's expectation that the King Air was arriving on Runway 10 biased their perception of incoming information, causing them to either ignore or misinterpret contradictory evidence—such as radio calls indicating the King Air was landing on Runway 28. This bias occurs as part of basic information processing, often without a person's active awareness.

With calm winds or a direct crosswind, there is often indecision about which direction to land or take off at a non-towered airport. Many pilots default to the runway most aligned with their inbound or outbound path, especially if there is no traffic in the pattern. One aspect the NTSB report did not address was the number of other aircraft in the pattern or approaching the airport, as noted in the Unicom operator's comment about "multiple aircraft inbound."

The JetBlue crew, upon hearing the King Air's initial call for Runway 10, mentally locked into that expectation. They seemingly did not process the subsequent calls announcing Runway 28. Additionally, they were occupied with engine start, FMS programming, and pushback. With so many tasks occurring simultaneously, even two highly trained pilots can only process so much information in a compressed time.

CTAF Procedure

The CTAF includes many numerical references—call signs, altimeters, mileage, and altitudes—so when discussing runways, it is essential to explicitly say "Runway 28" instead of just "28." The King Air crew was not as precise as they could have been in their transmissions. However, toward the end of the sequence, it became very clear what was happening.

At the runway entrance, the JetBlue captain stated they were going to wait for "...some planes to come in." The King Air had reported being 10 miles out. Thinking there was plenty of time, the JetBlue crew proceeded with their pre-takeoff checklist and picked up their IFR clearance, which included a two-minute void time. Meanwhile, the King Air rapidly progressed down the final approach toward Runway 28.

The short void time may have altered the captain's thinking from waiting to hurrying. Rather than requesting a longer void time, he decided to take off. He still did not understand that the King Air posed a direct conflict by landing on their departure runway.

Missed Opportunities for Deconfliction

Both flight crews could have taken steps to avoid the conflict. The JetBlue captain told his first officer, "Well, we're taking off now." The FO could have responded, "We'd better wait, that guy's inbound on 28." He didn't. The King Air crew, upon hearing JetBlue announce their departure from Runway 10, could have easily adjusted to the downwind for Runway 10 instead of continuing toward Runway 28. They didn't. Either crew could have deconflicted with minor inconvenience and better Cockpit Resource Management (CRM).

Regulations give priority to landing traffic, regardless of aircraft size or type of flight operation—whether air carrier, business jet, commercial operation, personal flight, or student pilot.

Entitlement:

The following discussion may not apply to this incident. In the introduction, I alluded to other human factors that may have contributed to this situation. Psychology is a messy business—we often attribute motives to others that may not exist, often reflecting our own biases.

Most airline pilots are unquestionably courteous and extremely risk-averse, a reflection of their operational environment. However, this is a good opportunity to discuss the sense of entitlement for operational priority that many of us sometimes feel.

Tail strike damage – more than skin deep, NTSB

Over decades of flying, I have observed the entitlement syndrome—and I've been guilty of it myself. As we gain experience and move up in aircraft size and complexity, some pilots develop a tendency to assume they have priority at non-towered airport runways. In the chapter introduction, we discussed that guidelines, rather than strict regulations, allow for flexibility to enhance both safety and utility.

For a large or fast aircraft on final, it's almost always safer to let them play through for several reasons, as mentioned earlier. Speed differentials and the need for larger aircraft to stabilize much sooner than light aircraft are key factors. A Cherokee can more easily go around, extend downwind, or wait to take off than a turboprop, bizjet, or airliner can. If the pattern is busy, sequencing issues arise, along with varying speeds and additional traffic to manage. It simply makes sense to get a fast mover down and out of the system efficiently. If someone gives you a break, publicly thank them—and pay it forward.

There's also the issue of fuel burn, which may not concern everyone, but larger aircraft often consume significantly more fuel during a go-around or extended wait time than a smaller aircraft does on an entire trip. That said, you'll sometimes hear an airline pilot announce, "We're going around because a light aircraft was on the runway." This can be a pejorative statement if the landing aircraft simply hadn't reached the exit yet. The burden is on the overtaking aircraft, and at a towered airport, ATC is responsible for sequencing. Pilots should clear the runway as quickly as safely possible, but they are not obligated to wear out their tires and brakes—or risk an incident—just because a faster aircraft cut it too close.

Always, it's often safer to yield to aggressiveness or cluelessness rather than risk a collision. Doing so avoids injuries, damage, paperwork, and possible violations. The only downside is a minor delay—a small price to pay for safety.

Check Six – Bogey on Your Tail

A C150 & C340 Get Together on Final

ABS-B Track, NTSB

The absolute worst place to be in the traffic pattern is in front of a faster-moving aircraft. This situation occurs daily at non-towered airports. It isn't the first time it's happened, and sadly, it won't be the last. Because we don't have eyes in the backs of our heads—even with Cessna's touted Omni-Vision—"see and avoid" is reduced to one pilot's ability to maneuver. They may not be as diligent as you or I. All fighter pilots agree.

The Flights

As with most midair collisions, this one occurred in good VFR conditions at a non-towered airport. A Cessna 152 was performing touch-and-go landings in the traffic pattern, making position reports on the common traffic advisory frequency (CTAF). A Cessna 340 pilot made an initial call at 10 miles out, stating that he would be descending for a straight-in approach to Runway 20. ATC VFR radar service had just been terminated with the advisory that other aircraft were in the traffic pattern. Shortly thereafter, the Cessna 152 pilot called crosswind, and the 340 pilot then reported he was three miles from the end of Runway 20.

The Cessna 152 pilot reported turning left base for Runway 20. About 19 seconds later, the 340 pilot reported he was on a one-mile final and looking for traffic on left base. Although the 340 is a moderately fast light twin, it doesn't travel seven miles in just over a minute. The Cessna 152 pilot said he saw the 340 and that it was behind him. Thirteen seconds later, he decided this was a bad place to be and called, "Going around because you're coming at me pretty quick, man."

The ADS-B ground track showed the Cessna 340 traveling at 180 knots ground speed until the collision at about 150 feet AGL. The landing gear and flaps were up. A witness overflying the airport at the time heard the calls on CTAF, recognized the collision potential, and saw the Cessna 152 turn final. He then heard the 152 pilot announce his go-around and saw the Cessna 340 turn right and collide with it. Neither pilot nor the C340 passenger survived.

The Aircraft

The NTSB found no anomalies with either aircraft.

The Pilots

The 32-year-old Cessna 152 pilot had just over 100 total flight hours, with 77 hours in make and model. He held a private pilot certificate. The autopsy showed a small amount of marijuana (THC) in his blood, as well as traces of ketamine and aspirin.

The 75-year-old Cessna 340 pilot had approximately 1,200 hours in total flight time, with nearly 750 hours in make and model. He held a private pilot certificate for both single and multi-engine land aircraft. He also tested positive for THC. Other prescription drugs were present in his system, but for both pilots, the drug concentrations were not considered impairing.

NTSB Probable Cause

"...*The failure of the pilot of the multi-engine airplane to see and avoid the single-engine airplane while performing a straight-in approach for landing.*"

Commentary

Three items in the crash report stood out as unusual.

First, this is the first incident I've seen where both pilots had measurable levels of THC in their blood. Perhaps it's a sign of the times, but permissive state laws do not override FAA regulations prohibiting marijuana use. Having seen too many failed appeals to the NTSB regarding drug violations—please, just don't use drugs, even if federal regulations change!

Second, it's highly unusual for a pilot flying a pressurized multi-engine aircraft not to have an instrument rating. This limits flight operations to below 18,000 feet and to good weather conditions. These aircraft are designed for cross-country travel, where weather encounters are inevitable.

Third, the final approach speed for a Cessna 340 is about 110 knots with gear and flaps down. On short final—where the crash occurred—speed should have been about 95 knots. We can only wonder what the pilot was thinking and what his training regimen was.

Traffic pattern choreography requires finesse and, definitely, some courtesy. Whenever a pilot states their distance and direction from the airport, take it with a grain of salt. Pilots often announce their positions late and are closer to the airport than reported. Some even misstate their location or their relationship to the runway.

The Aeronautical Information Manual provides guidance on non-towered airport procedures, encouraging pilots to use a standard traffic pattern and enter on the downwind leg. However, there are times when a crosswind entry or straight-in approach makes more sense. The FAA advisory circular AC 90-66B (or its successor) provides additional clarity—or confusion—on pattern behavior.

Maverick, if the pattern is full, don't barge in—take a little longer and join the flow. But slow movers, think about who loses in a right-of-way encounter and decide if your prerogative is really worth it.

Pilots flying high-performance aircraft prefer to make straight-in, stabilized approaches—and for good reason: inertia! It works to everyone's advantage to get the bigger players down and out of the system as quickly as possible. Stabilized approaches for larger aircraft are even more critical than for light aircraft. Slow down, and remember—size does not include entitlement.

CHAPTER 4
WEATHER – EVERYBODY TALKS ABOUT IT

"When all is said and done, the weather and love are the two elements about which one can never be sure."

Alice Hoffman, author, Here on Earth

Once the basics of handling the aircraft are mastered, weather becomes the biggest variable. It is ultimately one of the most interesting challenges of aviation. There's plenty to discuss. It's divided into sections except for turbulence, which is addressed below.

Light aircraft are vulnerable to weather and while we like to say they are "go-anywhere, go-anytime" machines, most are not. Even the airlines regularly stand down or delay. Ever been stuck in Atlanta, Dallas, New York, or Chicago on the airlines because of weather? Depending on the pilot, the geography, the aircraft, and the season, schedule reliability will range from about 98 percent to perhaps 30 percent.

Referring to these crashes as "weather accidents" is incorrect. They are pilot judgment tragedies - examples of trying to get somewhere on a schedule where the aircraft and/or the pilot just doesn't have the juice to do it under the existing atmospheric conditions. After the crash it's usually obvious what went wrong. The "why" is more elusive. The hard truth that safety advocates don't like to admit is that many of us get away with some egregiously bad weather judgment regularly. The last link in the accident chain just didn't snap into place. If you want to complete more trips and make better informed choices, understanding weather is essential. *It's also about being flexible in your plans.*

Mom was right!

Remember mom's hot stove warning as a child? We got smart quickly because the result was immediate and certain. That certainty is missing from VFR-into-IMC, flying near thunderstorms, or where icing is predicted. Often, the weather is not as bad as the dire predictions, but sometimes it's much worse. Smart pilots take the "almost" lessons to heart.

Risk is usually analog, not digital. It's seldom "either/or." If every time an IMC warning, thunder, or ice was in the forecast and the bad stuff really was there, pilots would stop messing with it. On the other hand, if we canceled every time there was some possibility of meteorological nastiness, a lot of trips would not be taken that could have been completed safely.

Learning weather - The weather is often better than forecast. The government has been on the losing end of too many "failure-to-warn" lawsuits so, rather than truth-in-forecasting, sometimes there is defensive forecasting. You can't blame them. The science is getting much better, but a lot of art remains. Many commercial operators use risk analysis matrices that guide the dispatchers and pilots to a go- or no-go decision. The pros are methodical about it and generally have highly experienced pilots and much more capable equipment. For light aircraft, it's tougher. We must identify, eat, and digest the whole weather enchilada ourselves.

In Alaska, much of NTSB's work centers on weather-related crashes. Low visibility, heavy icing, lots of vertical terrain, and the ever-present economic incentive create one of the most challenging flight environments on the planet. A second opinion can help tremendously and that's the role of dispatchers. For Part 135 operations, NTSB has recommended qualified dispatchers, so it splits the responsibility of deciding to launch between two people. Either pilot or dispatcher can turn down a flight. It doesn't work all the time, but the odds shift to there being at least one adult in the room.

Weather takes time to learn. One of the best books, *Weather Flying*, was written by Captain Bob Buck and revised by his son, Captain Rob Buck. Captain Buck wrote an article in the April 1972 issue of *Air Facts* magazine that laments that many forecasts are designed for meteorologists, not pilots. My take on that:

> **"Rather than get pilots to think like meteorologists, have meteorologists think like pilots."**
>
> *Author*

Flight Service is not how most pilots get weather briefings today. Now, it's mostly a do-it-yourself proposition for light GA without a dispatcher. 800-WX-BRIEF should have been 800-CAN-I-GET-THERE? Back in the heyday of flight service stations, briefings were a mixed bag – sometimes the crystal ball was clear when it should have been cloudy and in other cases the pessimist on duty would warn of destruction when the nearest convection was 100 miles away.

Sometimes, after a crash the view is "We told you so." So how do we stitch all the pieces together in a weather briefing, online or live, to make a smart decision? There is a lot of advice but the best protection, in my

166

view, is to be skeptical of any marginal forecast and there are plenty of those. Weather doesn't always fit into our neatly preconceived buckets. Time and place are variable. Study it, watch several days out, and always have a Plan B. Various models predict different outcomes, and they change. Are the Metars performing as the TAFs predicted? 50/50 predictions means the dartboard in the forecast office is getting a lot of use.

It's easy to make weather decisions when flying locally and there's no pressure to get anywhere. Most of the GA training system provides relatively little in the way of practical weather education. Early flying takes place in the confines of the practice area, traffic pattern or "long cross countries" in carefully curated weather. That's why student pilots and instructional flights have very few weather judgment crashes.

Such training doesn't prepare new pilots for taking family and friends on a trip that penetrates a weather system. Cross country training is limited at most flight schools to just what the FAA requires, a limited exposure at best. It's very expensive to fly for hours just to learn how weather develops and changes. But there are some opportunities.

"You can observe a lot just by watching."

Yogi Berra, previously cited

Watch the weather before the flight, during, and afterward. Did it behave as expected? One of the best suggestions from Captains Buck is to look in the direction from where the weather is moving from to get an idea of what it might do when it gets to your location or flight path. Are the timing and conditions conforming to the forecast? Is the METAR what was expected? If not, dust off the alternatives.

Take free online courses, attend seminars, and webinars. Fly with experienced pilots to see weather decision-making in action. Ask them why they made the choices they did. This is how the airlines and military pilots develop their weather savvy. It is also one of the main reasons why light GA will always struggle to match air carrier safety records. The GA mentoring system isn't that robust and realistically can't be.

Hangar Story: *On a dual cross-country flight, while prepping a VFR student for solo cross-country, we were headed toward a small airport in the foothills of Virginia. The forecast called for scattered snow showers with ceilings generally above 3,000 feet and visibilities 7 miles or better outside of the showers. The view ahead of the Cessna 150 was becoming progressively scuzzier (a technical wx term).*

Not sure if my body language had any influence, but just before penetrating a rather nasty snow squall, the student said, "We need to divert." Hallelujah and just in time. We'd discussed diversion tactics before the flight. Without assistance, he turned away from the weather and plotted a general course to get home. This was in the days when a well-equipped Cessna 150 included a single VOR receiver and a sectional chart. It was one of the best lessons I'd ever given (or rather the weather had) and from then on I looked for marginal weather to allow my students to see what the real deal was. The forecast might be OK but the weather might not. It was the experience that was important, not the destination.

Looking back over many seasons of flying, some of my decisions weren't all that good. It's gotten better over the years, but the game is never over until you hang up the headset for the last time. Whenever you cancel a flight because of weather, look back online to see how bad it actually was. The feedback loop develops weather knowledge. There's great satisfaction in calling off a trip for all the right reasons. A few times, I canceled a trip that was flyable. More often, it was a good call to not have to land in East Armpit and rent a car or hang out in the FBO for a day until it was safe to go. And several times, landing short was the smartest thing I ever did, albeit inconvenient. Live to fly another day. Occasionally, a flight turns out OK, but in retrospect maybe it wasn't a great call. Then it's time to ask, what did I miss and what would I have done differently knowing what the weather actually was? How much did the risk go up?

I had one of those recently with plenty of convective weather, although there were lots of escape routes. But strong winds aloft and on the surface made the ride uncomfortable. Glad I didn't have passengers. The windshear on landing with a mighty crosswind made the workload high. Used lots of runway. In retrospect, waiting a day would have been smarter and safer.

> **"VFR rules were written when there were fewer, slower aircraft, and obstacles to collide with. There was also much more open space in which to make an off-airport landing."**
>
> *Author*

"Weather" Airspace – Here's a secret that neither FAA nor flight Instructors discuss much. If you plan flights even slightly conservatively, a lot of airspace and regulatory complexity simply vanishes. Remember all that business about transition areas: 700 feet agl, surfaced-based class E airspace and all the magenta ink (electrons) on VFR charts with the 500 below/1000 above/2,000 feet horizontally cloud minutia? Then there are the visibility minimums of one mile versus three miles compared to five miles except when the phases of the moon coincide with months that end in R. Most of that disappears by not flying VFR in truly marginal conditions.

For check rides and training we need to be able to explain all the airspace details, but for safe and practical flight *on a regular basis*, going IFR eliminates most of them. If VFR is the only option, long cross-countries become speculative or leisurely. If you insist, learn about forecasts, weather reality, and plan on flexibility.

To develop skill without going anywhere, plan pseudo-trips. Look at the Prog charts and forecast discussions a few days ahead. On the "pseudo day" get an online briefing, look at the forecast and then decide if you would fly or not. Then a few hours

Surface -based Class E . Set your VFR minimums at 2000/5 and all that complexity goes away!

later come back and look at how the forecasts panned out. Did you make the right call? It's not real weather flying but it sharpens strategic planning skills. This technique should be taught to every VFR and IFR student . However, eventually we must fly.

> ***"Everybody has to be somewhere but there is nowhere that you have to be."***
>
> *Somebody said this sometime*

Have you heard this before? If you *just gotta* make that business meeting or family gathering, develop an alternate plan that doesn't involve traveling by light aircraft or has the flexibility to go earlier or later. The airlines will delay or divert when the weather doesn't cooperate. Safety first and then convenience. If the weather is looking iffy 24 hours before departure, buy an airline ticket. If you buy it directly from the carrier and later decide to fly, cancel within the 24-hour window, most will provide a full refund.

Did an accident pilot push weather on earlier flights? Previous bad weather encounters aren't usually logged. But with ADS-B and ATC tracking websites, prior transgressions show up. NTSB can now often look back at previous flights to get a view of the pilot's risk tolerance and weather savvy. Very few get nailed the first several or dozen times out.

Success emboldens. The Bill Gates' quote in Chapter 1 about success being a lousy teacher is on target.

"Weather crashes are aviation's version of Russian Roulette with infinitely more possibilities but the same ultimate outcome if you keep playing."

Author

VFR into IMC accidents will result in fatalities nine out of ten times. The odds of surviving thunderstorms are almost as bad. Icing encounters are roughly 50 percent fatal. Conditions that appear similar may not be. Change any one thing in the event chain and "an almost" becomes a statistic – or it doesn't.

For example:

- Terrain – mountains make weather worse. Mountains are reputed to hide out in clouds and after dark. Who knew?

- Airport availability – is there a nearby bail-out option? If not, what's the alternate plan?

- Weather is infinitely variable. It could be better or much worse than the last encounter. Fronts and low-pressure systems move faster or slower than predicted.

- The aircraft--may be more or less capable/forgiving.

- The pilot--may be more or less proficient than the last time.

- Time of day--humans don't see well at night and may be fatigued. Off-airport landing spots and nasty clouds are generally hidden after dark.

It's easier in Big Iron - Feeling a bit inferior to the high and the mighty? Having flown weather in both light aircraft and turbines, it's much easier in the bigger machines. It takes an accomplished light aircraft pilot to tackle weather that the turbine crowd can fly over, around, or even through, often just by flipping a few switches or with an easy course or altitude change. There is no substitute for more horsepower or, "In thrust we trust." High wing loading means turbulence is more likely to be just irritating, not profoundly uncomfortable.

The ability to reach the flight levels means a comfortable altitude can usually be found and thunderstorms are much easier to spot and avoid. Enduring bumps for 10 or 15 minutes is so much easier than looking forward to several hours of bounce.

DEPARTURES				
TIME	DESTINATION	FLIGHT	GATE	REMARKS
12:39	BERLIN	BA 903	31	CANCELLED
12:57	SYDNEY	QF5723	27	CANCELLED
13:08	TORONTO	AC5984	22	CANCELLED
13:21	TOKYO	JL 608	41	CANCELLED
13:37	HONG KONG	CX5471	29	CANCELLED
13:48	MADRID	IB3941	30	CANCELLED
14:19	LONDON	LH5021	28	CANCELLED
14:35	NEW YORK	AA 997	11	CANCELLED
14:54	PARIS	AF5870	23	CANCELLED
15:10	ROME	AZ5324	43	CANCELLED

The big guys cancel too

What about "Taking a look?" This is self-deception because human nature is such that once we've committed to something, it's much harder to change course. Once airborne, we've already driven to the airport, invested time and money to begin what may ultimately be a fool's errand. And the farther into the trip, the harder it is to quit. The gambler in us wants to keep playing and that psychology is well-studied.

Continuation Bias is hardwired into us, as mentioned in Chapter 1. Been guilty of it myself. It takes effort, backbone, and humility to quit while ahead. But the reward of survival versus the alternative can hardly be overstated. To help you decide, imagine the bad outcome or merely the inconvenience of starting out and having to land short or retreat.

We safety types know how to rig the scales because only crashes are counted, and those statistics tell only one side of the story. Nobody keeps score of the good, or lucky, decisions where trips are successfully completed, so the glass is always more than half empty.

> *"You've got to know when to hold 'em, know when to fold 'em, know when to walk away, know when to run..."*
>
> *Kenny Rogers, singer/songwriter*

Smart gamblers and pilots set their losing limits. The weather odds are far better than the famed one-armed bandits in Vegas. Although electronic slots lighten your wallet even faster than aviation, they only take your money, not your life. *Passengers need to know when a trip is first planned, not at the airport after they arrive, that light aircraft trips require flexibility.* Take the stress off yourself and always have a Plan B which may include canceling.

Alternate Planning – This isn't often discussed but can have a significant impact on diversion experiences. The alternate airport may have an adequate runway and fuel, but other aspects can make for "interesting" experiences. Diverting to North Nowhere may be the perfect exploration of Americana or it could be a "Deliverance-type" event.

Look for facilities that include rental cars (crew car, if you're lucky), some nearby hotels/restaurants, the right kind of fuel, and is it attended (or access is provided) so you can arrive or depart when desired (Gotten the T-shirt on this one.) It's irritating to have to wait until 0900 to gain access to the ramp and pay the bill.

Should have checked the attendance hours!

Turbulence & Wind - While often forecast, non-convective turbulence doesn't cause many fatal crashes. Moderate turbulence is uncomfortable for pilots, miserable for passengers and not so easy on the airframe either. It won't break anything, but moderate can change to severe in seconds under the right conditions. Many non-pilot passengers will not be flying in light aircraft again after experiencing "moderate" turbulence.

Lenticular Clouds – extreme turbulence possible, NOAA

At NTSB, we investigated quite a number of airline incidents where passengers did not heed seat belt warnings and got a very hard look at the ceiling, armrests, and other passengers – see Pauli's exclusion theory in Chapter 1.

I was on Go-Team when we launched to a medevac crash in Nevada. A Pilatus PC12 had just departed Reno, enroute to Salt Lake City. The departure was normal, and the flight reached about 19,000 feet. There was no distress call, but the aircraft entered a spiral dive before ultimately breaking up in flight. It was a dark night with moderate icing, strong winds, and moderate turbulence expected.

Other turbine aircraft were flying, but sometimes the wrong place at the wrong time can complicate things tremendously. At this writing, only the preliminary report is out but spatial disorientation seems like a solid possibility. Autopilots will tolerate only so much roughness before they give the aircraft back to the pilot, who may not be immediately prepared to take command. It's too soon to tell on this crash, but never underestimate mountain weather.

In mountainous terrain the wind will occasionally bring an aircraft down with wave activity and down drafts. Some questions to ask before flying in turbulence:

- How high can we reasonably fly and still retain some ability to climb?

- How strong is the wind aloft and in what direction relative to the ridges?

- How high is the terrain and can the aircraft climb above it with good margin?

- Will oxygen be needed to get above the turbulence?

- Is there lower terrain nearby to move away from the worst orographically generated effects?

- What effect will it have on takeoff and landing? Runway length and alignment are critical.

- Would it be better to delay the flight a day or two after a front passes or depart a day early, before it arrives?

Hangar Story - *Bonanza A36: March was in like a lion and out like a Tasmanian Devil this year. The winds were fierce and contributed to several crashes, as they often do. The airlines had several incidents with passengers and flight attendants injured. Tractor trailers were blown off the road.*

I had a flight from the south land up to the Washington, D. C., area at the end of March. The day before departure a stationary front dumped a lot of rain in the Carolinas leaving behind low visibilities, but it looked reasonable to fly through the IMC remnants of the front, albeit into a tight wind gradient on the north end. There were some Airmets for turbulence, standard fare in March, but I missed two subtle clues. First, the forecast top for moderate turbulence extended up to 15,000 feet—more than the usual 10,000 feet. Second, the winds aloft at 7,000 were booming out of the west at 45 to 50 knots as opposed to the standard 30 knots or so. There were no pireps of anyone complaining about anything at low altitude, so I expected nothing more than a few jolts on descent. That may have been because there were not many other fools up flying in light aircraft.

Airborne at the crack of 0930 and for the first two hours there wasn't a bump, but a 15-degree crab angle confirmed a ripping 90-degree crosswind. The Appalachian Mountains inconveniently interfere with the northern part of the route, rising to about 4,000 feet in place - paltry by Western standards, but enough to be a potent turbulence maker. Clouds are the signposts, and the fair-weather cumulus had that shredded look, which telegraphs a rough ride.

In perfectly smooth air, the autopilot gave the first indication that it might get lively soon. While maintaining 7,000 feet the indicated airspeed dropped from 140 knots to 115 knots. Mountain wave. While still 100 miles from the hills, ATC granted my request for 9,000. Should've asked sooner. Despite a light load and full power, the best the Bonanza could manage was about 200 feet per minute, when it was climbing at all. After several minutes of trying to go up the down escalator and watching engine temperatures climb I advised ATC that 9,000 might not be in the cards today. A block of 8,000 to 9,000 was granted.

Now the bumps started in earnest and could conservatively be described as "enthusiastic." I filed a pirep. ATC acknowledged there was a lot of that going around and handed us off to the next sector to start the descent to 5,000..

The perversity of weather never fails to disappoint, because now we were in the up part of a wave. Powered back gradually to the bottom of the green arc on manifold pressure and deployed speed brakes - the landing gear. Maneuvering speed (adjusted for weight) should be considered the upper limit of how fast to go in moderate turbulence, and slower is better. Never use flaps to manage speed in turbulence – the G-limits are typically cut by about half.

At 110 knots and comfortably in the airspeed white arc, we were coming down at a leisurely 200 feet per minute with periodic sucker punches to liven things up. Once more, I advised ATC that it would be a while before we could get to the assigned altitude, and I filed my second pirep about 50 miles from where the saga had begun. The controller, in a true act of charity, cleared us direct to destination, which shifted the route away from the terrain. It would have been even better to wait a day - less wear and tear on occupants and the aircraft.

Pireps (again) are a potential lifeline and the ATC guidance for controllers recognizes that as well. Their manual (FAA Air Traffic Control Handbook

Order 7110.65 and recommended review by pilots) requires that controllers solicit pireps whenever an Airmet is in effect. The system for getting these essential reports to the Aviation Weather Center in Kansas City, which issues and modifies forecasts and Airmets is, at this writing, cumbersome at best and it works only sometimes. In the above scenario, had there been pireps of wave action, shifting the route a hundred miles east would have resulted in a much better ride. ATC knew the flight conditions, but that info was not widely disseminated.

Airmets, by current necessity, are a blunt tool and they often over-warn of conditions because the forecast models just aren't that accurate yet. They cover too much altitude and geography for too long. Sometimes, though, they're bang on. Some pilots come to ignore them and go out to "take a look." With timely airborne observations, more flights will be safely completed, rerouted, or canceled. When forecasts are routinely verified, computer modeling will get markedly better. In a dozen or so cases a year, it's my belief that lives can be saved and thousands more flights will be safely completed or appropriately canceled. NTSB has made recommendations to industry, the National Weather Service, and the FAA to expand the number of pireps in the system. Maybe some technology will allow us to datalink them down on an EFB. That will make a huge difference.

Do it in the dirt - One action that saved many pilots in the early days of flight is the off-airport precautionary landing. If painted into a corner and faced with flying into a thunderstorm, going IMC when not competent, or getting totally iced up, there are records of highly successful off-airport landings without injury and little to no aircraft damage. The old tail wheel aircraft were better suited to rougher fields and there was much more open space in days of yore

Every pilot should keep this in the emergency toolbox--just in case. The Helicopter Association International (Now Vertical Aviation International) has a program called "Land and Live," encouraging helicopter operators to land off airport or off pad when things start to go bad (See Calabasas in this Chapter.) It's a lot easier with vertical takeoff and landing ability, but it works for all aircraft. It's better to just divert sooner.

Key Points:

- Unless only flying local VFR, become a weather junkie.
- Weather is what you find, not what was forecast.

- On longer trips, plan a weather window to be able to fly earlier or later-- sometimes it will only be a few hours and in other cases, a day before or after will make the trip much safer and more comfortable. Sometimes it may be longer.

- Risk is analog and each situation is different.

- VFR into truly marginal conditions is unlikely to kill you the first few, or dozen times. Some pilots get away with it for years, but many times the house wins.

- Turbine aircraft have it much easier - higher, faster, heavier.

- Mountains make weather worse.

- Give and get PIREPS. Even when the weather is good, let the system know. It will greatly improve forecasts.

- Plan an alternate where you can be reasonably comfortable as opposed to sleeping in the aircraft

- If cornered, an off-airport emergency landing is always better than the alternative.

Clouds and Fog – A big deal for the VFR pilot

Suitable for IFR on top- Not VFR, NOAA

Clouds are more prevalent than all the other weather challenges in aviation combined. For the VFR pilot they are deadly. A flight often starts out in great weather and then clouds start getting in the way. Should you go over the clouds, go underneath, or retreat? Some options to consider.

"What part of 'Cloud' don't you understand?"

Author

Over the top:

- It's tempting, with a smoother ride and no towers or terrain. How high are the clouds and how much higher will they get? For normally aspirated aircraft above 8,000 or 9,000 feet there isn't much juice left - Give it up.

- If the deck starts to go from scattered to broken - Give it up.

- Are you certain - bet your life certain - that it will be visual all the way from the cruise altitude down to the traffic pattern at destination? If not - Give it up.

Under the clouds:

- If the ceiling gets low enough, towers, trees and terrain intervene - Give it up

- If you're still using pilotage and the ceiling is 2,500 or below, it becomes much harder to navigate.(Does anyone still do that?)

- If the decision is made to quit - often there's an airport close by. In flat terrain, once the ceiling gets below 1,500 to 2,000 feet it's time to rethink VFR operation. *Because the flight rules require at least 500 feet below the bases tall towers should be of real interest* and it leaves many fewer options on that rare occasion of an engine failure. In mountainous terrain it's best add at least another 2,000 feet because the terrain varies, the weather reporting is sparse, and so are the forced landing/diversion options.

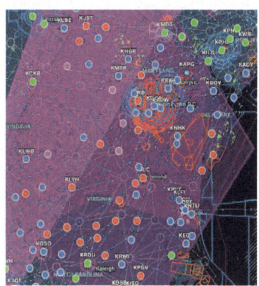

In the image here, the areas of low weather are abundantly clear. The light-magenta shows a general outline of instrument conditions. The colored dots represent airports with the latest METARs. Red is IFR, Magenta is Low IFR, Blue is marginal VFR and green is VFR. There are no hard decisions to be made here - It's an IFR day.

Cloudy Mid-Atlantic coast with mountains obscured to the west – Time for IFR!, Foreflight

Note the transition from instrument conditions to marginal to good VFR toward the north. There might be an optimistic Icarus who would make a run up the coast where the blue dots are. Could you get through VFR? Possibly, but is it worth betting your life? As Captains Buck noted in *Weather Flying*, weather is moving and constantly changing. Is this a transient morning condition that will get better as temperature and dewpoint spread or is it a stationary front that will get worse.

Visibility – Daytime, anything less than 4 to 5 miles has higher risk. Direction of flight and time of day complicates this. Anyone who's flown into the sun on a hazy day (morning or late afternoon) knows that forward visibility is zilch. At nighttime, outside the traffic pattern and especially in sparsely populated areas, mountains, or over water, visibility should be *at least* five miles or more. Instrument flying skills whether rated or not, will be needed. See Vineyard Spiral – this Chapter.

Special VFR – A seldom-used tool can save the day for the VFR pilot who waits too long before deciding to divert. It can be used at many towered airports, but not all--as identified on the charts. It allows limited VFR operations when the airport is operating under IFR. The reported visibility must be at least one mile, and the flight must be able to remain clear of clouds.

The pilot must request a special VFR clearance. ATC cannot offer it although occasionally a controller will attempt to guide a pilot by asking if there's "anything special" that they can do for them. At night, the pilot must be instrument rated and the aircraft IFR capable. Helicopters have different rules and IFR flights always have priority. If the weather is getting worse and/or you're starting to lose it, *declare an emergency* and you'll immediate help. There may be some questions and so what? File a NASA ASRS report and learn from the experience. Be cautious about using Special VFR outbound unless certain that it's just a local restriction and the weather is safe beyond the edge of the Class D.

Fog- Fog can form fast. Anticipate and have fuel to escape. It's most likely early and late in the day because there's not enough solar heat to increase the dewpoint spread. If VFR is the only option, a midday flight is often the best. But later in the day convection may fire up. It could be a narrow window.

Hangar Story - *As a young Air Force Officer, my first duty station was at Vandenberg AFB in Lompoc, CA (There's also a Federal Penitentiary there, but I had no occasion to visit.) The airplanes in the Air Force Flying Club fleet on the base were perfect for weekend jaunts. The club had nearly new Cessna 150s that a junior lieutenant could rent, even on government pay.*

One beautiful Saturday afternoon the sky was a bit hazy on departure, indicating a close temperature and dewpoint, but hey, it was good VFR to depart (I may have had some Rapture of the Sky as noted in the introduction.) After flying around and landing at various airports for about 1.5 hours it was time to go home.

On return, the ATIS was reporting the marine layer as 700 overcast and 2.5 miles. Since I was not instrumented rated at the time, there was no decision to be made and the flying club had a procedure for just such an occurrence. During the club checkout, the CFI noted that marine fog layers were common and if skunked, to simply land at Lompoc Airport next to town and farther inland, and they would retrieve the aircraft later – no penalty, no foul. It was good VFR at Lompoc, and I found a way back to base to retrieve my car. The club's planned alternative made the decision a no-brainer.

Key Points:

- FAA VFR minimums leave little margin at today's speeds.

- With automated weather reporting sites, it's a lot easier to check to see if the forecast became reality.

- Debate very carefully about going over the top versus staying low.

- Fog forms or dissipates at sunrise so anticipate that. If temperature and dew point spread increase with the warming of the day, it may get better. However, if there's a higher overcast that prevents the sun from raising the temperature, the fog could be there for a loooong time. Just the opposite happens as the sun goes down. Sunset with fog formation shortly afterward is not uncommon.

- Did you underestimate the potential for cloud presence and didn't escape sooner? Use the autopilot and declare an emergency if needed.

- Fuel is your friend in diversions.

- There are a lot more obstacles and aircraft to hit than in days of yore, but at least the terrain hasn't changed much (except maybe in California.)

Vineyard Spiral - (Loss of control over water at night)

A media frenzy and a challenging time for General Aviation

"Believe you can and you're halfway there."

Theodore Roosevelt, President of the United States

Being goal-oriented in life is a good thing. Every great leader, coach, successful businessperson will say so. But the desire to get to the destination must be tempered by real constraints. In the immortal words of

Yoda in Star Wars, "Do or do not--there is no try." However, with respect to President Roosevelt, sometimes believing just isn't enough.

The official NTSB report on the John F. Kennedy Jr. accident was released almost one year after the crash, an event that caused intense media and public scrutiny of general aviation--particularly regarding VFR flight at night. On July 16, 1999, at about 9:41 p.m. Eastern Daylight Time, a Piper Saratoga, N9253N, carrying JFK Jr., his wife, and his sister-in-law plunged into the Atlantic Ocean approximately seven and a half miles southwest of Martha's Vineyard, Massachusetts. Because of the exceptionally high profile, this report provides considerable insights into a tragic scenario.

The weather played an important part in this crash, although it appeared benign. The NTSB and National Weather Service determined that the weather at the crash site was VFR. It will become obvious, in hindsight, that this was not a good flight for a relatively

Piper Saratoga, Piper Aircraft

inexperienced VFR pilot to undertake. Even with considerable instrument training, Kennedy was unable to maintain control. The crash underscores the need for caution when flying over water and sparsely populated areas after dark.

Flight History – The flight originated from Caldwell, New Jersey (CDW), with a planned stop at Martha's Vineyard Airport (MVY), where Kennedy's sister-in-law was to be dropped off. The final destination was Hyannis, Massachusetts (HYA). Kennedy called the FBO in the early afternoon to say he planned to depart Caldwell around 6 p.m. Another pilot who was also heading out to the islands that night said that the auto traffic was the "second heaviest he had ever seen" and, as a result, Kennedy was delayed almost an hour beyond his own planned departure time. Witnesses at Caldwell saw Kennedy using crutches as he loaded luggage aboard the Saratoga. Air traffic control transcripts showed that the flight departed at 8:38 p.m. when it was nearly dark. After a quick discussion with the tower regarding the departure route, there was no further ATC communication.

A VFR transponder target, corresponding to the Saratoga, was observed a mile southwest of Caldwell at 1,300 feet. It exited the New York airspace and turned eastward to 100 degrees, climbing to 5,500 feet. The flight crossed the shoreline between Bridgeport and New Haven, Connecticut paralleling the Connecticut and Rhode Island coastlines; passed Point Judith, Rhode Island; and continued over the Rhode Island Sound.

About 34 miles west of Martha's Vineyard, radar data indicated a descent from 5,500 feet. The speed was calculated at about 160 knots, and the rate varied between 400 and 800 feet per minute. At approximately 9:38 pm, the flight turned right in a southerly direction. About 30 seconds later, the descent stopped at 2,200 feet and the target climbed for another 30 seconds.

The target then stopped turning and the airspeed decreased to 153 knots. At 9:39 pm, the target leveled off at 2,500 feet and flew southeasterly. Fifty seconds later, the target entered a left turn and climbed to 2,600 feet. As the left turn continued, it descended at about 900 feet per minute. On an easterly heading, it stopped turning with descent rate about 900 fpm. At 9:40:15 pm, while still in the descent, the target entered a right turn. By 9:40:25 pm, the bank angle exceeded 45 degrees, and airspeed increased to 180 knots. The last radar hit at 9:40:34 pm showed an estimated descent rate of more than 4,700 feet per minute. There were no survivors.

Wreckage Information - The Saratoga was located in 120 feet of water and struck the water right wing low, in a steep nose-down attitude. The attitude indicator showed a 125-degree right bank, and 30 degrees nose low. Data from flight and engine instruments showed that the engine was developing power at the time of impact. The tachometer indicated 2,750 rpm, above the redline of 2,700 rpm, and the airspeed needle was slightly above the

Airspeed indicator, note needle slap mark above 210 kts, NTSB

maximum 210 knot value shown on the instrument. The fuel-flow gauge indicated 22 gallons per hour. The landing gear was up with throttle and propeller controls full forward. Annunciator lights showed no evidence of filament stretching, indicating that the autopilot was not in use.

Aircraft information - The accident airplane was a 1995 Piper PA-32R-301 Saratoga II that Kennedy had owned since 1999. The annual inspection was completed in June 1999, with a total airframe time of 622.8 hours. There was no evidence of any pre-impact failures.

Pilot Information - Kennedy obtained his private pilot certificate in April 1998 and received a high-performance airplane sign off in his Cessna 182 in June 1998. His complex aircraft sign-off in the Saratoga was completed in May 1999. Estimated total flight time, excluding simulator training, was about 310 hours with 55 hours at night. Seventy-two hours were as PIC. Estimated flight time in the Saratoga was about 36 hours, with 9.4 hours at night. Only about three hours were without a CFI on board, and Kennedy had only about one hour of night solo in the PA32. In the 15 months before the accident, Kennedy had flown 35 flight legs either to or from Essex County/Teterboro, New Jersey, and the Martha's Vineyard/Hyannis, Massachusetts, areas.

The CFI who prepared Kennedy for his private pilot practical test observed that he had "very good" flying skills for his experience level. The pilot examiner who administered the check ride noted successful recovery from two unusual attitudes while wearing a hood. Kennedy flew approximately 179 hours in 1998 with 65 hours as PIC. He passed the instrument pilot knowledge examination in March 1999.

In April 1999, Kennedy went to a highly respected flight academy for concentrated instrument training, where he completed about half of the course. His instrument instructor noted progression was normal and that he "grasped all of the basic skills needed to complete the course." His instrument flying skills and simulator work were observed to be excellent. However, there was trouble managing multiple tasks while flying, which the CFI felt was normal for the pilot's level of experience.

Kennedy continued to receive flight instruction in the Saratoga from local CFIs. On one flight from Caldwell to Martha's Vineyard with an instructor, less than a month before the accident, an instrument approach was made into Martha's Vineyard through a 300-foot overcast.

Another CFI who flew with Kennedy for 39 hours between May 1998 and July 1999 accumulated 21 hours of night flight and 0.9 hour in instrument conditions. On July 1, 1999, the CFI flew him in the Saratoga to Martha's Vineyard, at night, with IMC at the airport. During the flight, Kennedy used and seemed competent with the autopilot. The CFI had to taxi the airplane and assist with landing because of Kennedy's leg injury. The instructor noted that Kennedy had the ability to fly the airplane without a visible horizon but was not ready for an instrument check as of July 1, 1999. The CFI would not have felt comfortable with Kennedy conducting night flight operations on that route and in those weather conditions. On the day of the accident, the CFI offered to accompany them that night but Kennedy "...wanted to do it alone."

A third CFI flew with Kennedy for nearly 60 hours between May 1998 and July 1999, including 17 hours of night flight and eight hours in IMC. This CFI had conducted the complex airplane sign-off in May 1999. On one or two occasions he noted a disparity in the airplane's autopilot where it turned to a heading other than the one selected but did not feel that the problem was significant. As noted earlier, no discrepancies could be found in what was left of the autopilot.

The CFI made six or seven flights to Martha's Vineyard with Kennedy in the accident airplane. Most were night flights, and Kennedy did not have any trouble flying the airplane. He was methodical about flight planning and cautious about his aviation decision-making. The CFI felt that he had the capability to conduct a night flight to Martha's Vineyard *if a visible horizon existed* – emphasis added.

In early June, Kennedy fractured his left ankle in a hang-gliding accident. The cast was removed, and on July 16, 1999, the day of the accident, he was given a straight cane. The orthopedic surgeon felt that, at the time of the accident, the pilot would have been able to apply the type of pressure with the left foot that would normally be required by emergency brake application in an automobile. It's speculative whether he would have been able to apply enough rudder pressure to recover from a high-speed spiral dive.

Meteorological Information - There was a wealth of weather data, but the warnings were subtle. *My apologies for all the detail but here's where a casual approach can lead to trouble.* Being suspicious is essential. Sunset was at 8:14 p.m. and civil twilight ended at 8:47 pm. When the accident occurred about 9:40 pm, the moon was 11.5 degrees above the horizon at a bearing of 270.5 degrees and provided about 19 percent illumination. Despite the relatively good weather report from Martha's Vineyard, several pilots reported considerable haze, which would have obscured what little moonlight there was.

Martha's Vineyard weather reported by automated surface observing system Automated Surface Observing Systems (ASOS):

- 7:53 p.m.--Clear; visibility 6 miles, haze; winds 240 degrees at 7 knots; temperature 23 degrees C; dew point 20 degrees C; altimeter setting 30.09.

- 8:53 p.m.--Clear; visibility 8 miles; winds 250 degrees at 7 knots; temperature 23 degrees C; dew point 19 degrees C; altimeter setting 30.09.

- 9:53 p.m.--Clear; visibility 10 miles; winds 240 degrees at 10 knots gusts to 15 knots; temperature 24 degrees C; dew point 18 degrees C; altimeter setting 30.10.

The ASOS could be edited and augmented by ATC tower personnel if necessary. Despite some assertions by the tabloid press, the NTSB found no anomalies regarding the ASOS. The tower manager reported, "The visibility, present weather, and sky condition at the approximate time of the accident was probably a little better than what was being reported. I say this because I remember aircraft on visual approaches saying they had the airport in sight between 10 and 12 miles out. I do recall being able to see those aircraft and I do remember seeing the stars out that night.... To the best of my knowledge, the ASOS was working as advertised that day with no reported problems or systems log errors."

However, the weather reported by the Nantucket ASOS was not nearly so good. Note the temperature-dewpoint spreads.

- 7:53 p.m.--Clear; visibility 3 miles, mist; winds 240 degrees at 12 knots; temperature 21 degrees C; dew point 20 degrees C; altimeter setting 30.09.

- 8:53 p.m.--Clear; visibility 4 miles, mist; winds 240 degrees at 11 knots; temperature 21 degrees C; dew point 20 degrees C; altimeter setting 30.10.

- 9:53 p.m.--Clear; visibility 4 miles, mist; winds 240 degrees at 12 knots; temperature 21 degrees C; dew point 20 degrees C; altimeter setting 30.11.

According to Weather Service International (WSI), a private weather service, Kennedy made two weather requests to WSI's Web site on July 16. The first was for a radar image at 6:32 p.m. and a route briefing request was made at 6:34 pm from Teterboro to Hyannis, with Martha's Vineyard as an alternate.

The 6 p.m. weather observations along the route indicated visibilities from 10 miles to four miles in haze at Caldwell. The lowest cloud ceiling was reported as 20,000 feet overcast at Providence, Rhode Island. At the departure point of Caldwell at 5:53 p.m., the sky was clear; visibility four miles in haze; and the winds were 230 degrees at 7 kt. Kennedy did not access the updated National Weather Service (NWS) area forecast (FA). Excerpts from the Boston area forecast, issued on July 16 at about 8:45 p.m. (after the flight had departed) and valid until July 17 at 2 a.m., included scattered clouds at 2,000 feet, occasional visibility 3 to 5 miles in haze, with haze tops at 7,000 feet for the area, including Martha's Vineyard.

Martha's Vineyard doesn't have a TAF but the forecast for Nantucket on July 16, valid from 2 p.m. July 16 to 2 p.m. July 17, was for clear skies, visibility greater than 6 miles, and winds from 240 degrees at 15 kt. A later forecast was not quite so optimistic. The 7:30 p.m. TAF, valid from 8 p.m. July 16 to 2 a.m. July 17, was for winds from 240 degrees at 15 kt; visibility 4 miles, mist; and scattered clouds at 25,000 feet. Temporary changes from July 16 at 9 p.m. to July 17 at 1 a.m.: clouds 500 feet scattered; visibility 2 miles, mist.

The terminal forecast for Hyannis also deteriorated; the 1:30 p.m. TAF called for clear skies and visibility greater than 6 miles, but the TAF issued at 7:30 called for winds from 230 degrees at 10 knots; visibility 6 miles, haze; and scattered clouds at 9,000 feet--with temporary changes from 8 p.m. July 16 to midnight July 17 of visibility 4 miles, haze. There were no Airmets, Sigmets, convective Sigmets, or in-flight weather advisories in effect along the route between Caldwell and Martha's Vineyard from 8 p.m. to 10 p.m.

Surface Weather Observations - The Nantucket weather was clearly marginal for VFR operations, especially at night. The one-degree temperature/dew point spreads at Nantucket and at Hyannis show the highly variable nature of weather in the islands. Even without clouds, especially at night, haze can be a significant obscuring factor. In this micro-climate, fog forms rapidly and can be localized. The Martha's Vineyard weather, *taken out of context,* would lead one to believe that VFR was a reasonable option.

Further evidence of that comes from some Coast Guard weather observations that would not normally have been available to pilots. At Point Judith, Rhode Island, the 5 p.m. and 8 p.m. reports were cloudy, with 3 miles' visibility in haze. By 11 p.m., however, it was cloudy with 2 miles' visibility.

Pilot Observations - One pilot who flew from Teterboro to Nantucket that evening requested weather for the islands. Visibilities were reported well above VFR minimums. He asked Flight Service "...if there were any adverse conditions for the route TEB to ACK. I was told emphatically: 'No adverse conditions. Have a great weekend.' I queried the briefer about any expected fog and was told none was expected and the conditions would remain VFR with good visibility. Again, I was reassured that tonight was not a problem."

The pilot departed Teterboro "...in daylight and good flight conditions and reasonable visibility. The horizon was not obscured by haze. I could easily pick out landmarks at least five [miles] away." Above 14,000 feet, the visibility was unrestricted. During descent to Nantucket, when GPS indicated that he was over Martha's Vineyard, he looked down and "...there was nothing to see. There was no horizon and no light.... I turned left toward Martha's Vineyard to see if it was visible but could see no lights of any kind nor any evidence of the island.... I had no visual reference of any kind yet was free of any clouds or fog." Upon contacting Nantucket Tower for landing, he was instructed to fly south about five miles; however, he maintained a distance of three to four miles because he could not see the island at five miles. Approaching the airport, he made a turn for spacing and "found that I could not hold altitude by outside reference...."

Another pilot flying from Bar Harbor, Maine, to Long Island, New York, crossed Long Island Sound at about 7:30 p.m. The preflight weather briefing indicated visual conditions, but the pilot filed IFR at 6,000 feet. Visibility ran two to three miles in haze throughout the flight. The lowest visibility was over water, but no clouds were encountered.

A third pilot departed Teterboro at about 8:30 p.m. destined for Martha's Vineyard. Climbing to 7,500 feet, the route took him over the north shore of Long Island. The entire flight was conducted under VFR, with a visibility of three to five miles in haze. Over land, he could see lights on the ground when looking directly down or slightly forward. Over water, there was no horizon; he encountered no cloud layers or ground fog during climb or descent. Near Gay Head, on the southwest corner of the island, he began to observe lights on Martha's Vineyard.

Another pilot at Caldwell canceled his planned flight from there to Martha's Vineyard because of the "poor" weather. "From my own judgment, visibility appeared to be approximately four miles--extremely hazy."

The interviewed pilots who flew that night, despite operating under VFR, were apparently experienced and qualified to fly IFR if needed.

NTSB Probable Cause -*"...Was the pilot's failure to maintain control of the airplane during a descent over water at night, which was a result of spatial disorientation. Contributing factors in the accident were haze and the dark night."*

Commentary - Kennedy had more than the average amount of instruction for his low total flight hours and flew more frequently than most non-career pilots. He could control an aircraft solely by reference to instruments and had done so before under similar conditions with a CFI on board. He was conducting a flight in an area with which he was very familiar and had been successful both at night and in IMC. Sadly, low time in the aircraft type and low pilot-in-command hours is a common predictor in many GA crash scenarios. The situation changes when a pilot must make all the decisions on his own and Kennedy had very little solo time in a high performance, complex aircraft in near IMC or IMC.

The reported weather could easily snare an unwary pilot, especially if the trip was compelling. Had the flight departed as originally planned, they would have arrived well before sunset. It's unknown how many non-instrument-rated pilots successfully made the trip that night and how many canceled. At night, over the water, hazy skies become dangerous. Decision-making must be based on what is seen (or not seen) through the windshield.

The flight proceeded normally until the descent when it appears the autopilot was disengaged. That was precisely the time, in retrospect, to leave it on. The fluids in the inner ear move in multiple directions, and the radar plot shows a classic disorientation spiral where the aircraft descended and climbed, made several turns, and then the pilot, when the aircraft was level, was tricked by his senses into rolling the aircraft into a gentle bank.

Ignore your senses. It is counter-intuitive, essential, and very difficult once vertigo sets in. It all unraveled in about two minutes from routine flight to impact. The unusual attitude training that most GA pilots get in training under the hood is almost worthless, in my view. While attempting to train it for instrument and private pilot students I've never had one totally lose it.

Kennedy had invested in quality training, was current, familiar with the area, and had the best equipment. The softer part of the analysis is the pilot's mindset. He was a successful entrepreneur, possibly under some business stress, pressured to make a schedule for a family activity, and must have experienced the self-imposed stress that most pilots feel to complete a trip. Couple that with his earlier successful exposure to the night/haze/IMC environment of the islands, and what seemed like a relatively low risk of spatial disorientation. What would you have done?

The Day the Music Died - (Buddy Holly crash)

VFR Pilot changes the course of music History

- *Pilot under tremendous pressure to fly top rock stars*

- *Weather moves faster than forecast*

- *Attitude indicator is different from what pilot is used to*

- *Inexperienced VFR pilot*

"Death is very often referred to as a good career move."

Buddy Holly, 1950s rock star (deceased)

Some accidents are burned into memory even decades after they happened. The sinking of the Titanic, the explosion of the Hindenburg, the

It was big news

collision of two Boeing 747s at Tenerife, and the space shuttle Challenger crash are all unforgettable tragedies.

Don McLean wrote a hit song, *American Pie*, in 1971 to commemorate the loss of singer Buddy Holly in an aircraft accident. The haunting line referred to "The day the music died." Charles Hardin Holley, better known as Buddy Holly, was and remains one of the giants in the music business. This may be the most-discussed music star aircraft crash in history. The impact on the music world still affects the public perception of

general aviation and generations of pilots later. His life and death have inspired numerous books, movies, and songs.

To say Holly was a star is an understatement. He has been described as the single most influential creative force in early rock and roll. He influenced countless musicians, including the Beatles, the Rolling Stones, and Bob Dylan. He exerted such a profound impact on popular music that Rolling Stone magazine ranked Holly number 13 on its list of the 100 greatest artists of all time. While only 22 at the time of his death, Holly was so prolific that new albums and singles were still being released years after his passing. The story of his crash and others like it, has been written many times but sadly, the outcomes don't change.

The Flight Plan--On the early morning of February 3, 1959, Holly and two other rising stars, Ritchie Valens, and J.P. Richardson (the Big Bopper), who were touring the country, had just finished a gig in Clear Lake, Iowa. They were scheduled to appear in Moorhead, Minnesota, the following evening but, because of bus trouble, the show headliners decided to go on to Moorhead by air. The group chartered a Beech Bonanza at the Mason City, Iowa (KMCW) airport to fly to Fargo, the nearest airport to Moorhead (KJKJ.)

The Civil Aeronautics Board (CAB), predecessor to the FAA and the NTSB, investigated the accident and what follows comes from their report.

Weather and Preflight - Around 5:30 p.m. Central Standard Time the charter pilot went to the Air Traffic Communications Station (ATCS—the equivalent of a Flight Service Station and Air Route Traffic Control Center) at the airport administration building, to brief the flight. He was provided current weather for Mason City; Minneapolis; Redwood Falls and Alexandria, Minnesota; and the terminal forecast for Fargo, North Dakota. The briefer advised that all stations reported ceilings of 5,000 feet or better and visibility of 10 miles or above.

However, the Fargo terminal forecast indicated the possibility of light snow showers after 2 a.m. and a cold frontal passage about 4 a.m. It all seemed reasonable for a VFR flight.

At 10 p.m. and again at 11:30 p.m. the pilot called ATCS to update weather. All stations had ceilings of 4,200 feet or better with visibility still 10 miles or greater. It was snowing in Minneapolis and the cold front that was previously forecast to pass Fargo at 4 a.m. was now expected to arrive at 2 a.m. At Mason City the ceiling was 6,000 overcast; visibility 15 miles plus; temperature 15 degrees F; dew point 8 degrees; wind south 25 to 32 knots; altimeter setting 29.96 inches.

At 11:55 p.m., the pilot, accompanied by the FBO/charter aircraft owner, a commercial/instrument-rated pilot, again went to ATCS for the latest weather update. With such important passengers on board, one couldn't be too careful. In the half-hour since the pilot had last checked, Mason City was now 5,000 overcast in light snow and the altimeter had dropped to 29.90. The weather was moving in.

The Flight - Holly, Richardson, and Valens arrived at the airport about 12:40 a.m., after the show, stowed their baggage and boarded the aircraft.

Although not noted in the CAB's report, I speculate the weight and/or balance might have been outside the limits with any kind of fuel load. That would have made the V35 a handful in the turbulence the flight would soon encounter.

The pilot stated he would file his VFR flight plan by radio when airborne. Taxiing to the end of Runway 17, the pilot called ATCS for a weather update. En route reports had not changed materially, but Mason City was coming down rapidly: The ceiling was now 3,000, sky obscured; visibility 6 miles, light snow; wind south 20 knots, gusts to 30 knots; altimeter setting 29.85 inches. The front had arrived.

The Bonanza was airborne at 12:55 a.m. and observed to make a left 180-degree turn and climb to approximately 800 feet. It passed east of the airport and turned northwesterly. Throughout most of the flight the aircraft's taillight was visible to the FBO/charter aircraft owner. About five miles from the airport the light gradually descended and disappeared. When the pilot failed to open his flight plan by radio soon after takeoff, the communicator (controller), at the owner's request, repeatedly tried to reach him but was unsuccessful. It was approximately 1 a.m.

The Crash- The Bonanza was reported missing at 3:30 a.m. The following morning the FBO owner flew the planned route and sighted the wreckage in an open field at 9:35 a.m. All four occupants had been killed and the wreckage was covered with four inches of snow. It's a given, even today, that accident investigations are usually done in decent weather, half a day later. Be patient with weather, it will get better. Had the group left at 10 o'clock that morning, they still would have arrived in plenty of time for the show.

The Bonanza struck the ground in a steep right bank, nose-low attitude at high speed. There was no fire and no evidence of structural or flight control failure. The landing gear was retracted, and the engine was producing cruise power at the time of impact. The attitude indicator showed a 90-degree right bank, nose-down attitude. The vertical speed indicator was pegged at a 3,000-feet-per-minute descent.

Pilot- The pilot, 21 years old, was employed by the FBO as a commercial pilot and flight instructor. He had been working with them about a year and had started flying in October 1954, with 711 hours total time and only 128 hours in the Bonanza. He had

approximately 52 hours of dual instrument training and had passed the instrument written examination, but had failed an instrument flight check in March 1958, nine months prior to the crash. His instrument training had been in several aircraft, all equipped with a conventional artificial horizon, but he had little experience with the Sperry attitude gyro that was installed in the Bonanza. These two instruments differ greatly in their

Sperry Gyro Advertisement

pictorial display, and the CAB believed that he would have had difficulty interpreting a completely different display.

The Aircraft - The Beech Bonanza, model 35, was manufactured in October 1947 and the engine had only 40 hours since major overhaul. The aircraft was purchased by the FBO in July 1958 and was well-equipped with high- and low-frequency radios, a Narco "Omnigator" (VOR), a Lear autopilot (recently installed but not operable), and a full panel of instruments used for instrument flying, including a Sperry F3 attitude gyro.

According to the CAB's report, "The conventional artificial horizon provides a direct reading indication of the bank and pitch attitude of the aircraft which is accurately indicated by a miniature aircraft pictorially displayed against a horizon bar and as if observed from the rear. The Sperry F3 gyro also provides a direct reading indication of the bank and pitch attitude of the aircraft, but its pictorial presentation is achieved by using a stabilized sphere whose free-floating movement behind a miniature aircraft presents pitch information with a sensing exactly opposite from that depicted by the conventional artificial horizon."

The Weather, Again - The weather was nastier than the surface reports indicated. The weather map for midnight February 3, 1959, showed a cold front extending from northwestern Minnesota through central Nebraska with a secondary cold front through North Dakota. Widespread snow shower activity was indicated in advance of these fronts. Temperatures aloft from Mason City to Fargo were below freezing at all levels with an inversion between 3,000 and 4,000 feet and abundant moisture present at all levels through 12,000 feet. Moderate to heavy icing and precipitation existed in the clouds along the route. Winds aloft were reported to be southwest at 30 to 50 knots.

A flash advisory (roughly equivalent to a Sigmet) issued by the Weather Bureau at Minneapolis at 11:35 p.m. on February 2, noted, "A band of snow about 100 miles wide at 2335 from extreme northwestern Minnesota, northern North Dakota through Bismarck and south southwestward through Black Hills of South Dakota with visibility generally below two miles in snow. This area or band moving southeastward about 25 knots. Cold front at 2335 from vicinity Winnipeg through Minot, Williston, moving southeastward 25 to 30 knots with surface winds following front north-northwest at 25 knots with gusts to 45 knots. Valid until 0335."

Another flash advisory issued by Kansas City, Missouri, at 12:15 a.m. on February 3 noted: "Over eastern half of Kansas, ceilings are locally below one thousand feet, visibilities locally two miles or less in freezing drizzle, light snow, and fog. Moderate to locally heavy icing, areas of freezing drizzle and locally moderate icing in clouds below 10,000 feet over eastern portion Nebraska, Kansas, northwest Missouri, and most of Iowa. Valid until 0515."

Neither ATCS briefers mentioned these flash advisories to the pilot indicating the virtual certainty that instrument weather would be encountered.

Civil Aeronautics Board (CAB) Probable Cause - *".... was the pilot's unwise decision to embark on a flight which would necessitate flying solely by instruments when he was not properly certificated or qualified to do so. Contributing factors were serious deficiencies in the weather briefing, and the pilot's unfamiliarity with the instrument which determines the attitude of the aircraft."*

Commentary - The CAB report noted that the flash advisories were not conveyed to the pilot. The weather briefing consisted solely of reading current weather enroute, and the current destination, and destination forecasts. Failure to "draw these advisories to the attention of the pilot and to emphasize their importance could readily lead the pilot to underestimate the severity of the weather situation."

The FBO owner said, he "Had confidence in the pilot and relied entirely on his operational judgment with respect to the planning and conduct of the flight." That confidence was sadly misplaced. It happens too often that enthusiasm and a strong desire to complete a flight overcome what little experience/judgment a new pilot has. This is especially problematic with a celebrity client as shown in the following case study on Calabasas. The desire to please should never take a back seat to suspicious, skeptical contingency planning. Easier said than done.

The CAB noted that with the obviously deteriorating weather at Mason City, which could be seen by all, and the fact that the charter company was "...certificated to fly in visual flight rules only...together with the pilot's unproven ability to fly by instruments, made the decision to go...most imprudent." That the pilot checked the weather so many times and that the owner went with him and then watched the flight depart shows that both of them probably had some serious misgivings. Listen to that inner voice--it's usually right.

The CAB's assessment was that shortly after takeoff the flight entered complete darkness with no horizon, falling snow, and moderate turbulence from the high winds. This required flight by reference to instruments. The pilot's unfamiliarity with the Sperry F3 gyro, noted above, because of its unique presentation, likely caused spatial disorientation.

This report could have been written last month, but the crash occurred more than half a century ago. If the weather is bad where you are, despite a decent forecast, the weather is bad. Period. The board's commentary: "This accident, like so many before it, was caused by the pilot's decision to undertake a flight in which the likelihood of encountering instrument conditions existed, in the mistaken belief that he could cope with enroute instrument weather conditions, without having the necessary familiarization with the instruments in the aircraft and without being properly certificated to fly solely by instruments." The lesson should not fade away.

Frontal Assault - (C210 pilot goes VFR into IMC)

A VFR pilot's first solo flight in a high-performance aircraft proves to be too much.

> **"The snow doesn't give a soft white damn whom it touches."**
>
> e.e.cummings, poet, artist. playwright

Exemplar C210, AOPA

Weather sometimes moves in fast and mysterious ways. Airplanes have always moved faster, but pilots must have the will to retreat. When the legal system gets involved, the aircraft is always suspect. What follows is the National Transportation Safety Board's analysis of an accident and an opposing viewpoint raised at the resulting trial.

The Pilot - The private pilot had accumulated 256 hours in about five years while progressing through the ranks of Cessna singles to the Cessna 210. This was his first flight in the 210 without an instructor although he had accumulated more than 30 hours of dual 210 time. He was not instrument rated.

The Flight - The flight departed Raleigh, North Carolina, at 8:45 a.m., for a VFR trip to Washington, DC. Shortly after takeoff, the pilot obtained VFR flight following from Washington Center. According to Raleigh Flight Watch, the weather did not appear threatening when the pilot called airborne, about 45 minutes before the accident. Although the flight was on top of a broken layer of clouds at 9,500 feet, the specialist seemed unconcerned: "...clear skies into the Richmond area with that scattered layer at 9,000, remains scattered through Washington at that altitude, so you should be okay."

The pilot replied that he was getting close to the cloud tops. The specialist provided the Washington National (now Reagan) terminal forecast of 4,000 to 5,000 scattered, with a possible 9,000 broken layer, and suggested that the pilot contact Washington Flight Watch in a few minutes. Just after the 210 returned to the Center frequency another aircraft requested weather for Harrisburg, Pennsylvania, which is about 80 miles north of Washington. The specialist responded with 1,900 scattered, estimated 5,000 broken, and towering cumulus to the east. A front had passed very quickly, dropping visibility in Harrisburg to only two miles in blowing snow, followed by rapid clearing. Had the pilot remained on the Flight Watch frequency to hear that information, it might have alerted him that the front was about to change the DC. area weather drastically. But it might not--Harrisburg was not near the destination.

Everything progressed routinely with a hand-off to Washington Approach at 10:02 a.m. EST. Descending to 3,500 feet, the pilot requested 2,500 and was cleared down.

Abbreviations: 01N is the Cessna 210, N6401N; Approach is a Washington Approach radar controller; and Dulles is Dulles Tower. Comments in italics.

10:09:51 01N: "Washington Approach, Zero-One-November; I'm getting into some snow here and low clouds."

10:09:55 Approach: "Zero-One-November, roger."

10:09:58 Approach: "Zero-One-November would you like an IFR...are you IFR rated?"

10:10:00 01N: "Negative."

10:10:04 Approach: "Zero-One-November, roger; maintain VFR."

10:11:44 01N: "Zero-One-November, I'm reversing my course." *[The frontal weather was approaching from the northwest. Once the pilot was inside the clouds a 180-degree turn would have turned the aircraft to the southwest and carried the flight deeper into the fast-moving clouds. A turn to the east or southeast would probably have been the shortest route to VFR.]*

10:11:49 Approach: "Zero-One-November, roger. Would you like me to check around to other airports and see what the weather is for you?"

10:11:53 01N: "Roger; I need to get out of this."

The controller called Dulles tower to check their weather.

10:12:24 Approach: "The Washington weather: 500 scattered, estimated ceiling, 1,500 broken, 3,500 overcast, visibility four, light rain and snow showers."

10:12:38 01N: "Roger, clear me to on top at 3,000 — ah — 5,500 I'm going on top."

10:12:41 Approach: "OK, Six-Four-Zero-One-November, fly heading 270. Climb and maintain VFR conditions on top. If not on top by 5,500, maintain 5,500 and advise." *[The controller apparently did not have a good picture of the weather, since the 270-degree recommended heading was putting the flight right into the front.]*

10:12:50 01N: "Roger; going up."

10:13:30 Approach: "Zero-One-November, what would you like to do now, sir?"

10:13:33 01N: "Roger; let me get on top and clear me to the nearest airport that's clear."

10:13:43 Approach: "All right, Dulles says they're VFR now, sir."

10:13:42 01N: "Roger. Do you have anything private, and I'll wait till this gets over, then come on into National?"

10:13:47 Approach: "I can't really say what the weather is at PG or College Park (general aviation airports in the vicinity), sir."

10:13:53 01N: "Roger; let me clear up on top, then I'll check back with you."

10:14:00 Approach: "OK. Don't leave the frequency, though. I've got traffic all around you down there."

10:14:02 01N: "Roger."

10:14:49 Approach (to Dulles Approach): Ten south of Davidson [another local airport], Zero-One-November is climbing VFR to 5,500 feet. He got caught IFR and he's not rated."

10:14:51 Dulles Approach (to Washington Approach): "My guy Two-Four-Lima [apparently another VFR aircraft whose pilot chose a lower altitude] can't stay VFR. He's been going down."

10:18:18 01N: "Zero-One-November needs a turn."

10:18:20 Approach: "Zero-One-November, roger. Where would you like to go now?"

10:18:23 01N: "Just turn me away from this stuff."

10:18:24 Approach: "All right, turn right heading 360, sir."

10:18:37 Approach (to Dulles): "What's the weather like at Manassas? You have any idea?"

10:18:56 Dulles: "It's pretty good, Joe. It's cleared out over here. They can probably get over there VFR if they're at 2,500 or below."

10:19:03 Approach (to Dulles): "OK; thanks, Jerry. I'll let you know."

10:19:07 Approach: "Zero-One-November, Dulles says it's clear around Manassas, if you'd like to go over toward Manassas."

10:19:56 Approach: "Zero-One-November, Washington."

There was no further contact with the airplane, despite repeated attempts. Recorded radar information showed that at 10:07 a.m. aircraft was about 26 miles southwest of Washington National when it descended from 3,500. At about 10:12, the aircraft was 16 miles southwest and had descended to 1,300 feet. This was well below the assigned altitude of 2,500. The aircraft went another mile, and at 1513 it reversed course and began climbing. It headed southwest until it was about 25 miles from the airport. The last radar hit showed 4,600 feet near the Marine Corps Air Station at Quantico, Virginia.

The Weather - The 10:20 weather observation at Quantico reported 500 scattered, 1,500 broken, and five miles in light rain and snow, with winds out of the northwest at 20 gusting to 29 knots. Visibility subsequently dropped to three miles.

National Weather Service radar showed a 90-mile-long area of snow showers (oriented southwest to northeast) centered near Quantico. The line was moving southwest at 40 knots, with tops to 13,000 feet.

Approx. crash location, Foreflight

An airline pilot flying a Boeing 737 had encountered "severe weather conditions" while inbound to Dulles about an hour before the accident. The pilot said he was unable to stay out of the clouds and requested an ILS. Once in the clouds, the air was very turbulent, and airspeed varied as much as 15 knots. He broke out over the approach lights at 300 feet agl and reported very heavy snow. "At no time during our approach and landing did approach control, tower, or other aircraft make any mention of these conditions or even the possibility of them," the pilot said. "Within 30 minutes, this storm was gone, and Dulles was VFR with scattered clouds. In 32 years of flying, with the exception of thunderstorms, I have never seen such a rapidly developing, moving, and severe weather condition move through an area."

There were at least five witnesses who saw the 210 descend from the base of the clouds in pieces. A flight instructor who had just landed at Quantico stated, "I elected not to go far from the air base, due to some very ominous-looking clouds that appeared to be approaching from the west northwest...with a base of 1,000 to 2,000 msl (roughly our altitude) with a layer of 'scud' underneath, perhaps 200 to 500 agl. The clouds were very dense, which I think caused their dark color.... The cloud cover to the east of the ominous clouds was scattered to broken at a fairly high altitude (estimated 3,000 to 4,000 or above)."

The wreckage was scattered over a one-square-mile area. The fuselage was generally intact and highly compressed. None of the three occupants survived. The rudder, elevator, upper vertical stabilizer, and horizontal stabilizer had separated. Both wings separated from the fuselage. The engine appeared to have been developing power, and the tachometer indicated 2,500 rpm.

NTSB Probable Cause - *"...Pilot-in-command — poor in-flight planning/decision, VFR flight into IMC—inadvertent. Aircraft control not maintained, spatial disorientation, exceeded design stress limits of the aircraft."*

Commentary - I would add one additional contributing factor that the forecast was substantially *incorrect*, but weather is what you find, not what you've been told (That may have been mentioned before.) Additionally, had the pilot used the autopilot, control would likely have been maintained. That was the same omission as in the Vineyard Spiral crash earlier

The pilot's widow sued both Cessna and the FAA. Cessna was charged with building a defective aircraft; and the government, with issuing a defective forecast. The litigation showed a considerable difference of opinion. The transcript of the trial filled more than 30 volumes and stood more than two feet high. The attorneys' views are summarized below.

Exemplar spar, Van's aircraft

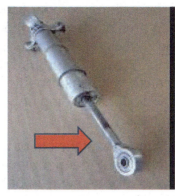

Trim tab actuator, Cessna

The plaintiff's attorneys contended that the Cessna 210 was improperly designed, and that the aircraft structure had failed because an aluminum spar had sawed through the steel trim tab actuator. Spars are formed from thin sheets of aluminum while the trim tab actuator is hardened steel, about the thickness of a pencil.

Lightening holes are made in the spar that are significantly larger in diameter than the actuating rod that passes through the hole enroute to the trim tab. For the rod to come into contact with a spar, the horizontal stabilizer/elevator would need to be severely deformed. The crash aircraft had undergone an annual inspection less than 200 hours prior to the crash and it's highly unlikely that a solid steel rod actuating rod would have managed to fail without being discovered in that time. Nor was there evidence that the spar was in contact with anything.

According to the plaintiff's attorneys, the "failed actuator rod" allowed the elevator to flutter. Under certain conditions, according to the plaintiffs, the entire empennage assembly would twist slightly, and the spar would start to chew into the rod. The contention was that Cessna had improperly tested the aircraft and had known about the problem as far back as 1963. It's creative theory. They cited multiple in-flight breakups without mentioning that none had crashed due to a failed trim actuator rod.

Cessna's attorney responded that these aircraft had tens of thousands of flight hours and they had taken video of an exemplar 210 loaded exactly as the crash aircraft which revealed no twisting or anything unusual. The aircraft was flown to Vne with no unusual results. However, the jury found Cessna guilty of improper design and awarded the plaintiff $5 million. There have been no Airworthiness Directives relating to Cessna 210 trim tab flutter to date.

It sounds reasonable to have a jury hear the case, with experts from both sides providing their views. The reality is seldom that pure. Over the past several decades, juries with no aeronautical expertise have passed negative judgment on an aircraft certification system that has delivered some excellent flying machines. To be sure, occasionally there's a flaw but in the vast majority of cases, a properly maintained aircraft is unlikely to be the probable cause, if flown properly. The NTSB's investigation is either vindicated or vilified, depending on your point of view and who's paying the expert witness.

Highly paid experts support a plausible or fantastical mechanical failure theory. Or they attempt to get the manufacturer off the hook. Objectivity on both sides *may* occasionally be swayed because massive amounts of money are on the line. Occasionally, an official investigation may miss something, and the system allows for reconsideration, if *new* evidence is found. This is called a Petition for Reconsideration. NTSB's role is to find probable cause - not defend or prosecute.

The FAA is responsible for the certification and safety in aircraft design, not juries or the courts, but we may have to disagree on that. If there is a problem with an aircraft, then it should be promptly dealt with through the appropriate recall systems. We shouldn't be waiting six years to find out that there's a defect - if there is one. NTSB will issue an urgent recommendation to FAA if they find a serious design defect and publish an alert to the aviation community to ensure timely notification.

The Beech Bonanza V-tail is an example where a change was dictated after multiple in-flight breakups occurring over several decades. The Cessna 441 needed the horizontal stabilizer to be strengthened after a design flaw caused a single fatal accident shortly after it was introduced. There hasn't been a tail failure since. The Boeing 737 MAX is perhaps the best-case study on a design/certification failure (See Automation chapter.) It happens, rarely though. In roughly 80-90 percent of the cases, the pilot has taken a perfectly functioning aircraft into the ground.

Plaintiffs are entitled to recover legitimate losses. Manufacturers are responsible for design and manufacturing rigor, and the FAA is there to see that it's done right. NTSB serves as a neutral arbiter. Finally, pilots are responsible for operating within the limits of their certificates and proficiency, and those of the aircraft. Maintenance techs have an equally awesome responsibility to maintain aircraft according to the regulations. *The pilot's decision* to fly VFR into IMC is a violation of both the FARs and common sense.

Pointing Fingers - (SR22 pilot tries to get to a hockey game)

Inadequate training or inadequate decision-making?

This crash is notable but not because of the event. Flight into marginal VFR or instrument meteorological conditions (IMC) by ill-equipped pilots is not as rare as it should be. There are constant reminders by the FAA, NTSB, AOPA Air Safety Institute, EAA, Flying, AVweb, and most aviation publications/websites that this is not a life-prolonging activity. The legal maneuvering afterward, however, nearly changed the training landscape.

This January 2003 crash also illustrates that isolated actions by individual pilots can have a major effect on general aviation when magnified by the legal system. There are similarities to the John F. Kennedy Jr. crash in June 1998 (see Vineyard Spiral) with marginal conditions at night, an almost-rated instrument pilot in a high-performance aircraft, and a strong desire to get to the destination. These factors have consistently proved to be lethal. The determination to press on, which sometimes serves well in personal and business life, can be catastrophic in aviation.

Pilot -The 47-year-old private pilot had logged 248 hours total time, with 57 hours of instrument time and nearly 19 hours at night. He had taken delivery of a new Cirrus SR22 six weeks prior to the accident and completed type-specific training. This resulted in a VFR-only completion certificate and a high-performance aircraft endorsement. At the time of the crash, the pilot had logged 19 hours in the Cirrus with only 0.3 hour of actual instrument time and 2.3 hours of night flight. Almost all his remaining flight time was in a Cessna 172 in which he was receiving instrument instruction. The pilot's CFII stated he was "nearly ready" for the instrument rating practical test.

> *"The door to safety swings on the hinges of common sense."*
>
> *Anonymous*

NTSB: "According to Cirrus Design/University of North Dakota records, the pilot completed the SR22 training course on December 12, 2002. The course consisted of 4 flights for a total of 12.5 hours of dual flight instruction and 5.3 hours of ground instruction." (The University of North Dakota was the training contractor for Cirrus at the time of the accident.)

NTSB: "The record indicates a ground lesson, which included 'Brief on VFR-into-IMC procedures,' was completed on the last day of the course. The flight lesson titled 'IFR Flight (Non-rated)' was not conducted." This was the probable basis for the ensuing lawsuit.

The Flight - At 4:55 a.m. (CST) on the morning of the accident, the pilot called flight service proposing a VFR flight from Grand Rapids, Minnesota (GPZ), to St. Cloud, Minnesota (STC), departing at 6 a.m. A friend was to accompany the pilot to a hockey game, where their sons were playing. The weather was marginal with Airmets for instrument conditions, scattered snow showers, and turbulence. The pilot called back again at 5:41 a.m. to recheck the weather, but it was essentially unchanged. The pilot mentioned that he was, "Hoping to slide underneath it and then climb out." The briefer didn't discourage him, saying, "The only problem you may have along the route, that I can see, is marginal ceilings." That was encouraging but it's not the briefer who is at risk. Low ceilings are deadly, especially in the dark when clouds are not so visible.

Weather can be deceptive and certainly not uniform. At the time of the crash, Grand Rapids, 20 miles north of the crash site, reported few clouds at 300 feet, a broken layer at 1,400, and 2,700 overcast with seven miles visibility and winds northwest at 17 gusting to 22 knots. Twenty-one miles south of the crash site at Aitkin, Minnesota, visibility was 10 miles, with scattered clouds at 2,500. There were two additional factors: It was still dark and morning twilight would not begin for about another hour, but there was a full moon low to the horizon.

The Cirrus struck trees at a 15-degree nose-down attitude, creating a 500-foot debris field indicative of a high-speed impact. The pilot and passenger died in the impact. The final flight path was estimated from radar data. Beginning at 6:30 a.m. the aircraft's altitudes varied between 1,700 and 3,200 feet. At 6:36 a.m. the aircraft began a descending left turn to 2,400 feet at about 1,200 fpm, followed by a climbing left turn of decreasing radius to 2,900 feet. The average true airspeed was estimated at 191 knots.

Several witnesses reported the aircraft between 75 and 100 feet agl. *One witness stated, "If he'd been two blocks east, he'd have hit the water tower.*)" The witnesses all agreed that the aircraft was moving fast, and weather was generally described as clear--except for one witness who lived a quarter mile from the accident site, and said it was snowing lightly and the atmosphere was "hazy." A question that arises from the witness statements was if the weather was "generally clear" why did the pilot feel compelled to fly so low? Weather phenomena are not necessarily uniform and while it may be good here, five miles away could be low IMC.

The Aircraft -The NTSB's examination of the aircraft and engine did not reveal any malfunction. The aircraft had logged 35.7 hours since new.

NTSB Probable Cause - *"…Spatial disorientation experienced by the pilot, because of a lack of visual references, and a failure to maintain altitude. Contributing factors were the pilot's improper decision to attempt flight into marginal VFR conditions, his inadvertent flight into instrument meteorological conditions, the low lighting condition [night], and the trees."*

Commentary - The plaintiffs' attorneys attempted to shift responsibility from the pilot to others outside the airplane. The complaint against Cirrus Design, as the aircraft manufacturer was known then, and the University of North Dakota Aerospace Foundation (UNDAF) was that the entities did not provide the contracted-for training. Transition training was included in the purchase price of the aircraft, and the pilot had paid for an additional 1.5 days of instruction to become more familiar with the SR22. The pilot's training records clearly stated that the training was to VFR standards only, and various autopilot procedures had been reviewed. However, one item in the course syllabus remained unchecked. It was use of the autopilot to escape from IMC. The plaintiffs' beliefs were that the crash would not have occurred if only that training had been received. It's an interesting theory, but to my knowledge, has never been tested in any scientific manner.

The Private Pilot Practical Test Standards at the time required applicants to demonstrate proficiency in basic instrument flight. Knowledge areas include aeronautical decision making and judgment. Also required is the ability to recognize critical weather situations from the ground and in flight, along with the ability to procure and use weather reports and forecasts. The required "training" is more for introduction to instrument flying than to develop any real proficiency.

As noted previously, the pilot had nearly finished a full instrument rating program but was inexplicably taking the training in his old aircraft, a Cessna 172. It's likely that the new aircraft was a bit overwhelming for this new pilot to manage basic flight in a high-performance aircraft and IFR procedures.

Instrument flight is the same for all aircraft, but the performance characteristics and avionics equipment between these two airplanes could hardly be more different. The pilot's CFII was not a defendant even though he had a lot of contact with the pilot and the ability to influence his decision-making process. However, many CFIs are in legal lexicon "judgment proof" because most have no assets worth recovering.

FAR 91.3 is unambiguous in declaring the pilot in command to be responsible for and the final authority as to the operation of an aircraft.

The FARs also are clear on certification, testing standards, and the requirement for VFR pilots to stay out of IMC. Even so, the jury found Cirrus and UNDAF jointly responsible at 37.5 percent each, and the estate of the pilot at 25 percent. It awarded combined damages of $16.4 million to the families.

Minnesota law prohibits lawsuits for educational malpractice because it's extremely difficult to determine if the teacher failed to teach or the student failed to learn. The plaintiffs and trial court felt that the defendants were not an educational institution and not entitled to such protection. (UNDAF hardly could be anything else, but legal logic is not always clear.)

Cirrus and UNDAF appealed, with AOPA filing a friend-of-the-court brief in support, because of the potential precedent. The Minnesota Court of Appeals reversed the trial court's ruling in a two-to-one split decision, thus finally settling the matter in favor of Cirrus and UNDAF. It said, "An airplane manufacturer's common-law duty to warn of dangers associated with the use of its aircraft does not include a duty to provide pilot training. A negligence claim against an aviation-training provider is barred under the educational-malpractice doctrine where the essence of the claim is that the provider failed to provide an effective education."

Additionally, the pilot's operating handbook was cited as explaining how to use the autopilot in great detail. There were explicit warnings about VFR pilots operating in IMC. The court held that the pilot had a duty to review such documents.

The trial court's verdict, had it not been reversed, would have been a significant disincentive for both manufacturers and flight schools to provide any training beyond that which is legally required. Unfortunately, expensive and emotional legal battles do not erase the pain for two fatherless families--or the associated negative perception that surrounds GA in the aftermath of crashes like this.

- Do you believe that this pilot fundamentally understood the risk he was about to undertake? The human psyche is incredibly complex, and the statistics of probability and possibility are arcane. The beauty of hindsight bias is that we can say, "Of course not."

- Did some outside impulse push him to take the risk, or did he honestly think he could make it? Is getting to a hockey game worth it?

- If more training in decision-making or autopilot management had been provided, would it have mattered?

- Is it possible to change someone's risk-taking profile?

- When is enough training provided, and who's responsible after a crash?

- Should high-performance aircraft be sold to VFR pilots?

As stated throughout, the pilot-in-command is responsible. There are many opinions, but the facts and the regulations are the basis of our safety system. Money distorts perceptions but it will not bring back loved ones. As stated throughout, the pilot in command is responsible for and the final authority as to the operation of an aircraft. We must recognize our limitations, those of our aircraft, and the vagaries of weather forecasting.

Aftermath - (Firsthand look at a crash investigation)

It wasn't the VFR pilot's first encounter with IMC

VFR into IMC – Unhappy ending , Author

This isn't something to share with passengers, but all pilots should read. Pilot-in-Command is an impressive title with commensurate responsibility.

With light winds and clear skies, the weather could not have been better as I flew out to the crash site. 20 hours earlier, a 500-foot overcast, and heavy rain obscured the view. The day prior to the flight, the forecast was a bit optimistic but the last forecast and latest Metars were accurate. A VFR pilot still chose to continue into instrument conditions.

Here in the heartland, the roads run in orderly fashion to cardinal compass points. Farms and homesteads are neatly kept in the wide-open spaces punctuated by lines of trees. It would've been a reasonably safe place to make a precautionary landing--had that choice been made.

The NTSB investigator was alone in the bean field when I joined him. We walked quietly around the wreck, mud sticking to our boots with the ground still damp from last night's rain. The conversation was clinical, not wanting to dwell on how these people had died. The aircraft hit steeply nose down, left wing low at over 180 knots according to the airspeed indicator, coming to a violent dismembering stop. No one walked away from this one.

The engine was partially buried with one curled propeller blade visible. Oil seeped into a crater filled with muddy water. The cockpit was completely crushed--a mass of shredded, jagged aluminum, upholstery, plastic parts, and avionics. Side forces ejected the right-side door and a few other parts about 20 yards. The windows had fragmented into hundreds of small pieces and were strewn about. Personal effects--a coffee cup, a pair of eyeglasses, and a small bottle of hand lotion were poignant reminders of human tragedy.

The first responders removed the two occupants last evening before our investigator arrived but some biomatter remained. The site was a brutal reminder of the forces involved and requires some mental processing for those of us not experienced with this environment.

The engine and airplane manufacturer representatives arrived. This team had worked together too many times and, after some pleasantries, started methodically measuring and documenting. Landing gear, flap, and trim tab position--check. Control continuity--check. Gyros--the case was damaged but looked normal otherwise.

The aircraft recovery crew arrived to load the wreckage onto a flatbed and move it to a nearby hangar. Next morning the wreck was pressure washed to clean it up as much as possible. Personal protective gear is always worn by teams as they perform recovery tasks--it's often uncomfortable but essential.

Was the engine running normally? It looked okay, with some superficial damage, but nothing is left to chance. The valve covers were pulled, and the mangled prop spinner cut away to see if the crankshaft could be rotated, looking for drive train continuity. A reciprocating saw, a tool not usually associated with aviation, shattered the morning calm ripping through an unrecognizable wing to inspect gear and flap actuators. 48 hours ago, this aircraft was someone's pride and joy. Now, it was just evidence in another fatal crash investigation that would eventually be discarded in a landfill.

From interviews and available evidence, it wasn't this pilot's first encounter with weather. He wasn't trained or certificated to fly in the clouds. There was no record of the pilot getting an official weather briefing, although there are many sources for weather these days. An Airmet was in effect for instrument conditions and the radar images shown here confirmed there was significant precipitation in the area.

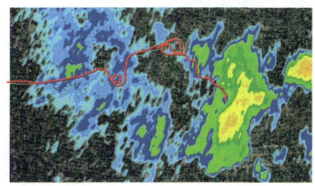

Flight path overlaid on radar plot, NTSB

Past flight tracks showed the horizontal and vertical meanderings of someone who wasn't used to taking NO for an answer even when clouds, prudence, and regulation strongly advised against a trip. Most of the time, when breaking the rules, the risk takers get away with it--until they don't. But it's easy to second guess.

What did we learn? Nothing new--sadly. Visual pilots must stay out of clouds. Forecasts aren't always accurate but were correct in this case. Sudden stops are lethal. An innocent person died trusting that the pilot would exercise great care. What's missing from this essay is the raw emotion of the families and friends, the massive complexity of transferring property, the legal fights, the empty holidays, and the dread that will come every year as the anniversary approaches.

It's so easy to be critical after the fact. None of us would fly in these circumstances if there was a chance it would end this way. Or so we say to ourselves. But there's almost always a history of success which emboldens. My hope in relating this tragedy is to revisit the realities of physics, aerodynamics, and human limitation. Would we bet our entire net worth; home, cars, aircraft, life savings, all future earnings on a card game where winning merely meant arriving at a destination on time? Perhaps it's a business meeting or just wanting to sleep in our own bed. When put in those stark terms few would take that bet. The risk just isn't worth the reward. Why gamble then with something so much more precious?

Icing, a descent below minimums, clouds, or a thunderstorm encounter-- there isn't a weather condition that cannot humble the best pilot. Sadly, NTSB will investigate a similar scenario again in a few weeks and the process will begin anew. It doesn't have to be this way.

Calabasas - (Well-known basketball star is lost to IMC)

Celebrity crashes always make news. Unfortunately, the cause of the crash is seldom new.

The NTSB does not refer to crashes by victim names but by location. While this doesn't help the public with immediate recognition, the objective is to provide a leveling effect that one person's life is just as valuable as another's. This incident, which included Kobe Bryant, retired LA basketball

Sikorski S-76B, Island Express, NTSB

superstar, and his daughter who were enroute to play at a Saturday morning game along with seven other friends, heightened the media attention immeasurably.

Like Buddy Holly in the Day the Music Died, and the John Kennedy Jr. in Vineyard Spiral, it shows that pilot motivations remain ever the same of trying to get somewhere on a schedule when the weather just isn't cooperating. This detailed analysis touches on much of what this book is about – learning from the past and remembering it for the future. It showcases failures at multiple levels even when all the safety holes are supposedly plugged.

Helicopters routinely fly in much lower weather than fixed wing VFR operations because of slower speeds and the ability to land off-airport almost anywhere if the weather shuts down. That bailout option is not used nearly as much as it should be.

Pre-departure - On January 26, 2020, a Sikorsky S-76B helicopter, N72EX was chartered to fly retired basketball star Kobe Bryant, his daughter, and several friends from John Wayne Airport (SNA) Santa Ana, California, to Camarillo Airport (CMA) Camarillo, California.

According to NTSB, the charter was requested several weeks in advance. Departure time was planned for 9 a.m. A flight risk analysis form was completed by the pilot showing a cloud ceiling of less than 2,000 feet agl and with all factors considered, he coded it a low-risk flight. Before departure the charter broker and the pilot discussed a color-coded flight map. It depicted marginal VFR for Long Beach, Santa Monica, and Malibu pass. Instrument conditions were shown over downtown Los Angeles,

Burbank (BUR), and Van Nuys. The pilot said he was going to go "up and around" and then "go east and north of the clouds."

The Flight - Upon leaving SNA 9:07 a.m., the visibility was 4 miles in mist with a 1,000-foot overcast ceiling. Automatic dependent surveillance-broadcast (ADS-B) data showed that it flew northwest. At 9:20 am about 8.5 miles southeast of BUR, the pilot contacted the tower requesting special visual flight rules (SVFR) clearance through the Class C airspace to follow US Route 101 west through the BUR airspace toward CMA. The flight was told to hold outside of the BUR airspace because of traffic. The pilot requested cloud tops and was advised that they were reported at 2,400 ft msl. They were subsequently cleared for transition.

At 9:32 am the flight was cleared to follow I-5 northwest through the BUR airspace. The BUR weather showed calm wind, visibility of 2.5 miles, haze, and an overcast ceiling of 1,100 ft agl. Van Nuys (VNY) reported nearly identical conditions.

At 9:35 am, the pilot contacted the VNY tower for SVFR transition and reported his altitude as 1,400 ft msl. The helicopter was flying about 1,320 ft msl (520 ft agl). The controller cleared the flight through the VNY class D airspace at, or below, 2,500 ft msl along State Route 118, as the pilot had requested.

At 9:40 am, the pilot contacted SoCal Approach and noted that they were transitioning in "VFR conditions" at 1,500 ft msl to CMA. *At that time, they were just 570 feet above the ground.* The SoCal controller asked if they planned to "stay down low...all the way to [CMA]," and the pilot replied, "Yes sir, low altitude." The controller advised that radar and radio contact would likely be lost and instructed the pilot to "Squawk VFR" and contact the CMA tower once closer to the airport, which the pilot acknowledged. At this point the helicopter was flying between 400 to 600 ft agl while remaining below 1,700 ft msl.

NTSB: "At 0942:45, the helicopter reached US 101 and began to follow it west toward CMA while flying about 1,420 ft msl (550 ft agl). Groundspeed was about 140 kts. A witness who saw the helicopter flying over US 101 reported fog and overcast cloud conditions that varied with "heavy low clouds" in some places and areas where the clouds were "quite high." She saw the helicopter flying "below or at the cloud line" before it "disappeared into heavy clouds" that she described as a "thick wall." Based on ADS-B data, the helicopter was at the location the witness described about 0944:32 and was at an altitude of about 1,370 ft msl (450 ft agl.)"

209

At 0944:34, the pilot advised SoCal that the flight was "Gonna go ahead and start our climb to go above the, uh, layers, and, uh, we can stay with you here." The helicopter immediately began climbing at about 1,500 ft per minute (fpm) and began a gradual left turn while remaining generally over US 101.

About 0945:10 it started a left turn away from US 101 reaching a maximum altitude of about 2,370 ft msl (about 1,600 ft agl). It then descended rapidly while remaining in the left turn. At 0945:17 (while the helicopter was descending), the SoCal controller asked the pilot's intentions and the pilot replied that the flight was climbing to 4,000 ft msl. The controller asked his intentions after that but received no response. The last ADS-B data point for the flight was recorded at 0945:36, and the accident site was about 500 ft east of this point at an elevation of about 1,100 ft msl in hilly terrain.

Flight path up to 0944:32 and location of ground

According to a ground witness near the accident site, the area was surrounded by mist. He described the "normal sound of a helicopter flying for about 20 seconds." He then saw the helicopter emerge from the clouds and roll left. He said the helicopter was visible for 1 to 2 seconds before it struck the ground and erupted into flames about 200 ft away. All nine people aboard the helicopter perished.

The Pilot - The 50-year-old pilot, held a commercial pilot certificate with rotorcraft-helicopter and instrument ratings and a second-class medical certificate issued on July 3, 2019. Total flight time was 8,577 hours with about 1,250 hours in Sikorsky S-76-series helicopters. He had about 75 hours of instrument flying time with 68.2 hours under simulated instrument meteorological conditions (IMC). In the 90 days, 30 days, and 24 hours before the accident, the pilot accumulated about 61 hours, 15 hours, and 1.5 hours flying time, respectively. There was one Class B violation in 2015 for which the pilot had received additional training.

The pilot was hired by Island Express in 2011 and flew smaller helicopters, transitioning to the S-76 in 2014. He became a flight instructor and the company's chief pilot in 2016.

NTSB: "The pilot's most recent Part 135 competency check was conducted in the event helicopter on June 21, 2019, and included use of the autopilot and maneuvers required for 'inadvertent IMC' (IIMC). In preparation for these checks, the pilot successfully completed training in the accident helicopter on May 8 through 10, 2019; the training form for the pilot showed satisfactory performance in a variety of flying procedures, including recovery from IIMC and unusual attitudes."

The pilot also completed the company's Safety Management System (SMS) course in June of 2018. In the days preceding the crash, there were no indications of fatigue or stress that would have impacted his performance. The pilot was viewed positively by his peers and management. No one reported any risk-taking behavior, and he was described as "proficient and always demonstrated sound judgment." The pilot's autopsy showed no evidence of impairing substances.

The Helicopter - The Sikorsky S-76B helicopter was manufactured in 1991 and equipped with two Pratt & Whitney Canada PT6B-36A turboshaft engines, which were manufactured in 1990. The helicopter, as equipped, was certified for single-pilot VFR and instrument flight rules (IFR) operations.

Island Express purchased the helicopter in August 2015, completed interior modifications in March 2016, and added it to the company's operating certificate in April 2016. It was equipped with a Honeywell SPZ-7000 digital automatic flight control system (DAFCS), which was a four-axis (pitch, roll, yaw, and collective) flight control system that combined autopilot and flight director functions. Depending on the level of automation selected by the pilot (by using the autopilot controller and flight director mode selector), *the DAFCS functions ranged from providing stability augmentation during manual flight to providing fully automatic flight path control* (emphasis added.)

Maintenance records indicate the airframe had accumulated 4,717 total flight hours, the left engine had 4,506 total flight hours with 1,009 hours since overhaul, and the right engine had 4,681 total flight hours with 1,226 hours since overhaul.

Post-crash analysis of the components and downloaded data showed no evidence of any pre-impact failure.

Training - An outside organization trained company pilots in the aircraft to prepare for FAA check rides. The instructor explained the inadvertent IMC procedure: "adjusting the pitch and roll to achieve straight-and-level flight, adjusting the power to about 70% to 75% torque and pitch attitude to establish a positive climb rate at an airspeed of about 75 to 80 kts, maintaining the current heading, and turning only to avoid known terrain or obstacles. He taught the pilot trainees that, once they had the helicopter established in a stabilized climb, they should transition to using the autopilot (by selecting heading, airspeed, and vertical speed) then communicate with ATC. For pilots who fly in the Los Angeles area, he taught them that the best choice was to climb above the cloud layer, fly visually to an airport, and land under VFR, if possible, or perform a precision instrument approach, which he practiced with them."

Both manual recovery and use of the autopilot were demonstrated and practiced The key points stressed: Slow down, fly level, and use the automation. *None of those procedures appear to have been followed in the crash scenario.* The instructor stated that they always discussed avoiding entering IMC so that the recovery and escape procedures were unnecessary.

Accident flight's final climb, left turns, rapid descent, and wreckage location, NTSB

According to the FAA Principal Operations Inspector who evaluated the accident pilot during check rides, appropriate adverse-weather-avoidance maneuvers included diverting, returning to base, or landing the helicopter.

Weather - The weather before takeoff, enroute, and at the crash site is not in question and called for an IFR flight plan. The image below is just one of several that showed the instrument conditions that morning.

NTSB: "At 0645, the National Weather Service (NWS) Aviation Weather Center issued airmen's meteorological information (AIRMET) Sierra advisories that forecasted 'IFR conditions' in mist and fog and mountain obscuration due to clouds and mist for areas that included the accident location and time.

According to the NWS forecaster at the local weather forecast office in Oxnard, California, the presence of a deep marine layer across the region near the accident site was uncommon for January. He said the near-surface relative humidity near Calabasas was still 100% at 1000, noting that it typically decreased earlier in the morning."

Organization and Management - Island Express's main office was in Long Beach, California, and they held an operating certificate for VFR-only day and night Part 135 on-demand passenger (charter) and cargo operations. The company had an FAA-approved training program as noted above.

51 °F | JAN 26, 2020 Calabasas Weather
09:47AM | Calabasas, CA, US

One of several wx cameras confirming that it was not VFR, NTSB

The Director of Operation (DO), the director of maintenance, and the chief pilot (the crash pilot) were the authorized management personnel for the operating certificate. NTSB: "The chief pilot reported to the DO and was responsible for, in part, supervising all pilots and their training activities, assisting the DO in formulating operations policies, and coordinating operations and training. The company also had a safety officer who reported to the DO and was, in part, responsible for scheduling safety meetings encouraging the reporting of incidents and following up on actions arising from the reports and encouraging the submission of comments and suggestions for improving safety procedures."

There were 25 employees, including 6 pilots and 2 mechanics. The company operated three Sikorski S-76-series helicopters and three Airbus AS350-series helicopters. In 2019, Island Express flew about 495 charter flights. Mr. Bryant had an established relationship with an air charter broker which included 13 flights prior to the crash. Bryant and the broker jointly decided that Island Express met their criteria for security, pilot experience, and the use of two engine helicopters.

Interviews with company management and the pilots showed that they mostly believed that the company had a good safety culture and that anyone could bring up a problem at any time. The Safety Management System (SMS) was a commercially produced program that was partially implemented.

NTSB Probable Cause - *".... was the pilot's decision to continue flight under visual flight rules into instrument meteorological conditions, which resulted in the pilot's spatial disorientation and loss of control. Contributing to the accident was the pilot's likely self-induced pressure and the pilot's* *plan continuation bias, which adversely affected his decision-making, and Island Express Helicopters Inc.'s inadequate review and oversight of its safety management processes."*

NTSB Abbreviated Findings and Recommendations - There were numerous findings and a few recommendations, not all of which are listed here. For more detail see the full report.

"The pilot's poor decision to fly at an excessive airspeed for the weather conditions was inconsistent with his adverse-weather-avoidance training and reduced the time available for him to choose an alternative course of action to avoid entering instrument meteorological conditions.

1. The pilot experienced spatial disorientation while climbing the helicopter in instrument meteorological conditions, which led to his loss of helicopter control and the resulting collision with terrain.

2. The pilot's decision to continue the flight into deteriorating weather conditions was likely influenced by his self-induced pressure to fulfill the client's travel needs, his lack of an alternative plan, and his plan continuation bias, which strengthened as the flight neared the destination.

3. Island Express Helicopters Inc.'s lack of a documented policy and safety assurance evaluations to ensure that its pilots were consistently and correctly completing the flight risk analysis forms hindered the effectiveness of the form as a risk management tool.

4. A fully implemented, mandatory safety management system could enhance Island Express Helicopters Inc.'s ability to manage risks.

5. The use of appropriate simulation devices in scenario-based helicopter pilot training has the potential to improve pilots' abilities to accurately assess weather and make appropriate weather-related decisions.

6. Objective research to evaluate spatial disorientation simulation technologies may help determine which applications are most effective for training pilots to recognize the onset of spatial disorientation and successfully mitigate it.

7. A flight data monitoring program, which can enable an operator to identify and mitigate factors that may influence deviations from established norms and procedures, can be particularly beneficial for operators like Island Express Helicopters Inc. that conduct single-pilot operations and have little opportunity to directly observe their pilots in the operational environment.

8. A crash-resistant flight recorder system that records parametric data and cockpit audio and images with a view of the cockpit environment to include as much of the outside view as possible could have provided valuable information about the visual cues associated with the adverse weather and the pilot's focus of attention in the cockpit following the flight's entry into instrument meteorological conditions."

Recommendations to the FAA - included requiring the appropriate use of simulation devices during initial and recurrent training for Part 135 operations to develop scenario-based training to improve pilot risk assessment, recognize the onset of spatial disorientation, and practice both avoidance and recovery.

Additionally, the Board reiterated that the FAA require Part 135 operators to establish a structured flight data monitoring (FDM) program that would identify when pilots where not following proper safety procedures and proactively monitor how operations were being carried out on a day-to-day basis.

Recommendations to Island Express (IE) and the industry - A new recommendation to IE was to install flight data monitoring equipment on all their helicopters and establish an FDM program as noted above.

To all the helicopter manufacturers the Board reiterated that they, "...provide, on your existing turbine-powered helicopters that are not equipped with a flight data recorder or a cockpit voice recorder, a means to install a crash-resistant flight recorder system that records cockpit audio and images with a view of the cockpit environment to include as much of the outside view as possible and parametric data per aircraft and system installation..."

Commentary - In theory, Part 135 commercial operations are supposed to perform at a higher level of safety because the traveling public is unable to determine what constitutes safe performance. The regulatory requirements for pilots, procedures, aircraft, and maintenance are more stringent than for Part 91 personal flight operations. NTSB has long asked for FAA to require all Part 135 operations to have a Safety Management System (SMS) in place. Island Helicopter, ironically, was one of the few that did. But as noted here and elsewhere, actively promoting, and performing safety, day in and day out is difficult.

Despite having considerable flight time, *the pilot's actual logged instrument time was less than 8 hours.* Also noted elsewhere, real instrument flight and simulated instruments are significantly different. The stress of flying in the clouds for a low time IFR pilot should not be underestimated along with the potential for vertigo.

Island Express, amazingly, did NOT have an authorization to fly IFR on their operating certificate Jim Viola, then president of Vertical Aviation International, noted that on the East Coast many Part 135 helicopter operators with larger helicopters like the S-76, all had IFR certificates but that was unusual on the West Coast. There's no shortage of fog on the West coast. FAA noted that revealed, "… that 476 certificate holders were authorized for Part 135 helicopter operations and that, of these, 411 were VFR-only operators." This seems like a significant safety opportunity despite the additional requirements. Many NTSB investigations of Part 135 and air tour operations involve VFR into IMC crashes.

Most corporate operators of large helicopters that serve transportation roles require two pilots who are instrument qualified and proficient in a fully equipped aircraft. Most train in motion-visual simulators that far exceed the capability of the view-limiting devices that are sometimes used by Part 135 operations. It certainly costs more but putting aside the human loss momentarily, think how much this one crash cost. Quality training is always a better alternative, or do you feel lucky?

The imperative to avoid clouds when VFR cannot be overstated, even for experienced pilots. It horrendous in fixed wing aircraft and worse in helicopters.

The PIC, upon seeing that the weather was substantially as forecast, could have decided that it was extremely unlikely and unsafe to be able to get out of the LA basin, over some higher terrain and back down to the coast at Camarillo under VFR. But the desire to perform for the client, or ourselves, is incredibly strong. Just suppose that after clearing the VNY airspace and moving toward the higher terrain over Highway 101, he had said to the passengers, "Folks, I'm really sorry but we're going to return to Van Nuys. We'll get a car to take you the rest of the way to Camarillo. This just isn't safe." There would have been disappointment but in most cases, people would understand. It's one of the hardest things a pilot has to do and it's the mark of a pro. An hour's delay (or a day) is such a superior tradeoff compared to the alternative.

It's For Real-(IFR)-Instrument Flight Rules

Flying under Instrument Flight Rules requires a much higher level of weather knowledge because we're no longer avoiding the weather but embracing it and some conditions are monstrously dangerous to the unwary. Here are some of the realities of cloudy flight. In Instrument

Meteorological Conditions (IMC) what was easy under VFR can become overwhelming.

Automation - We'll explore automation later but for now, understand it and use it *appropriately*. Autopilots are a big help with basic flight control. When operating single pilot Part 135 IFR, they are required. On most jets and turboprops, pilots are trained to use the automation most of the time. Air Traffic Control choreography can easily spike the workload so flying must become second nature. If you're sinking for whatever reason, turn on the autopilot if so equipped, and if VFR, turn around preferably *before* getting into IMC. NTSB crash reports regularly cite pilots getting into clouds that they were not rated or proficient to handle.

> *"The machine does not isolate man from the great problems of nature but plunges him more deeply into them."*
>
> Antoine de Saint-Exupery, French pilot, and author

Sometimes the temperature dewpoint can stay in lockstep for hours and remain VFR and in others, not so much. Here's just one instance I've encountered over the years.

Hangar Story - *Cessna 210: On a night IFR flight into Wichita, Kansas, the temperature-dewpoint spread was within two degrees, but visibility was unlimited. The ATIS was reporting clear and 10 miles visibility. Ten minutes later a special observation advised 600 broken and 4 miles—an IFR approach would be needed. At localizer intercept three minutes later, the tower reported 300 overcast and three-quarters of a mile. The weather was at minimums on landing, and it went below while I was taxiing in. In less than 20 minutes the airport had gone from clear to below minimums. I was the last flight in that night. It can change quickly.*

Two Great Regulations - Here are two of my favorite regulations. While we have greater flexibility under Part 91, personal flight rules, some of the Part 135 regs make a lot of sense.

FAR 135.219: No person may take off an aircraft under IFR or begin an IFR or over-the-top operation unless the latest weather reports or forecasts, or any combination of them, indicate that the weather conditions at the estimated time of arrival at the next airport of intended landing will be at, or above, authorized IFR landing minimums. What the lawyers made complicated: Don't takeoff under IFR unless the destination is forecast to be above (your) landing minimums at your ETA. Carry enough fuel to get to a solid alternate.

FAR 135.225: There are exceptions but what this reg says in essence: Do not start an approach if the weather is reported below (your) minimums before reaching the final approach fix. If after starting the approach you wish to continue, that's your prerogative, but beware that the temptation is often to slip just a little bit lower and there Icarus lurks. The temptation gets stronger when pilots come back around for a second or third attempt.

The record for tenacity, and stubbornness, occurred when a newly trained Cessna Conquest I pilot made seven attempts to land at a fogged-in mountain airport. Sadly, the final attempt resulted in fatalities to all aboard. If the weather condition is *transient and improving*, a second attempt might be reasonable. If the first attempt resulted in a miss because the aircraft was not properly positioned, a second shot might be warranted. The hard question to ask yourself, though, is why wasn't the aircraft lined up horizontally and vertically the first time?

Thought Exercise - Let's fly from Muscle Shoals, Alabama, to Myrtle Beach, South Carolina. A cold front is moving through the Southeast with convective Sigmets in the areas of heavier precipitation. (The Airmets for moderate turbulence, low-level windshear, and convective Sigmets *are not*

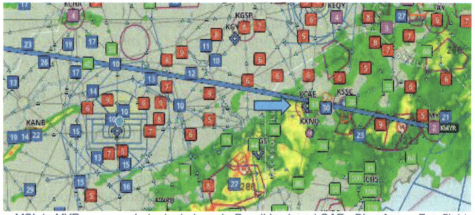

MSL to MYR, wx snapshot prior to launch, Possible stop at CAE - Blue Arrow, Foreflight

shown for clarity but on the original map there is plenty for pilots to consider.)

The colored squares show the ceilings in hundreds of feet--Red is IFR, Blue MVFR, Green is VFR, Magenta is Low IFR. (As luck would have it, that's our destination at the moment.) There's a band of IFR (all the red squares) behind the front but it's not real low and all the convection is along and ahead of the front. Flying from Northern Alabama Regional (MSL) to Myrtle Beach, South Carolina (MYR) is between 2.4 – 2.8 hours in a Cirrus, Bonanza, Cessna 210, or Mooney 201 and uses roughly half the on-board fuel depending on tankage and winds. Myrtle is currently showing low IFR at 2 OVC and the forecast for the next several hours isn't so good either; two miles in heavy rain/mist, 2 BKN but a few hours later improves to greater than 6 miles, 5 BKN and 20 OVC. Maybe.

Option 1—Delay departure until the forecast is above *your* minimums.

Option 2—if you're feeling optimistic, start out *but plan to land short.* That might be Columbia, South Carolina, (CAE) and where the forecast is two miles in mist and 800 OVC. If CAE fogs in, not far to the west it's already improving to MVFR or better and there's plenty of fuel left. Pack your overnight bag, however. Refuel at CAE so if the weather at MYR is still below your minimums as the front moves out, there's plenty of fuel to escape back to the west to better weather. This precautionary fuel stop creates an hour's delay compared to non-stop, but it opens up lots of options. If Myrtle improves to better than forecast while you're enroute to CAE continue on, but no wishful thinking. Stopping short makes it easy to reassess provided we are *honestly* willing to rethink the plan.

If MYR remains below your minimums, or there are too many thunderstorms, you're already an hour closer at CAE and can launch as soon as it improves, or tomorrow. It looks better south at Charleston (CHS), but that puts us back into the front, potential convection, with less fuel and well outside of the direct route. Bail out options aren't as good either with the Atlantic to the east and coastal fog always a possibility.

Option 3—Wait until tomorrow—Almost always a sure thing and a wise choice.

A similar VFR scenario could be constructed. The hypothetical ceilings should be at least 2,000 feet and visibility 4 to 5 miles. Temperature and dewpoint spreads should also be at least four degrees. With VFR there are more cautions and fuel is still your friend

Departing from KLRO or KJZI, it's much easier to divert to KCHS than to return to the departure airport, Foreflight

Takeoff Alternates - Takeoffs should have contingency plans too. As with approaches, under Part 91, unless there are specific minimums for takeoff it's up to the PIC to decide how much risk to take. It could be zero-zero—not smart but legal. It's better to have at least landing minimums at the departure airport, but circling minimums make more sense if a quick return becomes necessary. Pilots of multiengine aircraft can gamble a bit more provided single engine skills and aircraft performance are up to the task.

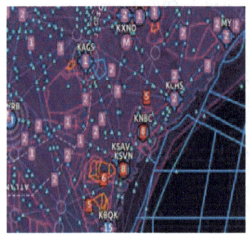

No good departure alternates here except down by Savannah. Magenta boxes show LIFR ceilings in hundreds of feet. Red boxes are "merely" IFR, Foreflight

A good rule of thumb, if you choose to take the risk, is to find a nearby airport within 15 minutes that's reporting a reasonable ceiling and visibility. That may be a better alternate than the departure field because there's no need to "pretzel" around to get realigned with the final approach course. Coastal fog tops are typically a couple hundred feet and clear above, but often there's no suitable nearby alternate. Waiting an hour or so, is much safer because in most cases, it will burn off or lift enough to allow for a takeoff alternate.

Immediate returns are high workload. It's much easier to intercept a straight-in final than maneuver for the procedure turn or extended vectoring to get re-aligned. Depending on the weather, the takeoff alternate should have an ILS or LPV approach that is easily intercepted. A little trick that could save a lot of FMS programming in a high workload situation is to put the departure airport (or departure alternate) at the end of the flight plan and then delete it once safely enroute. For example: KICT, EMP, KMKC, KICT. In the event of an immediate return, simply enter "direct to" for the final waypoint and load the associated approach.

A door opening, a mechanical malfunction, or an electrical/avionics problem complicate things. All this alternate planning is unnecessary until needed when it literally becomes a lifesaver.

As you read the crash reports in this, remember how easy it is to second guess. If the flights had ended successfully no one would have thought anything of it. What other alternatives were available?

Night Moves - Night flying in many countries must be done on instruments. After the JFK Jr. crash noted in Chapter 2, there was an outcry by some in Congress that the US should adopt a similar regulation.

I do not share that opinion, nor did the FAA. But pilots, especially VFR-only pilots, should understand that dark is different.

While the exposure denominator, as usual, is fuzzy, a disproportionate number of crashes occur at night. Humans don't see as well and cannot avoid clouds, obstacles, and mountains. Night flying is largely instrument flying. I spent much time learning and teaching instruments in VMC after dark. It's much more realistic than the hood, but it comes with caveats. The first several dozen times out, it's mesmerizing. The world looks completely different, but we've given up one of the most precious assets of flight–the ability to see well.

Temperature and dewpoint are a factor and depth perception on takeoffs and landings is more challenging. The currency and certification regulations attempt to address this. While the aircraft is not night sensitive, we are, and the nature of any emergency is compounded. Successful forced landings after a nocturnal engine failure are generally due to luck. Some say to go for lights, as in highways, while others say look for a dark spot.

My early night operations included some marginal airports, both in terms of length and lighting. Today, unless the airport has functional visual approach lighting (PAPI, VASI) it's off my list after dark unless it has a precision approach to the runway in use. Having an instrument approach with vertical guidance is a tremendous safety factor even when the weather is perfect. It also keeps you from landing at the wrong airport - ask a number of pilots about that sometime.

Fatigue is always a problem in flight and our ability to estimate our own alertness consistently and accurately is lacking. Older eyes do not see as well, and the night lighting in many legacy aircraft is atrocious. Finally, reduced oxygen affects night vision for everyone. Emergencies that arise will be compounded significantly, perhaps fatally. So, as the night whispers to come hither and enjoy the beauty with the silky-smooth air, remember Rapture of the Sky.

Key Points:

- The weather is what you find, not what was forecast.

- Temperature/dewpoint spread is important. Fuel and flexibility are the keys.

- FAR 135 has some great rules for Part 91 IFR operations.

  ◦ Use an autopilot for single pilot-operations.

- ◦ Don't start a flight if weather is forecast to be below your minimums *at time of arrival.*

- ◦ Don't start an approach if weather is below your minimums before reaching the final approach fix. If the aircraft just ahead of you missed–think about going elsewhere.

- ◦ Have a takeoff alternate if the weather is below circling minimums.

- Be conservative on fuel: 50 percent to destination, 30 percent to a solid (not paper) alternate, 20 percent absolute critical reserve.

- Minimums are just that. Beware the dark side of the temptation force.

- Weather window—Sometimes a few hours or going a day early or later makes for a much safer, more pleasant trip.

- Night flight is more complex and unavoidably compounds daytime hazards.

Cleared For the Approach - (TWA 514 crew is confused on descent)

One accident led to many changes

Controlled flight into terrain, or CFIT, in instrument conditions has been a leading cause of accidents for both the airlines and general aviation. An uneasy feeling that something's not right is a warning that, well, something isn't right

A classic CFIT accident that occurred years ago changed the way IFR flights are handled resulting in better charting and many other procedural changes. This crash was so disturbing that, even to this day, pilot examiners and flight instructors drill new IFR pilots on what the phrase "Cleared for approach" means. Simply, we

TWA B727, Wikipedia

must be on a published route with minimum altitude depicted before descending from the last assigned altitude unless on radar vectors. Even then it doesn't hurt ask what the minimum vectoring altitude is if you have any uncertainty

The Flight - On December 1, 1974, TWA Flight 514, a Boeing 727-231 was enroute from Indianapolis to Washington's National Airport (Now

Reagan.) An intense early winter storm system was moving up the East Coast. Crosswinds at National were too much for the north-south runway, and flights were being diverted to Washington Dulles International Airport's Runway 12. Weather observations at National showed a 1,200-foot overcast, with visibility better than five miles in light rain. Surface winds were from 070 degrees at 25 to 28 knots, with gusts as high as 49 knots. Winds at Dulles were substantially the same as at National, with a ceiling of 900 feet and visibility of three to seven miles in rain.

Weather radar showed large areas of precipitation, including some widely scattered embedded thunderstorms. Cloud tops at the accident site were 24,000 feet. There were Sigmets for thunderstorms with tops to 40,000 and moderate to severe mixed icing in the clouds. Numerous pilot reports for turbulence had been received, although none were specifically for the accident area. It was a grungy day to fly.

"You know, according to this dumb sheet, it says thirty-four hundred to Round Hill is our minimum altitude."

Captain, TWA 514

At 10:42 am Cleveland Center cleared Flight 514 to Dulles via the Front Royal VOR and to maintain Flight Level 290. At 10:43 am the flight was cleared down to FL 230 to cross 40 miles west of Front Royal at that altitude. The flight was handed off to Washington Center at 10:48 am During this time the captain turned over control of the flight to the first officer and they discussed the Runway 12 VOR/DME instrument approach into Dulles. The cockpit voice recorder reveals that the crew discussed the various routings they might receive from ATC, such as via the Front Royal VOR, Martinsburg VOR, or proceeding on a "straight-in" clearance.

At 10:51 am the Center controller provided a heading of 090 to intercept the 300-degree radial of the Armel VOR inbound, to cross a point 25 miles west of Armel at 8,000, and "...the three-zero-zero radial will be for the VOR approach to Runway One-Two at Dulles, altimeter two-niner-point-seven-four." The crew acknowledged. Cockpit voice recordings (CVR) showed that the VOR was tuned, and altimeters properly set. At 10:57 am the crew again discussed the approach, including Round Hill intersection, the final approach fix, VASI, runway lights, and the airport diagram.

At 11:01 am the flight was cleared to 7,000 feet and handed off to Dulles Approach Control. Dulles cleared it to proceed inbound to Armel VOR and to expect the VOR/DME approach to Runway 12. At 11:04 am the flight reported level at 7,000, and five seconds later the controller said, "TWA

Five-Fourteen, you're cleared for a VOR/DME approach to Runway One-Two." The captain acknowledged this.

The following is the Cockpit Voice Recorder (CVR) transcript. Captain— Capt, FO—First Officer, and FE—Flight Engineer. The NTSB did not quote all items verbatim except what is considered pertinent. The "non-pertinent" comments here are copied directly from the report. Edited for length.

11:04 Capt: "Eighteen hundred [feet] is the bottom."

FO: "Start down. We're out here quite a ways. I better turn the heat down [probably referring to cabin heat]."

11:06:15 FO: "I hate the altitude jumping around." Then he commented that the instrument panel was bouncing around.

11:06:42 FO: "Gives you a headache after a while, watching this jumping around like that."

11:07:27 FO: "...you can feel that wind down here now."

Capt: "You know, according to this dumb sheet [referring to the instrument approach chart] it says thirty-four hundred to Round Hill is our minimum altitude." The FE asked where the captain saw that, and the captain replied, "Well, here. Round Hill is eleven-and-a-half DME."

FO: "Well, but...."

Capt: "When he clears you, that means you can go to your...."

Unidentified: "Initial approach."

Unidentified: "Yeah."

Capt: "Initial approach altitude."

FE: "We're out of twenty-eight for eighteen."

Unidentified: "Right; one to go." [meaning 1,000 feet more before reaching the supposed level-off altitude of 1,800].

11:08:14 FE: "Dark in here."

FO: "And bumpy too."

11:08:25 Altitude alert horn sounds.

Capt: "I had ground contact a minute ago."

FO: "Yeah, I did too."

11:08:29 FO: "...power on this [expletive]."

Capt: "Yeah, you got a high sink rate."

FO: "Yeah."

Unidentified: "We're going uphill."

FE: "We're right there, we're on course."

Two voices: "Yeah."

Capt: "You ought to see ground outside in just a minute. Hang in there, boy."

FE: "We're getting seasick."

11:08:57 Altitude alert horn sounds.

FO: "Boy, it was--wanted to go right down through there, man."

Unidentified: "Yeah."

FO: "Must have had a [expletive] of a downdraft."

11:09:14 Radio altimeter warning horn sounds and stops.

FO: "Boy!"

11:09:20 Capt: "Get some power on."

Radio altimeter warning horn sounds again and stops.

At 11:09:22 TWA 514 struck the west slope of Mount Weather, about 25 nautical miles northwest of Dulles at an elevation of about 1,670 feet. Seven crewmembers and 85 passengers perished in the crash. There were no survivors.

The flight data recorder showed a continuous descent with little rate variation from 7,000 feet until about 1,750 feet. There were minor altitude excursions from 100 to 200 feet. Airspeed was fairly stable at 230 knots, with fluctuations between 222 knots and 248 knots. Gear, leading-edge devices, and flaps were up at the time of impact.

At the accident site, witnesses on the ground reported low ceilings with visibilities of 50 to 100 feet, drizzle, and wind estimated at 40 knots with stronger gusts. Possible altimeter errors caused by high wind speeds were calculated, with localized pressure changes, and the National Weather Service estimated that a worst-case scenario with 80-knot winds could result in an altitude indication 218 feet higher than the actual aircraft altitude.

NTSB Probable Cause *– " ...to be the flight crew's decision to descend to 1,800 feet before the aircraft had reached the approach segment where* *that minimum altitude applied. However, two of the five-member board dissented, identifying the probable cause to be the failure of the controller to issue altitude restrictions in accordance with the terminal controller's handbook. The pilot was also cited for failure to adhere to the minimum sector altitude depicted on the instrument approach chart and to request a clarification of the clearance."*

Commentary - Testimony following the accident indicated that ATC frequently vectored aircraft off published routes and cleared them to descend below altitudes published on the charts. Pilots and controllers had available published minimum sector altitudes (MSA- now minimum safe altitude within 25 miles of the airport.) MSAs provide 1,000-foot obstacle clearance within 25 miles of an airport and are considered emergency altitudes since they do not assure navigation signal coverage. Controllers also made use of minimum vectoring altitudes (MVA), which were not normally available to pilots.

The testimony also indicated that pilots had become so accustomed to receiving assistance from controllers that, unless advised by the controller, they did not know what type of ATC service they were receiving. Often, they were unsure of their position relative to terrain.

There was considerable debate as to whether the flight was being handled as "radar arrival," which could have put the burden for terrain separation on ATC. The term was not well defined, although in the controller's manual under "radar arrivals," the following guidance was provided: "Issue approach clearance, except when conducting a radar approach. If terrain or traffic does not permit unrestricted descent to the lowest published altitude specified in the approach, prior to final approach descent, controllers shall: 1) Defer issuance of an approach clearance until there are no restrictions or, 2) Issue altitude restrictions with approach clearance specifying when or at what point unrestricted descent can be made...."

FAA witnesses testified that Flight 514 was inbound to Armel by means of the pilot's own navigation, thereby relieving the controller of the responsibility cited above. The cockpit conversation is clear that the crew thought otherwise, although the captain had a nagging feeling that something wasn't quite right.

The NTSB also reviewed the handling of other arriving IFR flights at Dulles on December 1, 1974. A flight arriving about half an hour before the accident was cleared for the same approach. Because the aircraft was a considerable distance from the airport and was not given an altitude restriction, the pilot requested the MVA, which was provided. The controller then offered a surveillance approach, which the captain accepted.

Six hours after the accident another flight inbound from the southwest asked the controller its position relative to Round Hill intersection. The controller replied that he did not have that on his scope--and the captain, familiar with the terrain west of Dulles, stated that he did not descend until DME indicated 17.6 miles (Round Hill) and the aircraft was established on the final approach course.

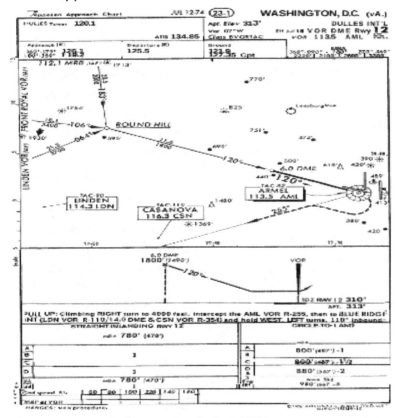

Accident approach chart, NTSB

The concept of crew resource management had not been formalized at that time, and the discussion between the captain and the flight engineer would probably raise some warning flags on today's flight decks. As noted above, several crews had some confusion and asked for clarification from ATC.

This is excellent guidance for all pilots--when in doubt, start asking lots of questions of yourself and ATC. Be very conservative regarding terrain clearance. Feelings of uncertainty are frequently a premonition of disaster.

As the NTSB conducted research on how the pilot community interpreted the concept of radar arrivals and when pilots were responsible for terrain separation, the only thing that was clear was that there were multiple interpretations.

Ironically, as early as 1970, TWA personnel were concerned about what they saw as conflicting information between the Airman's Information Manual (Now Aeronautical Information Manual) and the controller's manual.

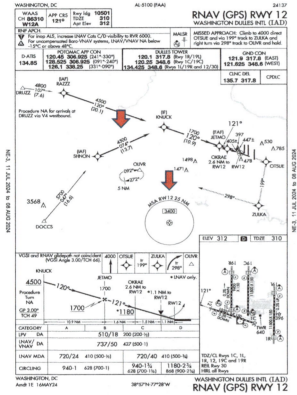

Both the U.S. Air Force and TWA pointed out to the FAA that "Cleared for the approach" terminology could be misinterpreted unless a specific altitude restriction was included in the clearance.

In its findings the NTSB assigned plenty of responsibility to all parties. The Board noted that procedures in the FAA's terminal control handbook were not clear and resulted in TWA 514's being classified as a "non-radar arrival," but the terms "radar arrival" and "non-radar arrival" were not defined.

A much clearer current chart. Note MSA and published route altitudes

Depiction of Round Hill intersection was not shown in the profile view, nor did it contain all the minimum altitudes associated with the approach procedure. The ATC system was not clear as to what services controllers should provide under the circumstances of this flight. Both military and civilian aviation organizations had shown repeated concern for the possibility of misunderstanding clearances that could lead to premature descent, but the FAA had taken no action.

Several things changed in the aftermath. The FAA directed that all air carrier aircraft be equipped with ground proximity warning systems (GPWS), and the FAA published a change to FAR Part 91, clarifying pilot responsibilities regarding operations on unpublished routes. An incident reporting system was established (NASA's Aviation Safety Reporting System), a pilot-controller glossary was published in the AIM, and chart depictions were improved. Procedures regarding ATC responsibilities during approaches were defined.

The legacy that grew out of TWA 514's loss is a staple of all IFR training and appears in the Aeronautical Information Manual: "When operating on an unpublished route or while being radar vectored, the pilot, when an approach clearance is received, shall, in addition to complying with the minimum altitudes for IFR operations, maintain the last assigned altitude unless a different altitude is assigned by ATC, or until the aircraft is established on a segment of a published route or IAP." It also appears in FAR Part 91.175(I). Note the significant change in charting and altitude depictions from the accident chart to current day charts.

With the widespread use of RNAV (GPS), terrain depiction, and synthetic vision, on electronic flight bags (EFBs) the number of Controlled Flight into Terrain (CFIT) accidents is dropping. But regardless of rules or protocol, no one should have greater interest in the height of terrain than the pilot.

Aspen Arrival - (GIII Captain is browbeaten into an approach)

A Gulfstream III and its overbearing passenger never make the runway

Aspen Airport--lots of hills, Aspen Airport Authority

A strong desire to reach a destination sometimes means never getting there. Mountainous terrain and deteriorating weather always complicate arrivals, and Aspen, Colorado, is known to be challenging. Over the past decade there have been about 15 accidents with several involving fatalities. It was late March 2001 when a Gulfstream III was hired to bring a party of 15 from Los Angeles to the resort area for an evening social event.

The captain and first officer reported for duty around noon Pacific Standard Time (PST) in Burbank, California. The first officer checked weather with flight service and while the forecast called for occasional IFR in low ceilings and restricted visibility with passing snow showers, there was nothing to cause major concern — other than the fact that they were flying into Aspen.

The FSS specialist noted that the circling minimums for the only instrument approach were no longer authorized at night. This would play a significant role in the events to follow. An IFR flight plan was filed, with Garfield County Regional Airport in Rifle, Colorado, as the alternate. Aspen's night landing restriction required the G-III to land within 30 minutes after sunset (6:58 pm MST) because of noise-abatement rules.

The G-III departed Burbank at 3:38 pm Mountain Standard Time (MST) (2:38 pm PST) for the positioning flight to Los Angeles International Airport, arriving 11 minutes later. At Los Angeles, the captain and a charter department dispatcher discussed the fact that the passengers were late and discussed the weather for the flight's arrival time at Aspen-Pitkin County/Sardy Field.

Overbearing Customers - Anyone who has flown charter knows that the desire to complete the mission is compelling but top operators focus on safety first and transportation second. The customer is not always right. The charter client's business assistant, in a post-accident interview, mentioned that "the boss" was hosting a party in Aspen. The charter company dispatcher called the assistant at 4:30 pm wondering where the passengers were and advising that if the flight departed later than 4:55 pm, it would have to divert to Rifle. The assistant located the passengers chatting in the airport parking lot.

One of the pilots mentioned the nighttime landing curfew and "the boss," upon hearing this, instructed his business assistant to call the charter company back and tell the crew to "Keep their comments to themselves." When informed of the possible diversion, the boss became irate and again, through the assistant, told the charter company that the airplane was not going to be redirected. He had flown into Aspen at night before and was going to do it again.

The G-III departed Los Angeles at 5:11 pm MST (4:11 pm PST), 41 minutes late for the one-hour-35-minute flight. The estimated time of arrival at Aspen was 6:46 pm, only 12 minutes before the landing curfew. The dispatcher recalled that during a subsequent conversation on the company frequency, the captain said the boss had pressured him to land at Aspen because "He'd spent a substantial amount of money on dinner."

The approach - At 6:37:04 pm the first officer called for the approach briefing. The captain answered, "We're...probably gonna make it a visual... If we don't get the airport over here, we'll go ahead and shoot that approach. We're not going to have a bunch of extra gas, so we only get to shoot it once and then we're going to Rifle." At 6:39:56, the crew received ATIS information Hotel.

Three-quarters of an hour earlier, at their originally scheduled arrival time, the 5:53 pm observation indicated wind from 030 degrees at 4 knots, visibility of 10 miles, scattered clouds at 2,000 feet, ceilings of 5,500 feet broken and 9,000 feet broken, temperature of 2 degrees Celsius, and a dew point of minus 3 degrees C.

Shortly afterward, a Canadair Challenger declared a missed approach. The G-III crew was probably feeling pressured about the night curfew, the deteriorating weather with a trip to the alternate and a very unhappy client. The Gulfstream captain asked the controller if the Challenger was practicing or had actually missed the approach. The controller replied it was a real missed approach. While in descent, the flight crew discussed the location of a highway near the airport and the chance to follow it in for a visual. A quick, easy arrival by visual reference was fading with the evening twilight and the snow showers.

At 6:47:30 pm ATC announced that a Cessna Citation saw the airport at 10,400 feet and was making a straight-in approach. At 6:47:41, the Gulfstream captain advised the controller, "I can almost see up the canyon from here, but I don't know the terrain well enough, or I'd take the visual." At 6:47:51, the first officer stated, "Could do a contact but...I don't know, probably we could not..." (A contact approach procedure allows an IFR aircraft to proceed by visual reference, with the pilot responsible for terrain avoidance. A flight must remain clear of clouds with reported ground visibility of at least one mile. Pilots must specifically request this procedure to use it as a shortcut to a full instrument approach.)

At 6:48:51, the captain said, "There's the highway right there." But the first officer replied, "No, it's clouds over here on this area. I don't see it." The captain advised the flight attendant that, if the approach was not successful, they would go to Rifle because "It's too late in the evening..." Shortly after this a passenger was allowed to sit in the jump seat to observe the approach. At 6:53:09 another Canadair Challenger missed the approach, to which the first officer remarked, "That's...not...good."

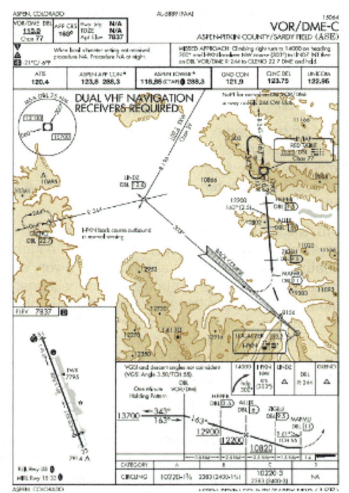

Aspen approach chart--Not so easy, NTSB

At 6:56:06 pm the flight was cleared for the VOR/DME-C approach. A few seconds later ATC reported that the visibility north of the airport was two miles (at minimums). The approach controller then switched the G-III over to Tower frequency. The Tower advised that the Gulfstream was following another Challenger, whose crew subsequently reported a missed approach at 6:58:00.

At 6:58:13, the jump-seat observer asked, "Are we clear?" The captain replied, "Not yet. The guy in front of us didn't make it either." There was some discussion with the crew about the approach step-down fixes, and at 7:00:08, the jump-seat observer said, "Snow." At 7:00:22, with the aircraft at 10,400 feet, 200 feet above minimums, the captain declared, "OK...I'm breaking out," and asked the Tower about 5 seconds later whether the runway lights were all the way up. The controller responded that they were on high. Eight seconds later the first officer said, "OK, you can go...10,200 [the minimum descent altitude]."

At 7:00:43 pm, there was more discussion about seeing the runway and then the Tower called to ask if the G-III had the runway in sight. The G-III was passing through 10,000 feet. The first officer advised the controller that the runway was in sight. At 7:01:13, the first officer stated, "To the right is good." According to radar data, the airport was to the left of the airplane at this time.

At 7:01:21, the airplane's configuration alarm sounded, noting that something was not set properly. This is not unusual and reminds the crew to make the necessary adjustments. At 7:01:31 and 7:01:34, the cockpit voice recorder (CVR) captured the sound of the aircraft's 900- and 800-foot callouts, respectively. The flight profile advisory (FPA) unit continued 700- and 600-foot callouts and at 7:01:42, the first officer stated, "To the right," which the captain repeated about one second later. Radar data indicated that the airport was still to the left of the G-III.

Three seconds later the ground proximity warning system (GPWS) sounded, indicating the flight was getting very close to terrain. Radar data showed the airplane started a left turn at 7:01:47 and two seconds later the GPWS announced a sink rate alert. This indicated that the flight path with a high rate of descent was not stabilized. At 7:01:52 there was another GPWS sink rate alert and the FPA 400-foot call-out. One second later the engines went to maximum power. The FPA 200-foot call-out and a GPWS warning of steep bank angle occurred almost simultaneously.

The G-III crashed into terrain at 7:01:58 in a steep left bank about 2,400 feet short of the Runway 15 threshold, 300 feet to the right (west) of the runway centerline and 100 feet above the runway threshold elevation. It was 34 minutes after official sunset. Post-crash analysis estimated that the G-III impacted at a 49-degree left-wing-down attitude, with a flight-path angle of minus 15 degrees. There were 18 fatalities: two pilots, a flight attendant, and 15 passengers. There were no survivors.

Weather - The automated surface observation system (ASOS) transmitted a special report at 7:12:26 pm (10 and one-half minutes after the accident) showing visibility down to one and three-quarter miles in light snow. Eight minutes later, the visibility had increased to three miles in light snow and mist. The ASOS five-minute observations surrounding the time of the accident (7:00:31 and 7:05:31) indicated that the visibility was nine and six miles in light snow, respectively. The ASOS one-minute visibility value about the time of the accident was eight miles. Weather observations at the alternate airport at Rifle indicated light winds, clear skies below 12,000 feet, and 10-mile visibility.

ASOS observations use an averaging algorithm and thus will not instantaneously reflect rapidly changing ceiling or visibility values. Obviously, the weather observations at the site, whether human or automated, may be quite different from what a pilot may see on an instrument approach a mile or two away. The augmented observation reported to the crew by ATC, of two miles visibility north, was reasonably accurate for operational purposes.

The Crew -The captain was type rated in the Gulfstream III in January 1990 and had accumulated 9,900 hours' total flying time, including 7,900 hours as pilot in command and 1,475 hours in the G-III. The first officer received his initial G-III type rating on February 9, 2001, and had accumulated 5,500 hours' total flying time, including 4,612 hours as pilot in command and 913 hours in the G-II and -III (110 of which were with the charter operator as a G-III second-in-command). According to family members, both pilots had slept about eight hours the night before the accident. Charter operation records showed the captain had been paired with the first officer for five months, and they had flown to Aspen twice before during daylight.

The pilots were trained in accordance with the company manual, which included initial qualification followed by annual recurrent simulator training in the G-III. Part of that training included instrument approach procedures, crew coordination on the approach, and a controlled flight into terrain (CFIT) training module. CFIT hazards, enhancement of situational awareness, recognition and evaluation of potential dangers, and a simulator profile that includes a ground-proximity escape maneuver are part of the training.

The Aircraft - The accident airplane was manufactured in 1980 and was equipped with an avionics unit which announces radio altitudes every 100 feet between 100 and 1,000 feet above the ground and deviations from a selected altitude. The airplane

Gulfstream III , Gulfstream Aerospace

was also equipped with Ground Proximity Warning System that announces the 500- and 200-foot radar altitude callouts and alerts for excessive descent rate, excessive closure rate to terrain, insufficient

234

terrain clearance, and excessive bank angle. According to the NTSB, the aircraft was maintained, loaded, and fueled properly. No pre-impact failures were found.

The Airport - Aspen's airport elevation is 7,815 feet msl and is surrounded on all sides by high terrain. There is only one runway, 15/33. The VOR/DME-C instrument approach does not include straight-in minimums because of excessive descent gradients. The terminal instrument procedures require that the maximum gradient for straight-in minimums between the final-approach-fix minimum altitude and the runway-threshold-crossing height cannot exceed 400 feet per mile. The descent gradient between ALLIX and the threshold-crossing height of Runway 15 is 700 feet per mile, hence the high circling minimums. Pilots are not required to circle provided they have the runway visually and have adequate time and altitude to make a normal landing.

Notams - Less than 10 days before the accident an FAA flight inspection crew working on a proposed GPS approach procedure to Runway 15 recommended that circling should not be allowed at night and a Notam was issued stating, "Circling NA [not authorized] at night" at Aspen. Since no straight-in minimums were published for the VOR/DME approach, the entire procedure was not authorized at night, but the wording was not clear.

The Notam was given to the FSS and included in the first officer's preflight briefing but not forwarded to the Aspen Tower as required. The approach never should have been authorized in the first place.

Air traffic control tower information - The Aspen controllers closely monitor the progress of airplanes executing the VOR/DME-C approach. The controllers pay close attention to minimum step-down fix altitudes and advise pilots immediately when they see an aircraft below a minimum altitude. During a post-accident interview, the Tower controller noted that she saw the G-III descend on radar below a step-down altitude but was unable to see the aircraft visually, although the crew called the runway in sight. It was at this time that she asked the crew if they had the runway in sight. When the G-III appeared around a minute later, it was pointed at Shale Bluffs (located northwest of the runway). From the Tower it appeared to be accelerating and was rolling rapidly to its left. The controller anticipated the crash and reached for the crash phone.

Radar profile - According to radar analysis, the G-III was not stabilized in the final approach and after passing ALLIX, it descended in excess of 2,000 feet per minute. It leveled momentarily at 10,000 feet for about 10 seconds and descended again at 2,200 feet per minute, hence the sink rate warning from the GPWS.

Landing gear and flaps were down but the wing spoilers were deployed. This is contrary to the G-III flight manual, which does not allow spoiler use with either landing flaps or gear extended. The pilot's intent appears to have been to bring the aircraft below the snow showers and gain visual contact while not allowing the speed to increase. This improper configuration created the very high sink rates.

***NTSB Probable Cause** –"... the flight crew's operation of the airplane below the minimum descent altitude without an appropriate visual reference for the runway. Contributing to the cause of the accident were the Federal Aviation Administration's (FAA) unclear wording of the March 27, 2001, Notice to Airmen regarding the nighttime restriction for the VOR/DME-C approach to the airport and the FAA's failure to communicate this restriction to the Aspen tower; the inability of the flight crew to adequately see the mountainous terrain because of the darkness and the weather conditions; and the pressure on the captain to land from the charter customer and because of the airplane's delayed departure and the airport's nighttime landing restriction."*

Commentary - There were multiple factors at work in this tragedy. Obviously, descending below landing minimums without a clear view of the runway or its environment is potentially lethal. This crew was committed to land despite all evidence that advised against it. Pressure from the charter customer contributed heavily to that mindset. Having a passenger in the jump seat and not adhering to a sterile cockpit rule (no irrelevant conversation below a predetermined altitude) were additional distractions.

Additionally, the FAA failed to write the Notam prohibiting night approaches clearly and transmit it to Aspen Tower. Both those oversights were corrected immediately after the accident. FAR Part 91 flights don't have the same external pressures that charter flights do, but in some respects the decisions involved may be tougher. When we decide not to fly or decide to divert, we're saying "No" to family, friends, and ourselves. Balancing the risk-and-reward equation correctly is impossible if we rationalize the risk. As pilot-in-command, tough love means occasionally having to say you're sorry.

Three Strikes - (PA32-260 pilot never gives up)

A low time instrument pilot attempts three very low approaches at different airports. The legal system determined a different outcome from what most pilots might expect.

When is it reasonable to decide that the weather just isn't cooperating and it's time to rethink your entire approach (pun intended) Weather changes constantly and requires both controllers and pilots to keep up. But sometimes it doesn't change, is widespread, and then it's time for Plan B.

How much experience should a pilot have before attempting approaches to minimums? What role does air traffic control play in leading a pilot down the primrose path. The first two questions pilots must answer for themselves, and a U.S. District Court provided its own answer to the second one. The court's decision may surprise you.

Piper Cherokee 6, Piper Aircraft

The Flight - On December 12, 2001, at about 3:35 pm Eastern Standard Time, the pilot of a Piper Cherokee Six called the Gainesville Flight Service Station in Florida for a weather briefing and to file an IFR flight plan from Fort Lauderdale Executive Airport to St. Augustine, Florida, with Craig Municipal Airport, Jacksonville, as the alternate. A stationary front stretched across the northern Florida panhandle to just south of the Jacksonville area before continuing off eastward into the southwestern Atlantic Ocean. Low ceilings and visibilities were widespread in the frontal area and to the north of the frontal boundary. Visibilities south of the front were mostly unrestricted.

At 4:50 pm N7701J departed Fort Lauderdale Executive Airport IFR for St. Augustine. At 6:09 pm the pilot called the Miami Flight Service Station to request St. Augustine and Craig airport weather. St. Augustine showed two miles in mist and 200 feet overcast (well below the landing minimums for the VOR 13 approach, which required a ceiling above 450 feet). Craig was hovering slightly above minimums with 1.5 miles visibility and a 200-foot broken ceiling. Temperature and dew point were 19 degrees Celsius.

Strike One - St. Augustine - At 6:35 pm as the flight checked in with Jacksonville Approach Control, the pilot inquired again about the Craig

237

weather. The controller replied that he'd had one flight miss the approach and advised the weather as one-half-mile visibility and 100 feet overcast. The pilot responded, "All right, I guess I'd like to go ahead and try the VOR in St. Augustine and see what happens. I'm not very optimistic, though."

The controller confirmed that the pilot had the current St. Augustine weather and cleared the Piper for the VOR 13 approach. Anticipating that Craig wasn't such a great option either, the controller suggested Jacksonville International Airport if the approach at Craig didn't work out. At 6:58 pm the pilot advised that he had missed at St. Augustine, was diverting to Craig, and asked if anyone had landed at Craig. ATC advised that two flights had landed using the ILS.

Strike Two – Craig - The 6:55 pm Craig weather was essentially unchanged from before: wind 010 degrees at 7 knots with one-half-mile visibility, overcast at 100 feet; temperature and dew point at 20 degrees C; and altimeter setting of 30.18 inches. Minimums for the ILS 32 approach were 200 feet agl and one-half mile. At 7:09 pm the flight was cleared for the approach and the pilot contacted Craig Tower around 7:11 pm At 7:16 pm the pilot called Jacksonville on the miss and was provided vectors for an ILS approach to Runway 7 at Jacksonville International Airport 14 miles to the northwest. Note that good VFR prevailed about 60 miles to the south.

Strike Three – Jacksonville - Behind the scenes there was considerable discussion at Jacksonville Tower about the weather. It was so marginal that at 7:14 pm ground control mentioned to the TRACON data controller on the phone: "These guys are getting lost out here on the taxiways...." At 7:16 pm the special weather observation was wind 050 degrees at 6 knots, visibility one-half mile in fog, clouds 100 feet broken, 500 feet overcast, temperature and dew point 19 degrees C, altimeter setting 30.20. Visibility was now at minimums, the ceiling was below minimums, and the altimeter setting had increased by three-one hundredths. The newly revised automatic terminal information service (ATIS) "November" was broadcast at 7:24:39 pm but there is no evidence that the pilot ever received it.

The flight was handed off to another approach sector at 7:19:50, and the new controller verified that the pilot had the current Jacksonville ATIS information "Mike," which was now more than an hour old. At 5:56 pm information "Mike" reported the wind was from 070 degrees at 7 knots, visibility one and one-half miles in mist, clouds 200 feet broken, 500 feet overcast, temperature 20 degrees C, dew point 19 degrees C, altimeter 30.17 inches. The reported visibility was three times better than what was needed to land but with the ceiling right at minimums. It was a roll of the dice whether the approach would be successful. The pilot contacted the final approach controller and at 7:29:06, was cleared for the ILS approach.

At 7:30 pm another special report was issued: wind calm; visibility one-quarter mile in fog; ceiling indefinite, 100 feet; temperature and dew point 18 degrees C; altimeter setting 30.20 inches. This should have been broadcast but does not appear that it was. The pilot may have been thinking "Mike" was still current.

At 7:31:26 the flight was handed off to the tower, which, at the time, was operating from a temporary facility while the main control tower building was undergoing renovations. The temporary tower had no audio recording capability. It was a significant oversight, as will be seen later, and there was no additional transcript of what information was, or was not, provided to the pilot of N7701J.

Two other single-engine airplanes on IFR instructional flights successfully made the approach to Runway 7 ahead of N7701J, seven minutes and about four minutes before the Piper made its approach. The pilots reported breaking out of the clouds about 50 to 100 feet above decision height and seeing the approach lights, but not the runway lights, at minimums. The tower passed these reports on to N7701J as the flight was passing the final approach fix.

Aftermath - Radar data showed that the Piper on the localizer course until about 7:39:04 pm, until about two miles from the runway approach end. At 500 feet the aircraft turned slightly to the right, continued to descend to 300 feet and turned back to the left. It appears that the flight never reached decision height. At 7:39:59 when about 2,000 feet from the end of the runway, the flight turned left and began to climb.

The Piper completed one 360-degree left turn, climbing to 1,000 feet, and then another 360-degree left turn, descending to 300 feet, where radar contact was lost at 7:41:23 pm The tower then heard the pilot report he was making a missed approach; this was followed by some crackling sounds on the radio and then the pilot said, "something about his instruments malfunctioning."

Witnesses at the approach end of Runway 7 reported seeing a red navigation light as the airplane passed overhead and hearing the engine noise increase as the airplane approached the end of Runway 7. The airplane then "appeared to start climbing and turn hard to the north. The airplane then continued to make several circles and descend, followed by the sound of the airplane crashing through trees." The pilot and three passengers were fatally injured. The wreckage was located in a wooded area about one mile north-northwest of the approach end of Runway 7.

Two subsequent aircraft on approach were sent around to clear the airspace but landed a few minutes later with weather reported at 300 feet overcast, visibility three-quarters of a mile.

Pilot and Aircraft - The 52-year-old pilot had received his private certificate more than 20 years previous to the accident but had earned his instrument rating only 10 months earlier. His logbooks and other information showed about 965 total flight hours, including 442 flight hours in the Piper Cherokee Six, 17 hours of actual instrument flight time, 24 hours of simulated instrument flight time, and 225 flight hours at night. It could not be determined if the pilot met the FAA recency-of-experience requirements for instrument flight.

Apparently the pilot was suffering from a cold. NTSB's toxicology report noted, "The tests were positive for .97 mg/L ephedrine/pseudo-ephedrine and 5.652 ug/ml acetaminophen. Examination of the pilot's briefcase ...contained Medic brand A-Phedrin pills, Acetaminophen gelcaps, Robitussin CF, and Medic brand 'Stay Awake' caffeine pills."

The 1968 Piper PA-32-260 had about 7,850 total flight hours. The last inspection was in November 2001, about 35 flight hours before the accident and the engine had 1,267 hours before the accident. No malfunction could be found with the aircraft.

NTSB Probable Cause -*"...was the pilot becoming spatially disoriented and losing control of the airplane during a missed approach descending uncontrolled and colliding with trees and the ground."*

Commentary - When temps and dew points are close or the weather is otherwise questionable, always ask for the latest report. With ASOS and AWOS, updates are often available up to the minute. In low conditions it may go above or below minimums constantly. Current pireps are extremely valuable — give them and ask for them.

The fact that other flights landed is a powerful inducement to continue but there are multiple factors to consider:

- The weather may have changed since their approaches.
- They have better equipment--a horizontal situation indicator, flight director, and autopilot for example.
- They have cheated and have a higher risk tolerance.

- Hard as this one may be to swallow--the pilot(s) may be more skilled.

After multiple attempts at different airports, the effects of fatigue and stress are not to be ignored, and it's time to exit the entire situation for a gold-plated alternate—not necessarily the "paper" one that's listed on the flight plan.

Most of us don't practice approaches to minimums and missed approaches very often in actual conditions. Get some training from an experienced instructor and set your personal minimums to a level appropriate to your comfort level, experience, and proficiency. Consider following the professional's rule of not starting an approach if the weather is reported below the minimums, or your minimums—just go to the alternate. Use the autopilot, again, like the pros—it will fly a consistently better approach than most humans. Caveats are that it's properly adjusted and it's not too windy on the approach for the more basic autopilot.

There was a strikingly similar crash that occurred about 15 years later where a relatively inexperienced instrument pilot, flying a Piper Arrow went shopping for approaches and finally ran out of fuel on the last attempt. Alas, Icarus!

You've heard not to fly with a cold or when taking cold medication, even a nonprescription drug. The effects on equilibrium can be powerful. According to AOPA: "Any of the over-the-counter drugs found in common upper-respiratory-tract-infection treatments, particularly pseudo-ephedrine, or diphenhydramine, could contribute to physical/mental impairment that could affect safety of flight. There are many variables in the equation, including the duration of use of the medication, the dosage, the frequency of use, the person's morphology (body type: obese, slim, fit, unfit), any underlying chronic/acute medical conditions, especially those for which the pilot is self-medicating, and the variety of different drugs being taken together (the cocktail)."

"For a severe head cold, the big concern in addition to the side-effect potential of the drugs taken is the symptoms themselves that could, alone, cause spatial disorientation and/or subtle, if not sudden, incapacitation. Barotrauma, or ear block, could affect the sensory receptors in the inner ear that help us stay straight and level, so if those sensors are upset by inflammation and/or fluid buildup, VFR flight, let alone flight on instruments, could be risky at best."

Court Case - This crash seems cut-and-dried until we collide with some convoluted logic of the U.S. District Court of Middle Florida. The pilot's family members sued the FAA for failing to exercise due care by not providing the pilot with current weather and altimeter information. According to the family's theory, the lack of a current ATIS caused the pilot to become disoriented and lose control of the aircraft on the missed approach.

The trial judge held, "All parties agree that the crash occurred because the pilot became spatially disoriented, causing him to lose the ability to control his airplane. I conclude that the plaintiffs have proven by a preponderance of the evidence that FAA air traffic controllers failed to give...the pilot...the current weather information on that night which would have alerted him that weather conditions were rapidly deteriorating, and that this failure contributed to the pilot's spatial disorientation.

"However, I further find that the pilot himself also contributed to creating his spatial disorientation by forgoing the other options available to him and attempting instead to make his third instrument approach landing of the flight [after two missed approaches] when he was fatigued, ill, and on medication."

"Applying Florida comparative negligence principles, I hold that the FAA's negligence was the legal cause of 65 percent of the accident and that pilot...negligence was the legal cause of 35 percent of the accident."

The judge went on to cite plaintiff expert witness testimony that, given the outdated weather information on ATIS Mike, "the pilot...would have reasonably begun to look for external reference points at an altitude of approximately 500 feet. At that elevation, he could expect one-eighth to three-eighths of the cloud cover to be breaking, *giving him obvious reference to the ground* (emphasis added)."

"As the pilot descended farther to 300 feet or less, he likely continued searching for visual references outside, which he could not see because of the deteriorating weather conditions that were known to air traffic controllers at the time. The medical and aeronautical experts agreed that by repeatedly turning his focus from his instruments to the outside environment and back, [he] likely began to experience spatial disorientation."

The lack of a current altimeter setting also was cited as a problem. ATIS Mike gave a setting of 30.17, and at the time of the accident the setting was 30.20 (which equates to 30 feet. Since the pressure was rising, although not by very much, the aircraft would have been approximately 30 feet higher than indicated — or on the safe side). On most light aircraft, altimeters without a digital Kolsman window three one-hundredths of an inch is almost unreadable and 30 feet above or below minimums will not result in a crash.

Granted, ATC may have been remiss in not providing current ATIS weather, but there was no way to conclusively prove that, given the absence of recording equipment. *The tower controller did provide two timely pireps that should have prepared the pilot for exactly what he was getting into.* Somehow that was overlooked in the judge's reasoning. The minor change in altimeter setting was irrelevant.

The judge's comment that "He could expect one-eighth to three-eighths of the cloud cover to be breaking, giving him *obvious* (my emphasis) reference to the ground" is remarkably ignorant of the realities of low instrument approaches. The fact that flights immediately before and after the accident aircraft were able to land shows that the weather was at or above minimums, assuming those pilots didn't cheat. With the right skills and equipment, it was flyable but certainly no place for amateurs. It also was not "...a rapidly deteriorating weather condition."

What this judge failed to understand and in my opinion added to some extremely bad case law, is that pilots, not air traffic controllers, are the final authority as to the safe operation of the aircraft. Landing expectancy out of an instrument approach is a dangerous mindset and is never an excuse for losing control. Following the judge's "reasoning," anytime a pilot misses an approach because the weather is somewhat worse than reported, it's somebody else's fault if the pilot loses it.

ATC is there to assist, but as pilot-in-command, the decision is yours. The weather is what is seen from the left front seat, not what anyone has told you it was or might be. The government settled the case for something more than $9 million and did not appeal. Recording equipment for the temporary tower would have been an excellent investment.

Snowy, Foggy, and Overloaded - (C172 CFI pushes beyond the limits)

Not only is the right equipment needed, but a realistic attitude.

This business flight took place on February 2006 in an aircraft that was ill-suited for the job. The plan was to fly from Warrenton, Virginia, to Mitchellville, Maryland, in a Cessna 172. Then, pick up a passenger before continuing on to Atlantic City, New Jersey. Moderate icing had been predicted along with snow, low ceilings, and visibilities. The ATP-rated pilot, a private pilot without an instrument rating, and a passenger departed Warrenton under IFR for Freeway Airport (W00) in Mitchellville, Maryland.

About 8:00 am EST a pilot called to ask the Freeway airport manager about the weather. The ceiling at the time was 500 overcast with one-mile visibility in rain. Published landing minimums for the approach were ceiling 700 and 1. The pilot said he would be arriving about 8:30. A few minutes later a passenger arrived at Freeway and said he was expecting a flight from Warrenton.

The trip proceeded routinely and was cleared for the RNAV (GPS) Runway 36 approach at Freeway Airport. Although the approach qualifies as a straight in, it is not aligned with the runway. This compounds the chore of finding the runway at the MDA since any wind correction angle must also be applied to the offset approach course.

The Approach - Radar data showed the Cessna approach from the south and track the final approach course inbound. About 9:30 am the pilot announced on the CTAF that he was 5 miles from the airport and inbound on the RNAV 36 approach. The airplane over flew the airport at about 500 feet MSL and the pilot then asked if the runway lights were illuminated. The airport manager responded that the lights were on but recommended that the flight continue to Baltimore-Washington International Airport (BWI), 14 miles north for landing, because the "visibility was only one-half mile in heavy snow." The pilot did not respond.

The Cessna executed a missed approach and was vectored for a second attempt. When the controller asked what the weather conditions were over the airport during the first attempt and the pilot responded, "(the clouds) were broken at 600 to 700 but we couldn't see the runway." During the second approach, the airplane descended and leveled at 500 feet MSL. The last radar hit was observed at 400 feet MSL, about 1/4 mile prior to the approach end of Runway 36.

Witnesses saw the 172 appear over the south end of the runway, between 200 and 300 feet above the ground, and it flew the length of the runway at low altitude. At the north end of the runway, the airplane turned west away from the airport, then circled to the right in a "dramatic" and "nose-high attitude" back towards the runway. It flew down the runway southbound and turned west again.

Engine power increased and the flaps were retracted. At 09:50 am the Cessna entered a steep left bank back towards the airport and "nose-dived" out of view. Both pilots were killed, and the rear seat passenger received serious injuries.

Pilots - The pilot held an airline transport pilot certificate and a flight instructor certificate for airplane single-engine, multi-engine, and instruments. He reported 2,900 hours total flight time on his last medical. The second pilot held a private pilot certificate, no instrument rating, and about 180 total hours logged.

Aircraft - The 2001 Cessna 172 had approximately 2,411 total hours with a 100-hour inspection completed a few weeks before. It was equipped with an IFR capable Bendix/King KLN 94 GPS receiver, and a Bendix/King

Approximate flight path- Foreflight

KMD 540 Multi-function display. The GPS navigation and multi-function database cards were expired. NTSB could find no malfunction relative to engine, instruments, or flight controls.

Weather - At 9:41 am Andrews Air Force Base, nine miles southwest of Freeway Airport, reported scattered clouds at 300 feet, an overcast layer at 500 feet, with two miles of visibility in snow and fog. The wind was from 140 degrees at three knots. At 9:42 am the weather reported at BWI, fourteen miles north of Freeway Airport, included broken clouds at 500 feet, an overcast layer at 900 feet with three-quarter mile of visibility in snow and fog. The wind was from 160 degrees at three knots. Witnesses at Freeway airport said the clouds were "….on top of the trees" and that visibility was about one-half mile due to snow and fog.

NTSB Probable Cause – *" …was the pilot's improper inflight planning/decision to attempt a landing in weather conditions below landing minimums, and his failure to maintain airspeed while maneuvering. Factors in the accident were the fog and snow."*

Commentary - This is an excellent case study in poor decision-making:

Error One - The weather alone should have caused the pilot to cancel. A Cessna 172, or any non-icing approved light aircraft for that matter, is just not safe transportation in a winter storm. If your travel profile requires regular winter trips outside of the sunny South, approved deicing equipment is needed. The airport manager's early report of rain is indicative that the weather was relatively warm and wet on front side of the low-pressure system or front. It will likely generate moderate to severe ice at some altitude and flights in non-deiced aircraft in February in the Mid-Atlantic region are best restricted to gentle IFR. The fact that the flight didn't get into severe icing is only due to luck.

Error Two – The choice of airport was poor. The minimum descent altitude for the GPS 36 approach at Freeway was 532 feet agl and a minimum visibility of one mile. Even before takeoff, the pilot knew the weather was "right at minimums." Weather does fluctuate, however, and under FAR Part 91, we're allowed to take a look even if it is reported below minimums. While we cherish that freedom, most pilots are better served to plan for a different airport. This report does not state whether the pilot filed an alternate that certainly would have been required under the prevailing weather condition. Logic and safety dictates that it would have been smart to have a Plan B. The pros flying in an icing-approved aircraft would have told the passenger to go to BWI where there's a full ILS, plenty of runway, snow removal and a complete lighting system to make finding the runway much easier.

Error Three - The aircraft did not have the weight-lifting capacity needed. The NTSB calculated that the Cessna weighed 2,604 pounds at takeoff; 147 pounds above the maximum gross weight of 2,457 pounds.

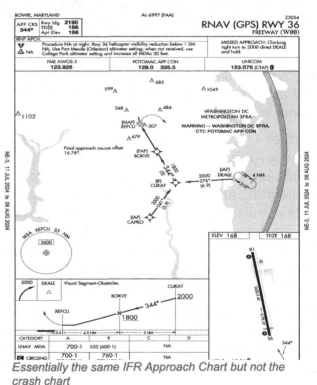

Essentially the same IFR Approach Chart but not the crash chart

At the time of crash, based on nominal fuel consumption, the airplane weighed about 2,526 pounds; 69 pounds above gross. They were about to pick up a second passenger at Freeway who weighed 175 pounds. Now, think this through. The airplane would have been nearly 240 pounds overweight, taking off from a short runway with obstacles and the runway was contaminated with snow that would have extend the takeoff roll. Even if the flight had successfully landed, the subsequent takeoff would have been in jeopardy. Wing contamination from snow could also become a factor.

Error Four - If the first approach was on speed and on altitude (it should be if you're flying in weather) and the runway isn't there, go to the alternate. Anecdotal evidence shows that many accidents occur on subsequent approaches as the pilot tries a little harder to get in. Simply, take another bite of the apple and hope there's no worm?

Error Five - Everyone knows that stall speed increases with weight, bank angle, and abrupt control input. The stall speeds chart in the 172R Skyhawk Information Manual shows that at maximum gross weight, the most rearward center of gravity, and with a 10-degree flap setting, the airplane would stall at 58 knots in a 45-degree bank and at 69 knots at a 60-degree bank. But, the flaps had been retracted, which raised the stall speed. Overweight and off the chart at a mere 300 feet above the ground is no place to be! Additionally, there may have been some wing contamination from icing, but that's speculative. A note at the bottom of the chart states: "Altitude loss during a stall recovery may be as much as 230 feet." You'll recall that that the witnesses put the aircraft at between 200 and 300 feet AGL.

A moderately experienced ATP/CFI was sure he could handle it. He'd been flying long enough to become complacent, but the list of errors is extensive and any one might have been enough to precipitate a crash. As stated before, with enough bad decisions, one can overcome any skill level. This pilot was trying to get too much utility out an unsuitable aircraft. It's plausible that the VFR pilot was hoping to learn more about how to fly in instrument weather. The ATP wanted to help, and a business trip seemed like a perfect example. The intentions weren't bad, but the outcome certainly was.

Hard Truths and Phantom Aircraft - (C206 pilot crashes in trees)

Wishful thinking does not alter the facts.

It is not in the natural order to lose a child before your own passing, even when they are already adults. It's the tallest mountain that any parent ever has to climb. It is with great respect and empathy that these facts are offered in response to a

Exemplar Cessna 206, Cessna Aircraft

"documentary" video that a businessman produced to exonerate his daughter. The purpose of the video was to show that another aircraft interfered with the event aircraft on a night instrument approach. The theory was that the intruder caused the crash while the intruder landed successfully in minimum or below conditions.

The daughter was, by all accounts, a gifted young lady and graduate student, aspiring to a career in opera. She was pilot-in-command in a tragic crash that took her life and that of four graduate student classmates on a night instrument approach into Bloomington, Indiana. The young people were returning from a musical practice session late on an April evening in a Cessna 206. Here is a recap of the final accident report.

Weather Briefing - The pilot called Flight Service at 10:13 pm EDT for a weather briefing from Lafayette, Indiana (KLAF) to Monroe County (KBMG). Flight time was estimated at 40 minutes with three hours fuel on board the Cessna 206. The briefer suggested an IFR flight plan due to an Airmet for instrument conditions throughout the night. KBMG weather at the proposed 10:30 pm departure time showed southwesterly winds at three knots, eight statute miles visibility, temperature, and dew point of 17 and 16 degrees C respectively and a ceiling of 800 feet with broken clouds.

The forecast for KBMG predicted light winds from the southeast, 600 broken and 5,000 overcast, visibility better than 6 miles with a chance for 5 miles in mist. It was not a daunting forecast for an IFR pilot, except for the temperature-dewpoint spread of one degree which foretold of potentially much lower conditions. The pilot filed an IFR flight plan but did not include an alternate which would have been both required and prudent under those uncertain conditions.

The Flight - The IFR flight proceeded normally and contacted Indianapolis Center at 11:19 pm. The KBMG Tower had closed at 9:30 pm and Indianapolis Center was handing approaches into the airport. The pilot advised she had the automated weather (ASOS) and requested the ILS to

Runway 35. The weather, just prior to the crash, was reported as wind 230 degrees at 5 knots, visibility one mile in mist and overcast at 100 feet. Landing minimums for the ILS 35 approach were 200 and one-half, so conditions had deteriorated significantly from the earlier forecast.

Ceiling is not considered controlling so by definition, the approach was legal since the reported visibility was one mile. (Part 135 or Part 121 will not even begin an approach if the weather is reported below visibility minimums and the flight has not reached the final approach fix.) In personal flying are we allowed to "take a look" even the though the risk goes up significantly.

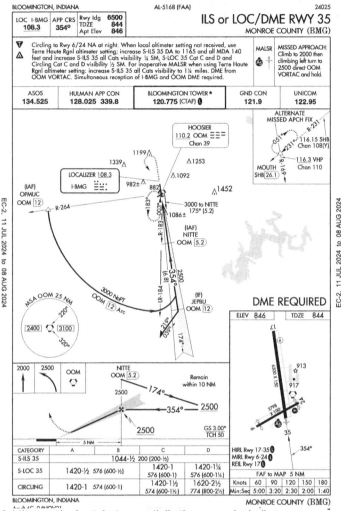

Not the crash chart, but essentially the same, including minimums, DME is added.

The controller provided vectors to the localizer but gave the incorrect CTAF, cleared the flight for the approach and terminated radar coverage. The radar plot tells the story even though radar was unavailable below 2,000 feet msl or about 1,200 feet above the airport elevation.

Simply, the aircraft never intercepted the localizer nor the glideslope, nor slowed to approach speed. The approach was unstabilized from the outset being about 30 to 50 knots above a normal final approach speed, well to the right of the localizer and well above glideslope. This could not have been safely resolved in the short time and distance remaining after the Cessna dropped below radar coverage. The pilot voiced no concerns about the weather and there was no discussion regarding a missed approach. Instrument students are taught that whenever an approach becomes unstable due to a poor vector or pilot performance, a missed approach is much the better choice. There is no shame in going around or in this case, going to an alternate. It would have been lifesaving.

The Crash - The Cessna was found between a quarter and half mile from the approach end of Runway 35, inverted, after colliding with trees. Six feet from the wreckage the nosewheel rim was embedded in a tree. The stop and aircraft inversion were brutal. There was no indication of any aircraft malfunction, and the engine monitor showed that full power had been applied just prior to impact, likely indicating that the pilot saw trees illuminated by the landing light. All five occupants perished.

Pilot - The 24-year-old private had single and multi-engine land ratings and instrument privileges. She had nearly 380 hours total flight time, 24.5 hours of actual Instrument time, and 51.1 hours of simulated instruments. She was instrument current and had flown 18 hours in the previous 30 days.

NTSB Probable Cause - *"…was the pilot's continued descent below decision height and not maintaining adequate altitude/clearance from the trees while on approach. Factors were the night lighting conditions and the mist."*

Petition for Reconsideration (PFR) - Accident investigations follow a well-defined process. Preliminary reports are typically available about two weeks after a crash. Factual Reports are completed from six months to about two years after the crash, depending on complexity, with the Final Report containing Probable Cause (PC) following shortly thereafter.

Any time there is new factual (and credible) evidence, interested parties can Petition for Reconsideration (PFR) which the NTSB must review. NTSB staff, other than the accident investigator of a particular crash, looks at the data separately along with the Office of Aviation Safety management team to ensure thoroughness and quality. It is then presented to the Board for their concurrence or disapproval. The facts control the outcome. If there was something that was missed in the investigation, that will be considered. If nothing credible or new is presented then the original report stands.

The father hired an expert witness engineering firm to review the data and investigate further. In such situations there is almost always considerable civil litigation. NTSB Probable Cause is not admissible in court because the Board's function is not to aid either party but to conduct a factual analysis. Probable cause is based on fact, but it is not certainty, although in most cases it's highly accurate. Additionally, PC is used to identify any systemic flaws in the system which should be addressed by FAA, industry, or companies. The factual report, however, is admissible in court.

In summary, the PFR was based on two witnesses claiming that they heard another unseen aircraft east of the airport that theoretically might have interfered with the accident aircraft. The engineering firm attempted to reconstruct the scenario using sounds described by the witnesses. It all seems plausible and very scientific until the layers of the proverbial onion are peeled back. NTSB audio experts found the testing flawed by not considering the tree foliage and ground cover, time of day, ambient road traffic noise, and the foggy conditions that existed at the time of the crash. Any of these factors can change sound transmission characteristics. NTSB's factual rebuttal addressed what the witnesses thought they heard and how that more likely corresponds to the unstabilized approach of the accident aircraft. *There were five other witnesses who did not hear any other aircraft.*

Commentary - Forecasts are never guarantees and what pilots see (or don't see) out the windshield always takes precedent. Stabilized approaches in all conditions, VFR and under instruments, are essential to success. Temperature and dew-point spread is critical, and weather can go from clear and 10 miles to below minimums in less than 15 minutes (My experience, as mentioned earlier.) New pilots, as talented as they might be, should be as suspicious and cautious as the gray heads are.

It is extremely unlikely that another aircraft would have intruded without

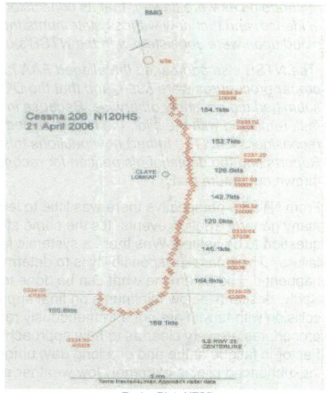

Radar Plot, NTSB

being seen on radar at some point, prior to descending below radar coverage, or upon executing a missed approach. No aircraft landed or were in the vicinity of Bloomington, Indiana, shortly before midnight according to data presented. *No evidence of a phantom was ever found.* With the weather that low, it's illogical that anyone would have opted for a non-precision approach without the benefit of an approach lighting system that the crosswind runway did not have. Nor would they have attempted a hard right turn at low altitude half a mile from the runway turn to join the localizer for Runway 35 just ahead of the accident aircraft.

The PFR noted that FAA settled the case against Air Traffic Control due to supposed procedural problems, which the FAA did not admit to except for the CTAF error. FAA legal and business decisions do not, and should not, bear on NTSB's review of the safety matters. The Board voted unanimously to deny the reconsideration.

The promo for the father's video mentions that the case went all the way to U.S. District Court of Appeals.

Not mentioned in the movie promo was the court's statement: *"The NTSB reviewed (the father's) materials, but found that the engineering firm's methodologies were flawed, that its conclusions were not supported by the evidence, and that new witness statements the firm had obtained and relied upon were consistent with the NTSB's original report.*

The NTSB also addressed the alleged FAA failures and concluded that proper procedures were used, and that the DOJ letter (the father) submitted did not show otherwise. Because in its judgment the probable fault remained with the pilot, the NTSB denied the petition for reconsideration. (The father) now petitions this court for review of the Reports and the denial of his petition for reconsideration." The case was thrown out before trial.

From NTSB's perspective there was little to learn from this crash, as in many general aviation events. It's the same story but different day. The question to be settled: Was there a systemic failure or an individual failure? The Board's responsibility is to determine what happened, why it happened, and determine what can be done to prevent a recurrence. Sadly, descents below minimums on final approach and subsequent collision with terrain are not a tremendously rare occurrence and the procedures are very clear as to how approaches should be conducted. Factor in fatigue at the end of a long day, unforecast weather, an inexperienced pilot in extremely low weather and prudence dictates leaving a significant margin for error. An unstabilized approach is a mandate for a miss.

NTSB has no vested interest in the outcome and is not subject to outside pressure, a rarity in government. It's essential to the Board's mission. Political, personal, or legal system interference are not the way to build a robust safety system, especially if competing views are thoroughly vetted. The PFR process takes this into account. We should admit that mistakes are occasionally made, and *a case is never closed when new and credible factual information is presented to the Board.* There are changes and reversal, but rarely. The NTSB has solid rules of evidence that have withstood the test of time.

> **"Facts are stubborn things; and whatever may be our wishes, our inclinations, or the dictates of our passion, they cannot alter the state of facts and evidence."**
>
> *John Adams. Second president of the U.S.*

The father deserves all possible solace and compassion in his grieving process in whatever way he chooses to pursue it. But there is no backing away from aeronautical reality. Facts to prevent future occurrences mean probable cause must be correctly identified. The Board must honestly review any Petitions for Reconsideration and must include NTSB's own admission of error in the original investigation or when new data are presented. That should be done as accurately, transparently, *and quickly* as possible, regardless of how fervently litigants or family members may believe in their cause.

Ice

There not as much ability to predict ice and then verify that it's actually there as with other weather phenomena because it's a bit more slippery to forecast accurately (sorry). It's less of a threat to properly equipped and flown aircraft, but nature can always beat us.

It's not going to fly well – if at all, NASA

There must be enough moisture *and* cloud *and* the proper temperature to get airframe icing. It's almost entirely an IFR problem because ice forms only when in clouds (we'll get to freezing rain and wet snow momentarily). In unapproved aircraft it just has to be avoided, which limits the usefulness of light airplanes during the colder seasons.

This puts a premium on pilot reports (Pireps) and verification of forecasts. It can be a transient condition confined to a narrow layer that almost any aircraft can penetrate, or it can build alarmingly fast and bring down even approved machines. Ask for and give Pireps. When ice is forecast and it's there, other pilots will find that very helpful. If icing is forecast but doesn't occur that allows others to fly and to pass along that the forecast models were in error.

But icing Pireps, or lack thereof, must be taken with caution. The pilot may have been too busy to give one, or they were flying a powerful, well protected aircraft and it just wasn't a problem. But if there are a several icing Pireps close to the area/altitude for the proposed flight, it's best to assume it's the real deal.

There are other variables such as the shape of the collection surface and the speed of the aircraft. Thin and pointy objects such as propellers, tail surfaces and fuel vents will collect ice more quickly than fat wings. You'll hear local "wisdom" such as "the unapproved Whizbang 400 with laminar flow wings will ice up in July at the equator while the Slingshot XZX will carry a load and has been flown around the Great Lakes every winter." This "advice" is suspect because the conditions in both cases may have been quite different, and the observer's veracity might be questionable. Neither have been objectively tested.

If an airliner or business jet ahead advises moderate ice, a light aircraft even with approved icing capability should go to Plan B. What's moderate for them may well be severe for light aircraft The official ice definitions also leave us out in the cold (my apologies again) on how it will affect a particular aircraft. *They also refer to the effects <u>with approved deicing equipment</u>, which most light aircraft don't have. Some new or bold pilots might believe that "light icing" won't be a problem as long as they escape in less than an hour, as detailed below.*

From the FAA's Advisory Circular (AC 91-74B or its successor) on Icing:

"Trace icing - Ice becomes noticeable. The rate of accumulation is slightly greater than the rate of sublimation. A representative accretion rate for reference purposes is less than ¼ inch per hour on the outer wing.

Deicing/anti-icing equipment is not utilized unless encountered for an extended period of time (over 1 hour).

Light Icing - The rate of ice accumulation may create a problem if flight is prolonged in this environment (over 1 hour). Requires occasional cycling of manual deicing systems to minimize ice accretions on the airframe. A representative accretion rate for reference purposes is ¼ inch to 1 inch on the outer wing.

Moderate Icing - The rate of ice accumulation requires frequent cycling of manual deicing systems to minimize ice accretions on the airframe. The rate of accumulation is such that anything more than a short encounter is potentially hazardous (emphasis added). A representative accretion rate for reference purposes is 1 to 3 inches on the outer wing.

Severe Icing - The rate of ice accumulation is such that ice protection systems fail to remove the accumulation of ice and accumulation occurs in areas not normally prone to icing, such as aft of protected surfaces and other areas identified by the manufacturer. A representative accretion rate for reference purposes is more than 3 inches per hour on the outer wing."

NWS's Aviation Weather Center's website is a great place to learn about icing. Shown here, an early spring storm moving up the east coast shuts down light aircraft flying for the day. The red splotches indicate potential for Supercooled Liquid Droplets (SLD)--no decisions to be made here unless you've got lots of power and hot wings.

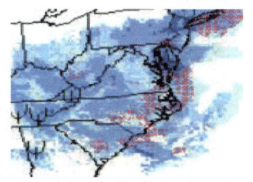

Moderate to severe icing and SLD at 7,000, Not flying today!, NWS

Advise ATC as soon as icing begins. From the Advisory Circular, based on a detailed study of icing accidents and incidents: "...conflicts with ATC were common when pilots take action to exit icing conditions after an inadvertent icing encounter. Very often, this was because the pilot deviated from an IFR clearance and failed to declare an emergency or otherwise clarify the situation with the controller."

"In a subset of these cases, the controller actually offered to declare an emergency for the pilot, but the pilot declined. In another subset, the frequency was too busy for communications, often because the controller was overwhelmed with traffic."

"A number of pilots expected an immediate response from ATC when they reported difficulties after encountering ice and expected a blanket clearance to escape icing without first declaring a state of emergency. In many cases, such assumptions proved to be not only false, but fatal."

Don't be hesitant to declare an emergency–that opens up lots of assistance. Controllers aren't interested in giving out violations, but they have to maintain separation. With luck, it will just be a powerful learning experience. But sometimes the potential for ice is so great that if it gets operationally complicated, it may also get legally complicated. In perspective, a discussion with the FAA is small potatoes to whacking the ground out of control.

SLD, Freezing Rain, and Wet Snow - Super-cooled liquid droplets (SLD) are the basis of freezing rain or drizzle. It can make an aircraft unflyable in literally minutes VFR or IFR. Fully approved airline aircraft have been lost to SLD (See Roselawn- Online.) Climbing to escape is often the best option for powerful aircraft but for light aircraft an immediate 180-degree turn may be the best, and perhaps only choice. It takes a lot of power to top the slush.

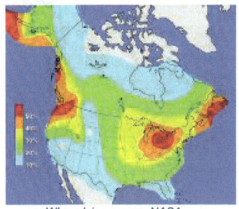

Dry snow is not a problem because in really cold clouds, the moisture is already frozen and ice crystals won't adhere. But if those conditions don't exist, wet snow can cause airframe icing because the snowflakes are only partially frozen and when the airframe or prop disrupts them, they'll take revenge and adhere. It can also block engine air intake filters and then the engine isn't happy.

Where Icing occurs, NASA

Below about minus 20 degrees C, the water has already crystallized and *generally* won't stick to the aircraft (note the disclaimer.) In the map to the right note that the icing potential is about the same in North Dakota as it is in the Southeast. It's much colder up there, but there's a lot less moisture. Note that Montana and Wyoming are like Florida–sometimes. *However, these statistics are averages that may not apply on the days you intend to fly.*

Warmer below-freezing temps are where things get ugly which can happen in the southeast in frontal passages. The clouds are juicier— cumulus clouds tend to be warmer and hold a lot of moisture. Periodically a warm but icy system will slide through Texas and shut down the Dallas-Fort Worth airports or Atlanta resulting in complete havoc on airline schedules.

Hangar Story – *It was over Kansas City in a fully ice-approved Cessna 310R at 8,000 feet when I encountered a cumulus deck behind a cold front that had passed through a few hours earlier. In about 20 minutes the airspeed had dropped 15 knots, and the tops were not within reach. I should have changed altitude sooner but had activated the anti and deice equipment appropriately. It was the exact definition of moderate ice, which had been properly forecast*

I was about to ask the controller how the BBQ was at Kansas City Downtown Airport when the skies opened up to clear blue above and a stratocumulus deck below. It took about an hour for the ice to sublimate off the unprotected parts of the airplane.

The real problems with Ice:

Weight isn't *usually* the problem. It's the aerodynamics that become royally fouled up in a variety of ways:

Wing stall – The shape changes with ice buildup and not for the better.

Tail stall – Ditto above but the recovery is exactly the opposite and sometimes it's hard to tell the difference. Not as nearly as common as wing stalls from what we know at this point.

Under wing plating – Behind de-ice boots. Most ice-approved aircraft have a posted minimum climb speed to ensure that any ice buildup stays on the protected part of the aircraft which is the front of the wing and tail surfaces. If ice builds behind that there's no way to remove it. Aircraft with fluid deicing systems may not have this problem but fluid supply is finite.

Propeller – If the blades ice up unevenly, the prop becomes unbalanced which leads to severe vibrations and may ultimately break engine mounts or worse.

Control binding – On unapproved aircraft, the control surfaces may bind with enough buildup.

Engine air intake blockage – If the primary airflow is blocked and there's no suitable alternate air source – power ceases.

Turbine engine flame-out – Ditto above or damage from ice going into the compressor blades. They are equipped with anti-icing equipment to use before ice starts to build.

Plugged fuel vents – Which leads to fuel starvation because a vacuum is created in the fuel tank. Non-approved aircraft may not have non-icing vents.

Inability to see through the windshield for landing - Kind of important to maintain alignment and depth perception.

Iced up windshield, NASA

Any *one* of these problems can make flight difficult or impossible. If several occur, you get the picture. Having a little deicing equipment can lead to trouble. Non-approved or "no-hazard" systems may provide short term protection, but they don't begin to cover the full potential of ice-degrading factors. Prior successful experience can lead to a bad situation.

NASA has researched this thoroughly: "Minute amounts of ice (equivalent to medium grit sandpaper) covering the leading edges or upper surfaces of wings can increase the stall speed up to 15 knots. Ice on the wing also can disrupt the airflow over the ailerons and cause the aircraft to behave in unusual ways. The aileron may deflect without pilot input and cause an uncommanded roll." Taking off with frost after the aircraft has been parked outside overnight turns it into an experimental machine. "Polishing the frost" is no longer recommended by the FAA. They used to say it was OK if you removed the roughness. Not anymore.

We need to know where the ice is not - There could be warmer air below, but it has to be well above the Minimum IFR Altitude. High terrain complicates things. It's colder in the hills and the ground is coming up as we're coming down.

Might not make the tops and the ground is rising, NOAA

Ice is bad in the tops - Escape could be to the clear air above the cloud tops. If the tops are low, climbing can work, but the climb has to be started before the aircraft has any appreciable ice. However, if you're already in moderately thin air and only capable of 50 to

60 percent power, better have another plan that doesn't involve getting above it all.

Just when we're sure to get above the clouds, the heavier ice in the cloud tops pulls us back down. That's where the atmospheric lifting stops, and all the moisture congregates. The blue sky is so close, visible through breaks in the overcast, but there just isn't enough power. Now it's forward and down and hope there's enough power and lift available to level off as we descend to warmer air. It is well above freezing down below, right? Suggest at least five degrees C or more.

Now, suppose the climb above the tops was successful, although the airspeed is down by 10 or so knots. The ice should start sublimating. But ahead the clouds are building. The aircraft may have only a small amount of climb capability left. It's time to descend to warmer air or return to where we were before the icing began. Don't wait too long.

The iced-up Landing – When flying fully deiced twins for Cessna in Wichita, the guidance was that if there was *any* ice on the aircraft don't try to land at Cessna field on the east side of town with a narrow 4,300-foot runway. Instead, go to the airline airport (KICT), on the west side of town, with a 10,000-foot runway and fly it on *with no flaps*.

Why no flaps? As you'll recall from basic aerodynamics, the horizontal tail normally exerts a down force to balance the aircraft. Using flaps changes the downwash and may upset a very delicately balanced aircraft because of ice. Normal approach speed across the fence at Cessna field in twins was about 95 to 105 knots and full flaps but with ice – at least 120 knots and no flaps even at the big airport. Many ice encounters *would have ended successfully* but the pilot stalled on final approach either from a wing or tail stall.

In cold country, runways are often icy as well, so high landing speeds and poor braking could mean a slide off the end. Obviously, higher speed translates to a *much* longer runway, but it's *always* better to slide off the end at 20 knots then plunge in short at 90 knots from several hundred feet.

While autopilots are a wonderful help, approved ice flight manuals often recommend disengaging the AP completely or perhaps every ten minutes to check the trim. If not, the AP will subtly adjust to the ice-induced increasing drag and will fly flawlessly right up to its limits. Then, it will hand the surprised pilot a badly out-of-trim aircraft with little lift left to keep flying. The early warning that more nose-up trim is needed is essential for an immediate escape.

Carburetor Ice - Another kind of ice goes back to when carburetors were prevalent. If you're flying an older model engine, it's still a threat. Every year there will be several engine stoppages (not failures.) resulting in a crash and the evidence will have melted before the investigators arrive.

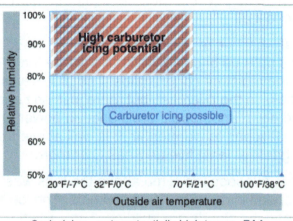

Carb. Icing - note potentially high temps, FAA

Carb ice can happen anywhere – including on takeoff and at cruise power even on warm days with high humidity. NTSB often investigates "engine failure" crashes where there was high humidity and no other obvious reason as to why the engine "just quit." Anticipate and use heat early and often. An NTSB study looked at crashes over a 10-year period (2000-2011) that showed about 25 per year. It's less now because many more engines have gone to fuel injection.

Thought Exercise – Icing - The end-run technique works with more than just convection. It's midwinter and we'd like to get out of snowy West Virginia and head to Hilton Head, South Carolina, but a low pressure is working up the coast. The icing Airmet is NOT shown. The ATC-preferred routing on the left is not safe for light aircraft (480nm, 3:20 and about 50 gallons in a high-performance single.) The long way around is only 60

Direct flight and probable icing, Foreflight

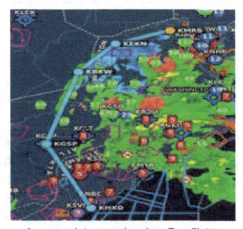

An easy detour and no ice, Foreflight

miles longer (540nm, 3:43 and about 5 gallons more). It's good VFR up north and out to the west to avoid icing. Down south, it's IMC but much warmer. Depending on the underlying terrain (mountains), even a bit farther west wouldn't hurt. Carry gas or stop and refuel, which will add about an hour but we're there in time for sundowners.

When a front is draped across the route, fly through at right angles rather than along it to minimize exposure. However, wishful thinking has no place without suitable escape options while penetrating.

Bottom Line – Avoid ice by going VFR if you can't get on top or stay underneath and the aircraft is unapproved. Unless there is freezing rain, freezing drizzle, or wet snow, staying out of the clouds means staying out of the ice. If IFR, advise ATC as soon as an altitude change is needed. If traffic is blocking a better altitude suggest an off-course vector, divert, or turn around. They need time to sort things out. If you're already at the minimum altitude for that segment of the trip, that's bad. Declare an emergency. Even if you just had a "great" learning experience, please file an ASRS report so the rest of us can learn and don't forget to file that Pirep. By the time 10 knots is lost, a 180° turn is often the best option to go back to where the ice wasn't. The total airspeed loss inbound will likely more than double during the escape. The aircraft is not the same old friend you're used to.

Key Points:

- If it's below freezing and there are clouds at cruise altitude, anticipate ice. It may not be there but at least you're prepared.

- If not ice-approved, advise ATC as soon as it starts to accumulate. If they can't respond in reasonable time declare an emergency. But ATC has to separate traffic which may require some coordination.

- Turn off the autopilot as it will mask poor handling qualities.

- Avoid routes over mountainous terrain when ice is unavoidable.

- VFR may be the only option – Is the weather good enough?

- When 10 knots are gone, depending on the aircraft, it's time to escape but start working on it before that. Another 10-20 or more knots may be lost during the retreat. Up, down, or turn around. Tops, bases, warmer air? What's the out?

- On approach, fly it on with no flaps and well above normal approach speed – a much longer runway is needed.

- Give and get Pireps.

Unpredicted, Unadvised, Unaware - (TBM 700 encounters severe ice)

Severe icing in the heavily traveled Northeast corridor goes unforecast, despite many reports to ATC. A relatively inexperienced TBM 700 pilot tries to climb through and loses control.

The loss of a TBM 700 turboprop over New Jersey surprised many pilots. The Pratt & Whitney-powered single-engine turboprop is certificated for flight into known icing (FIKI). For experienced pilots, all FIKI means is that when the aircraft starts to collect ice, it's time to go elsewhere. Turbine aircraft don't usually succumb to a hazard that catches roughly five to eight non-FIKI piston-powered aircraft every winter and although this accident occurred in some of the highest-traffic-density

TBM 700, AOPA

airspace in the world, surprisingly little warning was relayed to the pilot.

The Plan and the Forecast Weather - On December 20, 2011, at 7 a.m. EST time, the pilot filed an IFR flight plan from Teterboro, New Jersey (TEB), to Dekalb Peachtree Airport (PDK) in Atlanta, using the Direct User Access Terminal System (DUATS). No weather briefing was recorded through DUATS, but the NTSB noted that the pilot might have obtained weather data elsewhere.

The weather looked deceptively benign. The area forecast in effect until 10 a.m. called for a 7,000-foot overcast with cloud tops to 18,000 feet. After 10 a.m. the ceiling was forecast to improve to 15,000 feet broken. There was no discussion of icing hazards but there were numerous pilot reports, including several urgent ones that indicated danger aloft—although some came after the accident.

An Airmet issued at 6:45 a.m. advised of moderate icing between the freezing level of 3,000 feet and 9,000 feet and up to 18,000 feet. A subsequent Airmet for moderate icing between 2,000 feet and 8,000 feet up to 20,000 feet was issued at 9:45 a.m. that included the accident location. It's unlikely the pilot would have received that before departure.

On the ground, it looked like a good day to fly, with good visibility and high overcast. At Morristown (New Jersey) Municipal Airport (KMMU), near the accident site, the 9:45 a.m. observation reported wind from 360 degrees at 8 knots with gusts to 13 knots; visibility of 10 miles or greater; ceiling overcast at 20,000 feet; temperature 6 degrees Celsius; and dew point minus 2 degrees C. At Teterboro, just 20 miles from the accident location, skies were clear with unrestricted visibility.

The Flight - At 9:30 a.m. the pilot picked up an IFR clearance and departed Teterboro at 9:50 a.m. The TBM entered instrument meteorological conditions while climbing through 12,800 feet and was advised of moderate rime icing from 15,000 feet through 17,000 feet. The controller asked the pilot to advise him if the icing worsened, and the pilot said, "We'll let you know what happens when we get in there and if we could go straight through, it's no problem for us."

At 09:58:24, the controller directed the pilot to climb and maintain 17,000 feet and to contact New York Center (ZNY). While climbing between 12,800 and 12,900 feet, at 116 knots ground speed, the pilot acknowledged and advised that they were entering instrument conditions. *The minimum recommended climb speed, according to the POH, is 130 knots.*

At 10:02:17 a.m. the controller said that he was coordinating for a higher altitude. While at 16,800 feet, the pilot confirmed that, "Light icing has been present for a little while and a higher altitude would be great." The ground speed had dropped to 101 knots.

About 15 seconds later, the pilot stated that he was getting a "little rattle" and requested a higher altitude as soon as possible. About 25 seconds after that, the flight was cleared to Flight Level 200, and the pilot acknowledged. One minute later, at 10:04 a.m., the airplane reached a peak altitude of 17,800 feet "before turning sharply to the left and entering a descent." While descending through 17,400 feet, the pilot's last radio call was "and N-Seven-Three-One-Charlie-Alpha's declaring...." Ground speed was 90 knots.

The TBM came down very quickly from altitude, according to several witnesses, losing a wing and taking out part of the empennage before hitting the ground. The pilot and four passengers were fatally injured. The NTSB could find no mechanical problems with the airframe or the engine.

More than 80 pireps were received between 8 a.m. and 1 p.m., including an urgent report at 7:49 a.m. from a Cessna Citation reporting moderate to severe rime icing between 13,000 feet and 14,000 feet near the accident site. According to the NTSB, "an urgent pilot report was received at 8:08 a.m. from a flight crew operating an MD–83 airliner at 14,000 feet over

263

Morristown. The pilot reported moderate to severe rime icing between 14,000 and 16,500 feet...the worst he had seen in 38 years of flying experience, and that he had never seen ice accumulate so quickly."

A regional jet operating close to the accident aircraft reported that the wing anti-ice system could not keep up with the accumulation. The pilot estimated 2.5 inches of ice on protected areas of the wing, and four inches accumulation on some unprotected areas, in about five minutes. That critical information did not get into the system or was ignored.

Pilot and Aircraft - The 45-year-old pilot held a private pilot certificate with an instrument rating. He had more than 1,400 hours and held a current second-class medical certificate. His TBM flight time was not reported. The pilot had recently completed a simulator-based recurrent training program where icing procedures and escape were discussed. The TBM is prohibited from flight in severe icing conditions but can operate for some period of time in moderate icing.

According to the NTSB: "Impact damage prevented functional testing of the aircraft deice systems. The airframe deice, propeller deice, pitot heat one and two, and stall warning heater switches were found in the "On" positions. The inertial separator switches were found in the Off positions."

NTSB Probable Cause - *"....was the airplane's encounter with unforecast severe icing conditions that were characterized by high ice accretion rates and the pilot's failure to use his command authority to depart the icing conditions in an expeditious manner, which resulted in a loss of aircraft control.*

Commentary - This Pratt & Whitney PT6 engine has a switch-activated inertial separator on the TBM. When selected to bypass mode, it forces ice overboard rather than allowing it to enter the engine. Failure to deploy the separator could easily lead to a loss of power and a subsequent stall. The pilot's comment about "a little rattle" could indicate an engine problem because of ice ingestion, or possibly an airframe buffet warning of an impending stall.

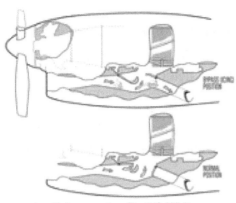

Inertial separator, Pratt & Whitney

The NTSB's analysis does not note if the engine was producing power at the time of impact, although there was rotation of the propeller and turbine rotors. The POH warnings about proper use of anti and deicing equipment are depicted in bold print and capital letters.

The Airmets warned of light to moderate icing between 2,000 feet to 9,000 feet and then all the way up to 20,000 feet. This is too broad to be of much use operationally. Research shows ice typically exists in bands of 2,000 feet to 4,000 feet. Moderate ice is nasty stuff and when jets with hot wings are reporting it, approved booted light aircraft should take that as a mandate to get through fast or stay out. Non-FIKI aircraft need not apply.

Icing Airmets do not distinguish severity based on aircraft type, but it makes a difference, and pilots should pay heed. Where moderate icing locations are well established, as they seemed to have been on this December morning, controllers should be providing unrestricted climb and descent clearances through those altitudes. Pilots, likewise, should ask for that if there is any doubt about their ability to remain in ice-impacted altitudes.

Be inquisitive about ice and ensure that aircraft anti-ice and deice systems are used early and often. In moderate to severe conditions, don't suffer in silence—immediately advise ATC that you've got a problem and if there's any doubt, declare an emergency.

Failure to maintain the minimum climb speed for icing can lead to rapid underplating on the underside of the wing and in moderate icing, that's a critical error.

It is troubling that in some of the busiest airspace in the world, critically urgent icing information was not passed to the pilot or to the forecasters. There were numerous pireps indicating potential danger lurking. Icing/instrument meteorological condition forecasts and Airmets can be verified by using pireps if there are enough of them. But sometimes you have to ask.

Ice Crisis - (SR22 pilot waits too long)

There are times when the PIC must advise ATC they are unable.

Anyone who has ever collected an appreciable amount of ice on their aircraft will avow the horrible degradation of flying qualities. The omnipresent Airmets from October to April in northern climates frequently miss the reality. The assumption is that ice will be present when the temperature, in clouds or precipitation, is below freezing. That's not necessarily so but it's far better to assume the ice is there until proven otherwise.

No icing reported or forecast -

Cirrus SR 22, AOPA

The Cirrus SR22 pilot contacted the Reno Flight Service Station (FSS) at 4:20 p.m. Pacific Standard Time on February 6, 2005. He received a standard weather briefing and filed an IFR flight plan from Reno/Tahoe International Airport, Reno, Nevada, to Oakland, California. The filed routing was from Reno, via the Mustang VOR (FMG), Victor 200 to Truckee intersection, Victor 392 to Sacramento VOR (SAC), then direct to Oakland at 12,000 feet. The briefing included current and forecasted weather for the Reno area, the intended route of flight, and the destination. An east-west stationary front was located in the Reno area. The briefer advised the pilot that there were no pilot weather reports (pireps) for the intended route of flight and no Airmets or Sigmets.

The freezing level in the Reno area began at 6,000 feet with no precipitation. There was an Airmet about 50 miles north of the Reno area that did warn of icing from the freezing level that began at the surface sloping to 5,000 feet, up to 18,000 feet. The pilot indicated he might request 14,000 feet once airborne. The minimum instrument altitude for the initial segment along V200 and 392 was 11,500 feet. The flight departed Reno about 5:50 p.m. The Reno weather at 5:56 was reported as: wind 020 degrees at 6 knots with 10 statute miles visibility and overcast at 5,500 feet with temperatures at 6 degrees Celsius, dew point -1 degree Celsius, altimeter setting 29.83 inches. A 7-degree spread on the temp-dew point indicated relatively dry air at the surface, but below-freezing conditions were only a few thousand feet aloft based on the cool surface temperatures. There was still no mention of icing by flights in the area or in the forecasts. The National Weather Service's Aviation Web site did show some potential for icing at 16,000 feet using Current Icing Product (CIP).

The Cirrus executed the Mustang Six departure and climbed on a westerly course to the assigned altitude of 12,000 feet. At 5:57 p.m. the departure controller asked whether N286CD could accept 14,000 feet to allow for departing traffic out of Truckee. The pilot acknowledged, and the flight was then handed off to the North sector of Reno Departure. The acquiring controller issued a 270 heading to join V392, and at 6:05 p.m. the Cirrus was handed off to Oakland Center.

Transcript - 6:07:46—N286CD: "Oakland Center, Two-Eight-Six-Charlie-Delta, any chance I could go up to one six thousand, see if I can get above these clouds?"

6:07:53—Oakland: "Cirrus Two-Eight-Six-Charlie-Delta, affirmative. Climb and maintain one six thousand."

6:07:57—N286CD: "One four thousand for one six thousand, Six-Charlie-Delta."

6:12:24—Oakland: "Cirrus Six-Charlie-Delta cleared direct Sacramento."

6:12:26—N286CD: "Direct Sacramento, Six-Charlie-Delta."

6:13:40—N286CD: "Oakland Center, Cirrus Two-Eight-Six-Charlie-Delta. Uh, I guess this isn't gonna work. I'm still in the clouds, any chance of lower?"

6:13:47—Oakland: "Cirrus Two-Eight-Six-Charlie, uh, Delta, Roger. Stand by one."

6:13:50—N286CD: "Six Charlie Delta."

6:15:00—N286CD: "Oakland center, Cirrus Two-Eight-Six-Charlie-Delta. If I could go up at, uh, another two, three hundred feet I could get above these clouds."

6:15:06—Oakland: "OK. Do you want to go up or down?"

6:15:08—N286CD: "Uh, let me go up first so I could build up some airspeed if that's OK."

6:15:12—Oakland: "November Six-Charlie-Delta, climb, uh, actually, November Six-Charlie-Delta, maintain the block one six thousand through one seven thousand."

6:15:18—N286CD: "Maintain the block one six thousand one seven thousand, Six-Charlie-Delta."

6:17:19—N286CD: "Uh, I'm coming down, Six-Charlie-Delta."

6:17:20—N286CD: (unintelligible) "I'm icing up."

6:17:39—Oakland: "November Six-Charlie-Delta, uh, say again."

6:17:42—N286CD: "I'm icing up. I'm coming down."

6:17:56—Oakland: "OK, uh last transmission was, uh, unreadable."

Aftermath - Oakland Center made several attempts to reestablish contact and contacted a nearby Cessna 210 to call the Cirrus, all to no avail. The Cessna pilot noted that another aircraft said he was icing up and descending, followed by static, but it could not be confirmed whether it was N286CD. Center acknowledged and then asked the Cessna pilot whether he was picking up any ice. The Cessna pilot replied, "Negative, I went through snow and I'm on top right here.... We just came out from underneath a layer ...garbled....about at we're between layers but all I got was snow. I did not get any ice the whole way out." The 210 pilot then attempted contact the Cirrus on another frequency. Three minutes later another aircraft reported picking up an emergency locator transmitter (ELT).

The Cirrus hit a steep rock and snow-covered slope. The pilot, who was the sole occupant, was fatally injured. The ballistic recovery system (BRS) parachute was deployed prior to impact and separated from the airplane almost immediately under extremely high loads. The maximum deployment speed was 133 knots, but the parachute was deployed at a much higher airspeed.

Pireps after the fact - At 6:40 p.m., 20 minutes after the crash, the Oakland Center Weather Service Unit issued "Numerous reports of moderate to severe rime ice between 13,000 and 17,000 feet. Conditions developing and increasing in the area." Ironically, two reports of severe icing came just prior to the crash, passed presumably by the controller to the Center Weather Service Unit.

Pireps included:

"6 p.m.—25 miles SE of Mustang VOR, Boeing 737, moderate rime during climb from 15,500 feet to 17,000 feet.

"6 p.m.—7 north CNO, Embraer 120, moderate clear 8,000 feet.

"6:15 p.m.—Over RZS (San Marcos on the coast near Santa Barbara), Cessna 182, light mixed, 9,000 feet.

"6:45 p.m.—25 miles NE of Reno, Turbo Commander 690, light rime 16,500 feet to 13,500 feet."

With the exception of the Cessna 182, all these turbine-powered aircraft were approved for flight in icing. Numerous other pireps started filtering in after the accident within 100 miles or so of the accident site.

The Pilot - The 54-year-old pilot held a private pilot certificate for single-engine airplanes with instrument privileges. He had started flying in July 2002, received his private certificate in April 2003, and received his instrument rating in February 2004, one year before the accident. He had logged 473 hours with more than 100 hours in the last 90 days. More than half his flight time was cross-country, with 43 hours of night and nearly 75 hours of instrument time with 11 hours of actual weather.

He had completed the Cirrus factory school, and it appeared that he had just returned home to the Oakland area two days prior to the accident with his new Cirrus. The logs estimated 69 hours in make and model with a current flight review issued December 29, 2004, six weeks prior to the crash.

The Aircraft - No pre-impact malfunction was noted during the investigation, and the aircraft had well under 100 hours of total flight time. The airplane was equipped with ice protection, that was certificated as a "No Hazard" system. The system was intended to allow escape from unexpected icing conditions. The tank held three gallons of deicing fluid that flows along the wing, horizontal stabilizer, and propeller blades.

The system did not allow flight into known icing, and the pilot's operating handbook (POH) made the limitations very clear that, "No determination has been made as to the capability of this system to remove or prevent ice accumulation." While pilots may be tempted to rely on this system, the guidance in the handbook is unequivocal. "Flight into known icing is prohibited.... At the first indication of icing, the most expeditious and safest course of exiting the icing conditions should be taken." To be fair, though, there was no indication that the flight would encounter icing.

We don't know from the investigation whether the system was serviced with fluid or it was operational. The pump can be airlocked if allowed to run dry. Pump priming must be done on the ground prior to takeoff. With fluid deice systems, it takes several minutes for fluid to fill the lines and be ready to flow onto the deiced/anti-iced area. If the pilot failed to prime the system, ice could easily have overwhelmed the aircraft.

According to the POH, "During simulated icing encounters, stall speed increases of approximately 12 knots in the clean configuration and 3 knots in the landing configuration were observed. In addition, cruise speed was reduced by 20 KCAS and the airplane's rate of climb diminished by at least 20 percent." Since this aircraft was not turbocharged and was flying close to its operational ceiling of 17,000 feet, the decision to climb, even

Flight Plan – Current chart - Truckee intersection is now Truck. Note MEA (in red box)is much lower and would have avoided ice. The red TFR is irrelevant to this event, FAA

without ice, was a non-starter. Going down was the only choice. The aircraft had a supplemental oxygen system, but it was impossible to tell if it was used.

The pilot was current and there was nothing in the preflight weather briefing that would strongly lobby against taking the flight. But many experienced mountain pilots will not fly single-engine aircraft after dark because there are really no good options in the event of an engine failure. The availability of an airframe parachute expands that debate.

NTSB Probable Cause *– "….was the pilot's in-flight loss of control following an inadvertent encounter with unforecast severe icing conditions. A factor in the accident was the inaccurate icing forecast developed by the NWS Aviation Weather Center."*

Commentary - The pilot should have been outspoken with ATC when he began to encounter the ice. Given his low total time, it is unlikely that he fully understood the nature of his predicament. Ten minutes elapsed from the time the pilot asked for a higher altitude to the time when the aircraft was so badly iced it could no longer maintain altitude. Had he deployed the parachute immediately, once the aircraft started losing altitude, the odds of survival would have been better.

Many more pireps might have helped. When the weather is poor or unexpected, pireps can literally be lifesavers. (I may have mentioned this before.)

The minimum enroute altitude would have avoided all the ice. Tell ATC you need a lower altitude "immediately, the code word" for "this is a big deal." If obstacles and traffic permit, ask for an off-course vector to get down faster. Declare an emergency if ATC balks and the ice is building. Certificate action is seldom taken. Be more afraid of gravity.

Roselawn, Indiana - A new respect for ice - (Simmons 4181 falls to

FZRA)

Certification authorities and flight crews learn more about freezing rain.

ATR -72, Simmons

Sometimes, even though flights have been lucky and soldiered through hundreds of encounters or more, the laws of physics and aerodynamics eventually assert themselves. So, it was with a Simmons Airlines ATR 72, Flight 4184, on October 31, 1994. The education was tragic and expensive.

NTSB conducted a detailed investigation that was subsequently modified by concerns voiced by the France's certifying authority, the Directorate General for Civil Aviation's DGAC. The report is nearly 300 pages for those who wish more detail. Here, we'll focus briefly on what was learned and the operational aspects. For ALL pilots, the message is simple – SLD is to be treated with great respect!

The Flight - The ATR 72, a Part 25 aircraft certificated by the FAA and DGAC, was operating on a scheduled flight between Indianapolis (KIND) and Chicago, O'Hare (KORD). During the late afternoon as traffic backed up into surrounding airspace, the ATR was giving a holding pattern at 10,000 feet with an expected 15-minute delay and reported entering holding at 1524 CST.

Maximum holding airspeed at that altitude was 175 knots. Because of a high deck angle the crew elected to deploy 15 degrees of flap to level the aircraft longitudinally. This did not help matters from an icing perspective. Deploying flaps usually exposes more unprotected surfaces to ice.

About 20 minutes into the hold the crew was advised to expect another 20 minutes in holding. At 1548, the First Officer (FO) as the pilot flying, noted some ice and the airframe deicing system was activated. At 1556, ATC advised it would be about another 10 minutes and to descend to 8,000 feet, which the crew acknowledged. There were no further radio contacts with the flight.

As the aircraft was descending through 9,130 feet at about 1557, the Angle of Attack increased and according to NTSB, ".... the ailerons began deflecting to a right wing down (RWD) position. About 1/2 second later, the ailerons rapidly deflected to 13.43 degrees RWD, the autopilot disconnected, and the CVR recorded the sounds of the autopilot disconnect warning .... The airplane rolled rapidly to the right, and the pitch attitude and AOA began to decrease. There were no recorded exchanges of conversation between the flight crew during the initial roll excursion, only brief expletive remarks followed by the sounds of "intermittent heavy irregular breathing."

"Within several seconds of the initial aileron and roll excursion, the AOA decreased through 3.5 degrees, the ailerons moved to a nearly neutral position, and the airplane stopped rolling at 77 degrees RWD. The airplane then began to roll to the left toward a wings-level attitude, the elevator began moving in a nose-up direction, the AOA began increasing, and the pitch attitude stopped at approximately 15 degrees nose down."

At 1557:38, as the airplane rolled back to the left through 59 degrees RWD (towards wings level), the AOA increased again through 5 degrees and the ailerons again deflected rapidly to a RWD position...."

Ultimately the aircraft pitched down and with the crew attempting to recover well above the maneuvering speed, the outer wing panels separated. This was an unrecoverable situation, unlike the Colgan crash cited in Chapter I. The four crew members and 64 passengers perished.

The Weather – According to NTSB, "At the time of the accident, there was no significant meteorological information (SIGMET) indicating the existence of icing conditions, and stations along flight 4184's route of flight

272

were not reporting any freezing precipitation. The only relevant in-flight icing weather advisory (AIRMET "Zulu") indicated, "light to occasional moderate rime icing in clouds and in precipitation freezing level to 19,000 feet."

A surface low pressure in southern Missouri was pumping out the usual instrument conditions, with plenty of moisture and moderate wind conditions. In short, it was a typical mid-western late fall storm. At 10,000 feet, where the flight was holding, temperatures were just below freezing with weak to moderate radar echoes in the area. There were a few pireps in the area reporting mostly light but some moderate icing. There was some rain and one aircraft reported sleet. The freezing level was estimated between 2,000 and 5,000 feet.

Loss of control - The NTSB noted that there was a sudden autopilot disconnect, uncommanded aileron deflection, and rapid roll of the airplane consistent with airflow separation near the ailerons caused by a ridge of ice that formed aft of the deice boots, on the upper surface of the wing.

Control wheel force data from icing tanker tests and the subsequent flight tests with artificial ice shapes indicated that the freezing drizzle ice shapes caused trailing edge flow separation and subsequent aileron hinge moment reversals.

NTSB: "The data did not show other airplane models to have a similar incident/accident history involving uncommanded aileron excursions in the presence of freezing drizzle/freezing rain. One possible reason for this is that other model aircraft use hydraulically powered ailerons, smaller mechanical ailerons with larger hydraulically powered spoilers, or different balance/hinge moment control devices to provide adequate roll control with less propensity for aileron hinge moment reversals."

Icing certification is one of the most complex areas to design and test. It has evolved somewhat since this crash and much more attention is now paid to such esoterica as droplet size, temperatures, boot area coverage, etc. Some Flight-Into-Known-Icing (FIKI)–approved GA aircraft that were once given Carte Blanche to operate in icing now have much more stringent limitations. Caveat Emptor! We're still learning.

Not Entirely Unexpected - NTSB: "The Safety Board determined that 13 of the 24 roll control incidents were related to icing conditions. (This indicates that there were other roll control incidents NOT related to icing.) Of these 13 icing-related incidents, the following 5 occurred in weather conditions consistent with freezing drizzle/freezing rain, and involved varying degrees of uncommanded aileron deflections with subsequent roll excursions:

• AMR Eagle/Simmons Airlines at Mosinee, Wisconsin, December 22, 1988;

• Air Mauritius over the Indian Ocean, April 17, 1991.

• Ryan Air over Ireland, August 11, 1991;

• Continental Express at Newark, New Jersey, March 4, 1993;

• Continental Express at Burlington, Massachusetts, January 28,1994"

The Continental Express incident in Newark, New Jersey, on March 4, 1993, was reviewed by NTSB on March 5, 1993. The pilots provided the following ASRS report regarding the events: "Apparently our problem was caused by ice formation on top of the wing in an unprotected area...Ice was noted accumulating on the side windows. The outside temp was fluctuating between 0 and minus 3 degrees C throughout the descent..."

"Passing approximately 7 NM and approaching the final fix the FO [first officer] began a power reduction in order to reduce speed so that the aircraft could be configured in the normal landing profile. It was at this time during the speed reduction the autopilot disconnected, and the aircraft immediately rolled to the right."

"...Both pilots immediately grabbed the controls to bring the wings level and nose back up. It took full aileron travel to do so. The aircraft returned to normal flight and was now being hand flown by the FO. Shortly after, the same flight characteristic was observed, and the aircraft once again was recovered. At this time, the trims were checked and were found to be normally positioned. The same flight characteristics were then observed for a third time."

"The captain took control of the aircraft. The trims were checked a second time along with the spoiler lights on the overhead panel, again found to be normally positioned. On the fourth roll, it was observed that prior to the roll, the flight controls became spongy and rough air disturbance could be felt over the ailerons. The aircraft was recovered again, and the captain observed that there was approximately 3 inches of ice aft of the leading-edge boots spanning the entire length of the wing. The ice extended back as far as could be observed..."

All the above were investigated/re-investigated by the Board and ATR, after the Simmons crash. As indicated, the design of lateral control seems to have lent itself to a rapid rolling event under certain icing conditions.

ATR made some design changes including vortex generators to respond to an Airworthiness Directive (AD 92-19-01). Even after redesign, problems remained in certain icing regimes.

Crew Expectation - This next section from the analysis accurately captures the crew's mindset with some emphasis added. NTSB: "Because there was no prohibition against flap extension in icing conditions, and no published information explaining the potential consequences of extending the flaps in icing conditions, the crew of flight 4184 would not have had reason to believe that the extension of the flaps would result in an adverse ice accumulation in front of the ailerons. In addition, the flight crew's training was such that the only performance degradation they would expect from ice accumulation would have been a continuous loss of airspeed and subsequent stall condition with stick shaker activation, rather than an aileron hinge moment reversal at an airspeed well above stall speed, that would suddenly overpower and disconnect the autopilot and cause the ailerons and control wheel to move uncommanded to near their full travel limits with no stick shaker activation."

"The flight crew's apparent lack of concern regarding the prolonged operations in icing conditions may have been influenced by their extensive experience of safely flying commuter aircraft in winter weather conditions, especially in icing conditions that are prevalent in the Great Lakes region. In addition, they were probably confident in the ability of the airplane deicing system to adequately shed the ice that had been accumulating on the wings and in their ability to perform safely under the existing circumstances."

"The flight crew was operating in icing conditions that exceeded the limits set forth in 14 CFR Part 25, Appendix C, resulting in a complete loss of aircraft control. However, the insidious nature of these icing conditions was such that the ice accumulation on the observable portions of the wings, windshield and other airframe parts was most likely perceived by the flightcrew as nonthreatening throughout the holding period. Moreover, the flight crew was *undoubtedly unaware* that the icing conditions exceeded the Appendix C limits and most likely had operated in similar conditions many times prior to the accident, since such conditions occur frequently in the winter throughout the Great Lakes and northeastern parts of the United States."

We are creatures of habit and training.

Findings – There were 43 findings – the critical ones, in my view, are listed here. To review the entire list, see the full report.

"5. Flight 4184 encountered a mixture of rime and clear airframe icing in supercooled cloud and drizzle/rain drops. Some drops were estimated to be greater than 100 microns in diameter, and some were as large as 2,000 microns."

"6. The forecasts produced by the National Weather Service (NWS) were substantially correct, and the actions of the forecasters at the National Aviation Weather Advisory Unit (NAWAU) and the meteorologists at the Chicago ARTCC's Center Weather Service Unit (CWSU) were in

accordance with NWS guidelines and procedures."

"8. The flight crew's actions would not have been significantly different even if they had received the available AIRMETs."

"9. The flight crew's actions were consistent with their training and knowledge."

"10. PIREPs [pilot reports] of icing conditions, based on the current icing severity definitions, may often be misleading to pilots, especially to pilots in aircraft that may be more vulnerable to the effects of icing than other aircraft."

"16. ATR 42 and 72 ice-induced aileron hinge moment reversals, autopilot disconnects, and rapid, uncommanded rolls could occur if the airplanes are operated in near freezing temperatures and water droplet median volume diameter (MVDs) typical of freezing drizzle."

"17. At the initiation of the aileron hinge moment reversal affecting flight 4184, the 60 pounds of force on the control wheel required to maintain a wings-level-attitude were within the standards set forth by the Federal Aviation Regulations. However, rapid, uncommanded rolls and the sudden onset of 60 pounds of control wheel force without any warning to the pilot, or training for such unusual events, would most likely preclude a flightcrew from making a timely recovery."

"22. The 1989 icing simulation package developed by ATR for the training simulators did not provide training for pilots to recognize the onset of an aileron hinge moment reversal or to execute the appropriate recovery techniques."

"31. The nearby air traffic control facilities were aware that light icing conditions were forecast for the area of the LUCIT intersection. Nonetheless, the release of flight 4184 from Indianapolis was proper because there were viable options for pilots who chose to avoid holding in icing conditions."

"34. Because there were no PIREPs [pilot reports] provided to the Boone sector controller by other pilots, and because the crew of flight 4184 did not provide a PIREP of icing conditions at the LUCIT intersection, it was reasonable for the controller to conclude that there were no significant weather events in that area and that the crew of flight 4184 was not experiencing any problems that would have warranted precautionary action by the controller.

The French DGAC filed a dissenting view on the probable cause and some of the findings, which caused NTSB to modify the report. For those interested go to the full report. It also provides some insight into how international investigations are conducted. You'll see reference to keeping an open mind in Chapter 5 where the loss of the second Boeing 737 Max is discussed.

NTSB Probable Cause (as amended) - *"....determines that the probable cause of this accident was the loss of control, attributed to a sudden and unexpected aileron hinge moment reversal, that occurred after a ridge of ice accreted beyond the deice boots while the airplane was in a holding pattern during which it intermittently encountered supercooled cloud and drizzle/rain drops, the size and water content of which exceeded those described in the icing certification envelope. The airplane was susceptible to this loss of control, and the crew was unable to recover."*

"Contributing to the accident were: 1) the French Directorate General for Civil Aviation's (DGAC's) inadequate oversight of the ATR 42 and 72, and its failure to take the necessary corrective action to ensure continued airworthiness in icing conditions."

"2) the DGAC's failure to provide the FAA with timely airworthiness information

developed from previous ATR incidents and accidents in icing conditions."

"3) the Federal Aviation Administration's (FAA's) failure to ensure that aircraft icing certification requirements, operational requirements for flight into icing conditions, and FAA published aircraft icing information adequately accounted for the hazards that can result from flight in freezing rain."

"4) the FAA's inadequate oversight of the ATR 42 and 72 to ensure continued airworthiness in icing conditions; and..."

"5) ATR's inadequate response to the continued occurrence of ATR 42 icing/roll upsets which, in conjunction with information learned about aileron control difficulties during the certification and development of the

ATR 42 and 72, should have prompted additional research, and the creation of updated airplane flight manuals, flightcrew operating manuals and training programs related to operation of the ATR 42 and 72 in such icing conditions."

Commentary - Icing is bad, freezing rain is much worse. It now goes by the description of Supercooled Liquid Drops or SLD. For anything other than the most capable aircraft, it's a mandate to avoid completely. What makes this accident unique is that we thought we understood icing and the certification needed to guard against it. It's difficult to know where to draw the line for "outlier" events both for regulators and manufacturers. The certification systems are usually extremely reliable; however, they are not perfect.

For any aircraft, the ability to climb to ultra-cold altitudes quickly is essential. Conventional wisdom is if the air temperature is colder than about -20C (-4F), the already frozen ice crystals will not stick to the aircraft flight surfaces. Back when lots of piston aircraft were being built in the late 1970s, manufacturers would only certify turbocharger-equipped models for any kind of icing. Normally aspirated need not apply because they didn't have the climb capability, especially when encrusted. That wasn't the problem in this accident but in any light aircraft it's always a good idea to coordinate with ATC on climb and descent strategy when icing is forecast and especially if reported. Get up or down through the freezing layer as quickly as possible. Retreat is sometimes a better option and, on some days, don't fly!

Cold Realities

Air Florida's Palm 90 succumbs to reduced power and wing contamination.

If you're feeling that something's wrong, you're probably right.

Cold weather makes aircraft wings and engines perform better because the air is denser. But adding moisture makes life tough for pilots and aircraft alike. Most pilots are aware of icing's dangers but snow can be a hazard as well. Any type of winter moisture can ruin performance. While this discussion is about an air-carrier accident, the lessons apply to all aircraft.

Delay, Delay

"Palm 90" was the call sign for Air Florida's Boeing 737-200 flight from Washington National Airport to Fort Lauderdale, Florida. Its 74 passengers and five crewmembers were scheduled to depart at 2:14 p.m. EST on January 13, 1982, but moderate snow delayed departure. The airport was closed for snow removal at 1:38 p.m.

Air Florida B737 - Courtesy, Airliners.net

with a scheduled reopening at 2:30 p.m. Palm 90's captain requested the deicing crew to spray down the aircraft with deicing fluid in time for the scheduled reopening. Half an inch of wet snow covered the aircraft. The captain stopped deicing upon learning that 11 other aircraft had departure priority and that the airport would not reopen until 2:53 p.m.

Around 2:50 p.m. deicing was resumed as heavy snow continued to fall and was completed by 3:10 p.m. At 3:23 p.m. pushback was approved by ground control, but the tug couldn't move the aircraft because of the snow, ice, some deicing fluid on the ramp, and a slight incline. The aircraft engines were started and reverse thrust was applied, but this was unsuccessful. Witnesses stated that snow and slush were blown toward the front of the aircraft; after the engines were shut down while a mechanic inspected the engine inlets. No ice, snow, slush, or water was found.

A second tug was brought up, and by 3:35 p.m. the Boeing was pushed back for engine start. At 3:38 p.m., after accomplishing the after-start checklist, the captain responded "off" to the first officer's callout of "anti-ice." Despite this, the crew was thinking about the snow while they sat in the line of departing aircraft. Nine air carrier and seven GA aircraft were awaiting departure ahead of Palm 90.

At 3:40 p.m. there was some further discussion about getting deiced again. The first officer (FO) made several comments about the elapsed time since the aircraft was deiced. At 3:46 p.m. the FO commented, "Well, all we need is the inside of the wings anyway — the wingtips are gonna speed up on 80 anyway; they'll shuck all that other stuff."

Reluctant Crewmember

At 3:48 p.m. the FO asked, "See the difference in that left engine and the right one?" presumably referring to an engine gauge reading. The captain replied, "Yeah." The FO continued, "I don't know why that's different unless it's hot air going into that right one, that must be it — from his exhaust — it was doing that in the chocks awhile ago...." The FO was referring to the jet exhaust from the preceding aircraft. At 3:53 p.m. the FO said, "Boy...this is a losing battle here on trying to deice those things. It gives you a false sense of security — that's all that does."

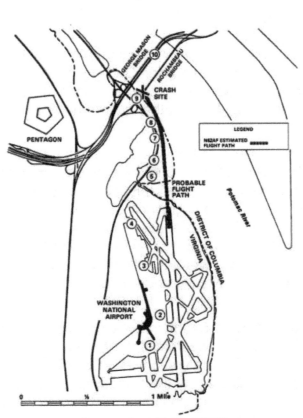

A very short flight, NTSB

At 3:57 p.m. the crew set the airspeed bug settings to 138 knots for V_1, 140 knots for V_R and 144 knots for V_2. The FO asked the captain, "There's slush on the runway — do you want me to do anything special for this or just go for it?" The captain replied, "Unless you've got something special you'd like to do." The FO then said, "Unless just take off the nosewheel early like a soft-field takeoff or something. I'll just take the nosewheel off and then we'll let it fly off."

At 3:59 p.m. Palm 90 moved into position on Runway 36 to hold for takeoff and then was cleared for takeoff with no delay because of landing traffic on a two-and one-half mile final. *Fifty minutes had elapsed since deicing was completed, and snow had been falling continuously.*

The FO began takeoff at 3:59:46 p.m. At 3:59:56 p.m. the captain commented, "Cold, real cold." The FO replied, "God, look at that thing, that doesn't seem right, does it?" At 4:00:05 p.m. the FO stated, "That's not right," to which the captain responded, "Yes it is, there's 80 [knots]."

The FO replied, "Naw, I don't think that's right." About nine seconds later the FO said, "Maybe it is," but then two seconds later, after the captain called "one-twenty" (an airspeed callout), the FO said, "I don't know."

Eight seconds after the captain called V_1 and two seconds after calling V_2, the stick shaker (stall warning) activated. At 4:00:45 p.m. the captain said, "Forward, forward. We only want 500." At 4:00:50 p.m. the captain said, "Come on. Forward, forward — just barely climb." At 4:01:00 the FO said, "Larry, we're going down, Larry," to which the captain responded, "I know it."

Palm 90 struck the northbound span of the Fourteenth Street Bridge and plunged into the ice-covered Potomac River three-quarters of a mile from the departure end of Runway 36. Heavy snow continued to fall, with visibility at the airport varying between one-quarter and five-eighths of a mile. 70 passengers and four crew-members died. There were nine survivors. Four persons in vehicles on the bridge were also killed.

Ground witnesses agreed that the aircraft was flying at an unusually low altitude, with the wings level and a nose-high attitude of between 30 and 40 degrees nose up.

NTSB Probable Cause

"The National Transportation Safety Board determines that the probable cause of this accident was the flight crew's failure to use engine anti-ice during ground operation and takeoff, their decision to take off with snow/ice on the airfoil surfaces of the aircraft, and the captain's failure to reject the takeoff during the early stage when his attention was called to anomalous engine instrument readings.

Contributing to the accident were the prolonged ground delay between deicing and the receipt of ATC takeoff clearance during which the airplane was exposed to continual precipitation, the known inherent pitch-up characteristics of the B-737 aircraft when the leading edge is contaminated with even small amounts of snow or ice, and the limited experience of the flight crew in jet transport winter operations. "

Analysis

The findings of the National Transportation Safety Board show the complexity of factors involved. This accident also illustrates how little information is available regarding safe flight operations for light aircraft in similar conditions.

While many light aircraft will not be operating under these circumstances, the information is instructional. According to the accident report, the aircraft's takeoff roll and acceleration were slower than normal. Under these conditions a B737 should normally accelerate to a liftoff speed of 145 knots in 30 seconds while using about 3,500 feet of runway and then climb at about 2,000 feet per minute. Palm 90 took 45 seconds and used about 5,400 feet of runway to reach 140 knots. There was an initial climb of 1,200 feet per minute, with the aircraft approaching a stall angle of attack shortly after liftoff. The maximum altitude gained was about 200 to 300 feet agl before the Boeing descended into the bridge.

Although there was contamination on the wing and slush on the runway that would have retarded the takeoff somewhat, the NTSB felt that the engines were not developing the appropriate power. By doing a sound analysis on the cockpit voice recorder (CVR) it was determined that the engines were not set to takeoff power. The appropriate engine pressure ratio (EPR) should have been 2.04; but the sound level on the CVR matched an engine output at 1.70 EPR.

The crew's failure to turn on engine anti-ice resulted in the freezing of the probe and an erroneous EPR, leading to a reduced thrust lever setting and significantly less thrust.

The report analyzes the effects of wing contamination and the Boeing 737's tendency to significantly pitch up in conditions where the wing was presumably contaminated by snow, sleet, or rain in near-freezing conditions prior to takeoff. Excerpts from the flight manual concerning proper engine operation and aircraft deicing are extensive. The pilot operating handbook for light aircraft, even those approved for flight in icing conditions, provide little, if any, guidance on takeoff performance under these conditions.

The NTSB concluded that neither the low thrust during takeoff nor the presence of snow or ice on the aircraft alone would likely have led to the crash.

Simply put, if the crew had firewalled the thrust levers, chances are good that Palm 90 would have flown. This was contrary to most crew training practices governed by concern for engine longevity.

The effectiveness of the anti-ice protection afforded by spraying down the aircraft was also questioned. A pre-departure spraying will buy some time before the wings are re-contaminated but there are many factors as to how long it will remain effective. The NTSB found that the deicing practices were not uniform. There were no fewer than three recommended procedures: one by Union Carbide, which made deicer fluid; another by

Trump, which built the deicing vehicle; and a third from American Airlines, which had the contract to deice Air Florida aircraft at DCA.

The NTSB reminded pilots that even though there may be residual anti-icing effect after having been properly deiced, the only way to ensure that the aircraft is ready for takeoff is to observe the wings and tail just prior to beginning the takeoff roll. Analysis of the deicer vehicle found that a nonstandard nozzle had been used and was dispensing about half as much deicer fluid as the mixture controls indicated. While other aircraft successfully departed under the heavy snow conditions, the NTSB was unable to reach a conclusion on whether the nonstandard deicing was a factor in the accident.

Since Air Florida procedures have changed a lot. Now, rather than deicing at the gate and then sitting in line awaiting takeoff, many busier airports have deicing pads where a flight is decontaminated just before takeoff. It's a much better system.

Commentary

Flight crew performance, according to the NTSB, was not optimal. At no time did anyone leave the cockpit to check the condition of the wings. The use of reverse thrust to back out of the gate was contrary to the operations manual, yet the crew tried it anyway.

The flight manual is very clear about the need for engine anti-ice under these conditions. The crew failed to appreciate this, although they constantly discussed the snow on the wings. Additionally, a Boeing operations manual bulletin recommends that aircraft should maintain a greater distance behind other aircraft when taxiing in areas of ice or snow. This is to prevent the blow-off from melting and refreezing on the following aircraft. Palm 90 did not follow this advice.

The decision to take off with visible snow on the aircraft was in violation of the FARs (91.527) and good operating practice.

The least bit of uncertainty during a takeoff is a mandate to abort. The first officer's comments regarding engine performance should have been a signal to both crewmembers that something was amiss . That Palm 90 was offered an immediate takeoff because of landing traffic may have predisposed the crew to continue rather than reject and cause another aircraft to go around. The departing aircraft owns the runway until becoming airborne and rejecting a takeoff for cause or suspected cause is always good form.

It was noted that the captain had relatively little experience in jet transport operations and very little in winter operations. An investigation showed that he had made only eight arrivals or departures where conditions were conducive to icing. While his flight hours were significant (8,300 total, with 2,322 while employed by Air Florida), they clearly were not in northern climates.

It's easy to blame the crew who is often the last link in the accident chain but as noted there were several other factors. This pilot upgraded to captain in fewer than

Aftermath - Note ice on top of vertical stabilizer - Courtesy, NTSB

three years, so there was little time to acquire winter operations experience. Local climate makes a difference and South country pilots who aren't familiar with winter hazards will be ill-prepared. The opposite applies to polar types.

Learning about cold-weather procedures in a classroom is not a good substitute for real-world experience to emphasize winter's hazards.

Forward and Down – an Icing Story - (PA32-260 sort of escapes ice)

Cherokee 6 pilot learns that icing can be a problem even in springtime. He saves the day by landing off airport.

Cherokee 6, AOPA

Springtime is a transition period where winter has not quite given up. A Cherokee 6 (PA-32-260) made a successful off-airport landing in Grantsville, Maryland, landing at dusk in low Instrument conditions on April 16, 2022. There were three minor injuries and the aircraft was heavily damaged. The reported use of 5-point front seat harnesses in an older aircraft was obviously a very worthwhile investment. While the pilot did a good job of managing an ice-crippled aircraft, there was also an element of luck that prevented this from becoming a tragic encounter. The ceiling at the crash site was approximately 300 to 500-foot overcast in mountainous terrain with moderate rain at dusk. A little daylight helped tremendously.

A detailed description of the weather shows both the strength and weakness of our current system and how an optimistic aviator could find themselves in dire straits. As noted in the human factors , the forecasts are sometimes just vague enough to lure the unwary into a bad situation. The flip side is that sometimes we are lucky in prior encounters and apply that thought process to a situation that appears similar but is not. Read on, knowing we have the benefit of hindsight.

The Flight – The pilot received an online weather briefing, and departed Wabash Municipal Airport (IWH), Wabash, Indiana, destined for Martin State Airport (MTN) in Baltimore, Maryland, on an instrument clearance. About 2 hours into the flight, at 11,000 ft they approached some clouds. The pilot increased power and

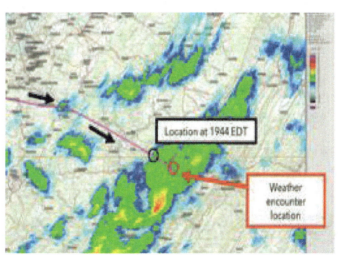

Radar Plot showing flight path and time, NTSB

turned the pitot heat on, planning to climb above the clouds.

According to the pilot, "Shortly after entering the clouds, I noticed the airspeed slowing and that the autopilot was pitching the nose up to maintain altitude. I disengaged the autopilot and pitched the nose down. At that time, I noticed that it was difficult to maintain pitch control. I began to suspect the presence of tail ice. However, I was not able to locate any visual signs of icing after studying each wing's leading edge, the fuel caps, and the windscreen.

Despite attempting multiple control and power settings, it became increasingly difficult to maintain altitude and directional control. I still suspected the presence of tail icing and informed ATC (Washington Center) of my difficulty in controlling the aircraft. While I struggled to retain control in the descending aircraft, ATC advised they were declaring an emergency, offered a block altitude, and advised that I might be able to go lower."

"The aircraft continued to descend, albeit with marginal directional control. Having descended out of the clouds, I was now in hard rain with no visibility of ground structures. At some point, perhaps a few hundred feet AGL, I was able to make out trees and what appeared to be level terrain. In that space, I located a small road next to a building within the trees and was able to maneuver to it. I initiated a landing on the road and did my best to use the trees on my wings to dissipate energy and preserve the aircraft cabin. We came to stop near the building (which was a home)." The crash occurred at 1952 EDT, just before dark.

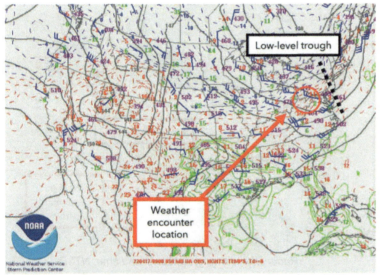

Bad flying over the Applalachians at low altitude!, NTSB

The Pilot – The 62-year-old Commercial pilot was instrument and helicopter rated. Total flight time was 955 hours and 48 hours in type. He had 17 hours of actual instrument time. Additionally, he had 18 hours in the preceding 30 days.

The Aircraft - The 1966 PA32-260 (Cherokee 6) had a total flight time of 3280 hours and showed no mechanical problems. The aircraft was equipped with an engine monitor which helped tremendously in determining that power was available. An icing-prohibited placard on the left cockpit sidewall was missing but likely did not enter into the pilot's decision process.

The Weather – On this day, the weather was worthy of considerable study. A cold front had just passed through the area with Sierra (IFR), Tango (Turbulence), and Zulu (Icing) Airmets valid for the weather encounter location.

The AIRMETs at 1645 EDT forecast obscuration due to clouds, precipitation, and mist, IFR conditions, moderate turbulence below 10,000 ft, and moderate icing between the freezing level (3,000 to 8,000) and 17,000 ft.

The radar showed a solid band of precipitation and on cold days that often means ice.

There were not many pertinent pireps. A Piper Arrow about 40 miles to the north had reported below freezing temperature, moderate turbulence, and moderate rime ice. Everything else was at higher altitudes. There were probably ice-certificated aircraft flying in the area so a couple of well-placed pireps might have helped the pilot decide this was not a good day to fly.

Performance Study - NTSB completed a performance study which sheds considerable insight on how the pilot, the wing, and the ice were interacting in the last 10 minutes of flight.

NTSB: "While a complete lift and drag relationship for the PA-32-260 was not available, the estimated aerodynamic model should accurately reflect the trends in the airplane's lift and drag. The loss of lift event at 19:48 occurred at a high lift coefficient, high angle of attack, and at a calculated calibrated airspeed lower than the reported stall speeds.

While ice accumulation on the airplane may have contributed to making it difficult for the pilot to control airspeed and attitude, it did not appear to cause the wing to stall prematurely.

However, the loss of lift at 19:51:30 occurred without an increase in the lift coefficient and a speed above 100 kts. Additionally, there was a marked increase in drag before the loss of lift This would be consistent with an accumulation of ice on the wing increasing drag and reducing the amount of lift it could produce before stall.

The airplane continued to descend for more than another minute and drag, again, began to increase before the airplane descended below 150 ft above the terrain and was no longer recorded by ADS-B."

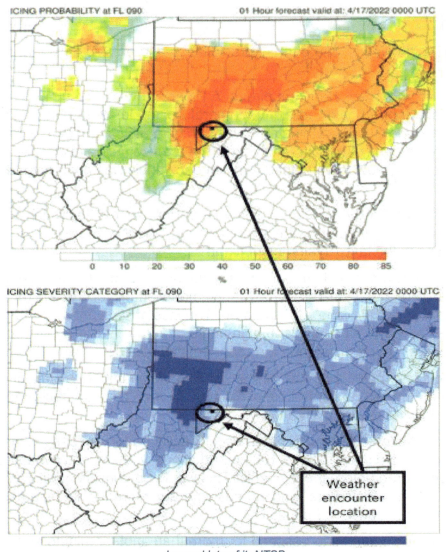

Ice and lots of it, NTSB

NTSB Probable Cause – *"…was the pilot's decision to continue the flight into an area of moderate to heavy icing conditions, which resulted in a degradation of airplane performance and subsequent loss of control."*

Commentary - This is a classic icing encounter in a light aircraft with a happy ending. The pilot's statement that he did not see ice on the air frame or windshield is somewhat suspect. The possibility of clear ice (SLD) might explain why he claims not to have seen it but in my very few encounters with SLD, the windshield will become coated, and it was visible on the wings. The description that the aircraft was unable to climb, losing airspeed and the autopilot pitching up to maintain altitude is a perfect description of a losing ice battle.

The text Airmets made it clear that a light aircraft was not going to climb over the ice. Going underneath was unlikely in the mountainous terrain over the Appalachians. The low-altitude IFR enroute charts show a Minimum Enroute Altitude (MEA) of 5,400 feet and a Minimum Off Route Altitude (MORA) of 7,100 feet.

Temperature forecasts at 6,000, 9,000, and 12,000 show +03, -01-, and -08-degrees C. Some optimism might start to creep in by thinking that at 6,000, we'd be above the terrain and still above freezing. Sadly though, the forecasts are not accurate to that degree. The winds aloft forecast, behind the front, showed the wind out of the west at 28 knots at 6,000 and at 49 knots at 9,000. That portends a bad ride even if the groundspeed is enthusiastic.

What about an end run to the south? It would have taken a pretty large deviation. The next day would have a been a far better choice.

Remember that the icing description in the forecasts are for aircraft with known icing certification. In mid-April, ice is still to be reckoned with.

Convection

"Thunderstorms are never as bad as they look – they're much worse!"

Anonymous

Mother Nature's Big Gun, NWS *The Bomb, The Brookings Institution*

Ever notice the similarity between a fully developed thunderstorm and the mushroom cloud of a thermonuclear weapon? It's been estimated that that the energy release between the two is comparable. Who does those calculations? Here's some guidance that's helped me stay out of the tiger's clutches, so far. But every flight in thunder country is a new opportunity to get into trouble.

Trip profiles determine what avoidance skills are needed. Local flights are easy but longer ones require more assessment. Thunderstorm crashes have a 70 to 90 percent fatality rate. There are no statistics on encounters where the tiger merely takes a nip. Many pilots experiment as they gain experience and either learn quickly from Icarus or become one.

Slow growing or bad stuff? At the most basic level, understand what's generating the weather system and the terrain beneath it. Is the atmosphere just slightly unstable or explosively so? Is the air juicy and hot? If the dewpoints are in the upper 60s to 70s, the juice is there. If the cells are moving faster than about 20 knots or the tops are above 25,000 feet, expect really bad things. But bad stuff often comes in smaller packages--it's hard to know in advance.

With severe weather, the conventional wisdom is to avoid all convective precip by 20 miles--not just the red that shows on radar. That means a 40-mile gap between cells, but that guidance was developed decades ago for the airlines at flight level altitude—especially downwind of a big cell. The closer you get, the more likely an unhappy encounter. Your mileage, literally, will vary depending on the situation.

 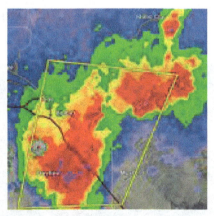

Shallow gradient, "Probably" OK to fly through or under, Author *Tight gradient – No Way Jose!, NOAA*

Inside a "small" but vigorously growing cumulus where there's no radar return it will be a ride to remember. There can be extreme turbulence outside the clouds as well. How do we know? Anecdotally and by T-storm penetrations done by research aircraft prior to entry.

What's tough about convection is that past experience is not a guarantee. The saddles (a low spot in the clouds) can fill in at 4,000 feet per minute and few aircraft can outclimb that, especially at altitude. If the temperature spread across a front is more than about 10 degrees there's lots of energy. The so-called "garden variety" thunderstorm may be circumnavigated more closely but nobody has come up with a consistent way of determining "GV" versus severe.

What starts out as a simple individual cell may grow into a Mesoscale Convective Complex, or MCC in weather jargon. They can last for hours, generate hail, tornadoes, derechos (straight line winds) and royally foul the airspace for as long as they're alive. A single cell may just build and die – easy to work around

> **"A good hockey player plays where the puck is. A <u>great</u> hockey player plays where the puck is going to be."**
>
> *Wayne Gretsky - Hockey Hall of Fame*

Timing is everything – Mornings tend to be less active because daytime heating hasn't gotten underway so it's usually a better time to fly. As shown below, in less than an hour the convection has grown from turbulent but probably flyable inland, to flying down the coastline. Caveat One – watch how the system is moving. Don't want to get chased out to sea. Caveat Two – As stated several times, radar only shows the precipitation, not the widespread growing towering cumulus that could easily have severe turbulence. If the destination is inland, wait for things to cool down – perhaps around 7 p.m. A noon arrival or before would have been a much nicer ride, and we could have enjoyed the fireworks from the ground.

1:55 pm – Flyable but lots of growing CU – a bouncy ride, Author

2:10 pm – It's growing! Time to think about Plan B, Author

2:25 pm – Not Good! Not flying inland, Author

2:40 pm – Game Over! Fly the coast if you're passing through, otherwise – land, Author

Sometimes there's activity early in the day that dies out before noon, perhaps as early as 0800, and regenerates a couple of hours later. This happens a lot in the Midwest. Late in the day can also work but darkness works against us to see the big clouds.

"Forecasts are like fruit – always better when fresh."

Author

Hangar Story *- April 26, 1991: Mooney M20E: The pre-departure IFR briefing for a mid-afternoon flight from St. Louis to Wichita (ICT) predicted scattered thunderstorms and indeed, there was a big one already on the route. The NWS Storm Prediction Center had forecast a high risk of severe weather, but the current route was good VFR. The main event was still a hundred miles west of Wichita and wasn't scheduled to arrive until early evening. I moved my departure time up by two hours. This would bear close watching. This was long before weather in the cockpit.*

The big cell at the half-way point was isolated and had drifted well north about an hour after takeoff. It was nice VFR, but surface and winds aloft were picking up from the southwest. Crossing the Missouri/Kansas line, the sky changed. There was energy – lots of it. The cumulus clouds were building and the ride at 6,000 feet became moderately turbulent. I slowed down and asked for a block altitude to ride the waves. The controller, after being queried several times, reassured me (again) that he was showing no precip into Wichita and the nearest storms were still about 70 miles to the west.

The airport was good VFR, but the sky had a coppery color from blowing dust with towering cumulus not far to the west. The surface winds were 28 gusting to 42 knots. Fortunately, it was down the runway so the Mooney may have rolled 200 feet. Taxiing in was almost as sporty as the landing and my sense was something big was about to happen. It did.

Two hours after landing half a dozen tornadoes touched down around ICT including an F5 that killed 17 people. A total of 55 twisters were confirmed. The forecasts were accurate, and flight was possible earlier in the day – not later.

Convective weather moves, ebbs, and flows - What's unflyable now may be good in 20 minutes and vice versa. On some days, a big chunk of the atmosphere is just not going to cooperate. Normally, following the herd is a good plan because smart pilots will avoid. This complicates things for ATC because everyone wants the same airspace. Following a lucky Icarus, however, could be bad. If it's good when the lead aircraft goes through, ten minutes later may be disastrous. Convection is guaranteed to either get better, get worse, or stay the same. The big question is what will the storms being doing when we get there per Gretsky?

Nexrad radar is a magnificent tool for seeing what's happening and makes the decision-making easier. Airborne datalink mostly keeps up with the major developments, but we've been told many times that it's not adequate for tactical maneuvering - for working in close and "picking" a way through the bad stuff. "Latency" is the problem - it can take several minutes for the radar scan to be updated and then more time for the images to be processed and sent up to the aircraft.

With active weather systems latency is a problem. At this writing, the Nexrad displayed in many light aircraft may show a storm that's moving at 30 knots about 5-8 miles from where it actually is. The tools are evolving and will get better but airframe-bending bounces may be miles from the nearest radar return. In clear air around the backside of a cell a rough ride can await - Note to self, add another 5 miles.

The eyes are the best avoidance tool - provided you can see. Confidence is much better when above or below the veiling cloud decks that often develop on thundery days. Even between layers is better than nothing. If it's mostly solid IMC, we're deprived of our best assessment tool. For normally aspirated piston single engine aircraft, the game is over when the deck grows above 10,000 to 12,000 feet outside the storms. Unless there's oxygen on board AND the aircraft is capable of at least 400- to 500-foot per-minute rate of climb, give it up.

Sometimes, lower gets us VFR beneath the bases and that can work well. A few pros – a few cons: The ride will be bumpy and hot, possibly with tall towers and terrain complicating things if you're really low. The ability to bail out to an alternate airport is much better, assuming you're not in sparsely populated country. With passengers, carry sick sacks.

VFR works sometimes - Go left or land soon, NOAA

If the storms are air mass type, avoiding the rain shafts may be a good plan but downdrafts and shear are a consideration. With fronts and lines of storms where there may be few places between storms, have an escape

Bow Echo - Beware!, NOAA

path and if the bases are getting below about 3,000 feel agl and visibility less than 5 miles best to give it up.

A bow echo is a sure tip-off of high winds. When landing, try to leave *at least* 30 miles between you and the leading edge of the weather. The gust front in a strong system may blow out well in advance of any precipitation. The last place to be is approaching the airport or taxiing in when 40 to 90- knot winds arrive. Radar doesn't always show wind or gust fronts.

Allow time to tie down and if hail is a possibility, buying hangar space even for an hour or two is far better than having the aircraft look like it was attacked by a ball peen hammer. Are the tie-down ropes are healthy? Better to carry your own. Install gust locks

End Runs—In a no-wind flight up the coast between New Smyrna Beach, Florida, and Mt. Pleasant, South Carolina, the distance is 267 miles. A 150-knot aircraft will take one hour and thirty-eight minutes with fuel burn of about 27 gallons. The deviation around the west end of the line is 332 miles, two hours and seven minutes and consumes 33 gallons. From my perspective, 65 miles, half hour and an extra six gallons is a brilliant trade-off

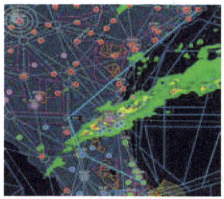

A nice end run, Foreflight

to avoid a tactical convective encounter. If the southwest end of the line builds, landing short and waiting an hour has us arriving safely at our destination perhaps two hours late. The airlines do that all the time. For VFR, the map shows a lot of IMC so best to cancel.

ATC Assistance - ATC radar has improved tremendously in the last several decades. Terminal Radar with a good controller can be a big help - assuming there's enough maneuvering room. *TRACONs can provide more detailed help because their weather is real time.* Enroute Centers, get the same weather that we do in the aircraft with some delay and less granularity. If it's an active weather day, flying in TRACON airspace

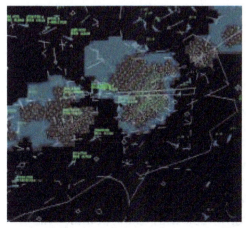

TRACON – More detail and real time, FAA

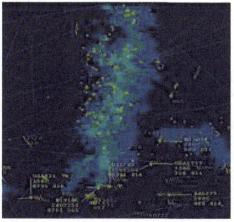

Center Radar - 3 levels of precip and some Lag, FAA

(lower) will often allow controllers to provide more detailed weather avoidance but you may not be high enough to get over the lower cloud shields.

The tradeoff is that TRACONs don't have as much airspace and often more traffic. Controller skill is variable, like pilots, so if there's some hesitancy on the other side of the radio, react accordingly. If ATC insists on putting you someplace that looks bad visually even though their radar says it's okay to respond "unable" and offer an alternate course of action

Racing a storm into or out of an airport is a bit risky and it's done regularly. I've done it myself, but if you decide to play, be ready to bail out immediately. A tragic crash occurred when a family leaving on a camping trip in a loaded Bonanza rushed the takeoff from Jabara Airport in Wichita, Kansas. A line of storms was approaching the airport from the west, but it was clear to the east. It would have worked had they departed just two minutes earlier but that's cutting it way too close. The gust front arrived just as they were becoming airborne--no survivors. Had they waited half an hour, all the weather would have been to the east.

"Go the left of the Good 'Ole Big Ones and to the right of the Big 'Ole Good Ones."

Richard L. Collins, aviation author and editor

Always ask ATC for deviations even around smaller buildups when operating IFR. Other aircraft may be close by and separation still has to be coordinated. By asking early, ATC will have time to sort things out. It's easy: "Charlotte, Bumpmaster 80X would like 10 degrees left for weather."

If you're VFR and not talking to ATC, now might be a good time to join the party and get flight following if they're not too busy with instrument flights. If the traffic gets heavy you may get dumped – "Squawk 1200 and have a nice day." ADS-B is invaluable in these situations, because all the traffic will be heading for the same area but you still have to stay visual.

No rain, yet but a very rough ride, NOAA

Often you'll hear this clearance while deviating: "Bumpmaster 80X cleared direct FIGGS *when able* and advise." There was a crash some years back where the pilot thought that meant to go direct to FIGGS right then. He turned into an extreme cell with fatal results. This clearance means you can decide when to turn, when clear of the weather, and let them know. If it's going to be a while, say so and if uncertain, ask for a vector to stay clear.

Slow Down - *Slow to well below maneuvering speed before getting into bouncy clouds.* Halfway between stall and V_a is good for light aircraft because in serious weather the speed and altitude will jump around a good bit. We remember that V_a varies by weight and is lower when lighter. If in a retractable and the ride is getting rough, put the gear down remembering to add power to maintain desired speed. This keeps airspeed down in case of an upset and provides more lateral stability. In any light aircraft, NO flaps. It weakens the structure. If it's getting that rough, I'm questioning why I got there and looking for a quick exit. Fly attitude, not altitude and let the aircraft ride the waves.

Sometimes down low and suffering through the bumps is really the only way. Trying to stay on top of a building cumulus deck means it will be IMC soon. Then there's no visibility and it's turbulent. If caught VFR, advise ATC, declare an emergency and get on the ground. The FAA is probably not going to hassle you but file a NASA safety report and vow not to get caught again.

Hangar Story: *- Flight from Washington, DC, area to Columbia, South Carolina (CAE): A strong line of thunderstorms passed through the DC area at 0700 preceding a cold front. The scheduled departure time was 0830 and the aircraft was a new Cessna P210 equipped with onboard digital radar. The weather was clear behind the line as we departed and climbed into the mid-teens. I was confident that with all that hardware that we'd be able to work through the line with minimal disruption. The storms were moving about 25 knots and lay in a northeast-southwest line across North Carolina by mid-morning.*

Approaching a line from the back side usually isn't as impressive as the front side. No classic towering CBs, just a benign looking mass of gray with rain - shouldn't be a problem for a good IFR pilot. Not so. The little pod radar antenna under the right wing was soon showing rain but no cells, just a band of water. Attenuation occurs when all the radar energy is being reflected back by significant precip without looking more than a few miles ahead. We were flying blind. (In the image below, using a far better (color) radar than I had, there's attenuation behind the heavy rain (red) band.)

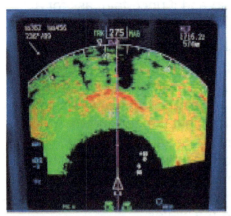

Radar shadow at 1230, attenuation caused by heavy rain, NOAA

I asked the center controller how it looked ahead and if there were any ride reports. This was long before the days of Nexrad (WARP) being depicted on their scopes. He replied that I'd be the first one through in his sector. It's best not to be a pathfinder under those circumstances.

Decided that Rocky Mount, North Carolina (RMT) was a great place to stop and after an uneventful ILS in moderate rain and low ceilings, we landed and waited about three hours. The flight into CAE had light IMC with no convective turbulence or rain. It was a good choice, and the coffee was uncommonly good.

What's it like to fly into one of the beasts with a professional weather researcher? Here's Bob Maxon's tale:

Hangar Story *- Capt. Bob Maxon-Director, NWS Aviation Weather Center, Prior Dir., NOAA Aircraft Operations Center (Courtesy, ahf.com)*

On March 2, 1984, my aviation career entered into a new phase of reality when we unintentionally flew our NOAA Shrike Commander (AC-50) into a thunderstorm near Lake Charles, Louisiana. In what can only be described kindly as an "E" ticket ride, our aircraft suffered numerous severe updrafts and downdrafts coupled with extremely heavy rain and lightning.

As a crew, we were very busy keeping the aircraft in a level attitude and trying not to "chase" an airspeed indicator that was oscillating from red line to stall. Believe me, when you hear an aircraft wind up like that and you see your airspeed indicator approach VNE it's hard not to pull back on the throttles. Every object that was not secured in the cabin went weightless, moved forward, and eventually landed in the flight deck. The aircraft went through several iterations of positive and negative "G" loads that even further disorientated us.

We had entered this small, embedded cell at 13,000 feet and were ultimately ejected from the storm at 4,700 feet, in spite of the fact that we were trying to maintain our IFR assigned altitude of 13,000 feet! Some of our navigational equipment failed because of turbulence, but we managed to make an emergency landing in Gulfport, Mississippi. I felt like I had been given a second chance at both life and flying – we could have easily been killed…

"What you see is what you get"

Comedian Flip Wilson

Key Points:

- Hot and juicy air masses may develop thunderstorms as the day wears on

  ◦ Temperature/dewpoint spread will help predict moisture aspect.

- Early in the day is usually much better than later – Plan to arrive by noon at the latest. However, a front, trough, or low can trigger convection any time.

  ◦ Evening from about 7 pm and later can also work BUT visually avoiding storms at night, even with lightning, will be much more difficult. Consider fatigue and hypoxia.

- Vision is the gold standard, regardless of onboard equipment. Night makes it a lot harder.

  ◦ If you can't stay visual, especially without onboard radar, Plan B needs to be close at hand.

  ◦ Recommended to have at least 50 percent of the sky clear of storms for escape. It's best if the weather is all off to one side of the projected flight path.

- Cumulus clouds with tops above 25,000 often give an awful ride. *But much smaller ones in growth mode may be nasty.*

  ◦ Anything moving faster than about 20 knots probably has a lot of punch to it.

  ◦ Temperature differentials of more than about 10 degrees across a front indicates strength.

- Mountains make everything worse. Orographic lifting, less Nexrad coverage, and fewer airports to escape to.

- What's worked before may not work on the last flight because the weather is different, or perhaps the pilot has become bolder. An airline pilot once referred to thunderstorms as "treacherous." I'm not sure about the anthropomorphic characteristics but completely agree with the sentiment.

- Beware the pathfinder who went through just 10 minutes ahead and said it was fine.

- End runs are good but sometimes stopping and waiting is a better option.

- Gust fronts - allow 20 miles from the nearest echo and the airport-- maybe more. Bow echoes on radar are bad - high surface winds are in the offing.

- Even though you ARE avoiding the Good 'ole Big Ones, slow to below maneuvering speed BEFORE getting into growing cumulus if there's no way to avoid. When it gets rough - put the gear down and NO flaps. Just concentrate on keeping wings level and don't chase airspeed or altitude. Advise ATC that you need a block altitude and ride the vertical waves.

- If ATC tries to put you into a nasty looking area, the simple term is "unable." As PIC you are responsible. Declare an emergency if you're unable to work out a satisfactory route. File a NASA report. However, give both yourself and ATC time to work out other options.

Bowling Alley Blues - (American 1420 runs into weather)

An airliner gets caught between a rock and a hard place

Summary: After a long day and an evening of convective weather, an MD82 lands in the middle of a thunderstorm and slides off the runway. The crew loses the psychological battle of trying to deliver the goods in an impossible situation.

MD82, American Airlines

There are few airline thunderstorm accidents today, but the storms disrupt schedules and cause innumerable flight cancellations. This air-carrier accident is exceptionally well-documented and holds great lessons for all pilots.

The airlines have a logistical problem with convective weather because of the interlocking nature of the hub-and-spoke system, which requires aircraft and crews to be at certain places at certain times.

The strategy of flying thunderstorms is simple--avoid them. But the tactics of how far you want to stick your head into the lion's mouth is always a judgment call. It's not just the pilot and a few passengers who are inconvenienced, but a whole planeload of customers, company teammates who are depending on you to get the job done, and the need to deliver the aircraft to the gate for the next leg. The captain is under significant pressure, much of it internal, to deliver the aircraft safely.

In the high flight levels, it is a rare day that a safe hole cannot be found to slide through or around a line. But during takeoff and landing there is little maneuvering room, altitude, or airspeed to trade with a storm.

The Flight - On June 1, 1999, at 10:40 p.m. Central Daylight Time (CDT), a McDonnell Douglas MD-82 operating as American Airlines (AA) Flight 1420, with 144 passengers and crew, departed Dallas on a routine flight to Little Rock, Arkansas. The flight was delayed for about two hours awaiting the inbound aircraft. The flight crew had been on duty since about 10:30 a.m. and had flown two legs, from Chicago to Salt Lake City and then to Dallas.

The National Weather Service surface analysis chart at 0600Z showed a low-pressure system on the Illinois-Wisconsin border with the cold front stretching across western Illinois southward into southeast Missouri and northern Arkansas, and then southwestward across west-central Arkansas and southeastern Oklahoma into north-central Texas. A squall line extended from Michigan through Indiana and into southern Illinois. A second line was depicted over Arkansas.

Convective Sigmets and severe thunderstorm warnings covered the Little Rock area at the time of 1420's arrival. At 10:54 p.m. the AA flight dispatcher sent the following printed message to the crew in the cockpit: "Right now on radar there is a large slot to Little Rock. Thunderstorms are on the left and right, and Little Rock is in the clear. Sort of like a bowling alley approach. Thunderstorms are moving east northeastward towards Little Rock, and they may be a factor for our arrival. I suggest expediting our arrival in order to beat the thunderstorms to Little Rock, if possible." At 11:11 p.m., flight dispatch advised the crew of a revised fuel load. Additional fuel had been added before departure for enroute weather deviations and diversion to Nashville or a return to Dallas if needed.

The cockpit voice recorder shows that the crew was clearly focused on the weather as they watched the storms moving in from the west and northwest. The captain was flying the aircraft. CAM-1 (cockpit area microphone) is the captain; CAM-2 is the first officer; RDO is a radio transmission from the aircraft; and APR is Little Rock Approach or the tower controller. An asterisk (*) indicates an unintelligible word and # denotes an expletive. Non-pertinent remarks have been edited out and my comments are in italics.

11:25:43 CAM-1 We got to get over there quick.

11:25:48 CAM-2 I don't like that...that's lightning.

11:25:56 CAM-1 Sure is.

11:26:48 CAM-1 This is the bowling alley right here.

11:26:50 CAM-2 Yeah, I know.

11:26:55 CAM-1 In fact, those are the city lights straight out there.

11:26:57 CAM-2 That's it.

11:29:44 CAM-2 Yeah, that alley's getting big...closing to the west.

11:29:48 CAM-1 Yeah, it is.

11:29:49 CAM-2 * be OK.

11:29:52 CAM-2 I say we get down as soon as we can.

11:31:52 CAM-2 Whoa. Looks like it's movin' this way, though.

11:31:54 CAM-1 Yeah *.

11:32:05 CAM-1 Just some lightning straight ahead.

11:32:11 CAM-2 Think we're gonna be OK. Right there.

11:32:28 CAM-1 Down the bowling alley.

11:34:09 APR American 1420, Little Rock Approach. We have a thunderstorm just northwest of the airport moving through the area now. Wind is 280 at 28, gusts 44, and I'll have new weather for you in just a moment, I'm sure. *[The crew acknowledged and said they could see lightning.]*

11:34:36 CAM-1 Right near the limit. *[The crew then discussed the crosswind limits of the aircraft and decided that 20 knots was the maximum allowable on a wet runway.]*

11:39:05 APR American 1420, your equipment's a lot better than what I have. How's the final for [Runway] 22 Left lookin'? *[The controller is referring to the aircraft's radar being much better for seeing weather than ATC radar. Today, that is not the case for most TRACONS.]*

11:39:11 RDO-2 OK, we can see the airport from here. We can barely make it out, but we should be able to make *[Runway]* 22. That storm is moving this way like your radar says it is, but a little bit farther off than you thought.

11:39:22 APR American 1420, roger, would you just want to shoot a visual approach? *[The captain declined the visual because of poor visual conditions. The controller advised of a wind shear alert at Little Rock and reported the current winds as "center field wind is 340 at 10, north boundary wind is 330 at 25, northwest boundary wind 010 at 15."]*

11:42:26 APR American 1420, it appears we have the second part of this storm moving through. The winds now, 340 at 16, gusts 34.

[The crew acknowledged. The gust factor exceeded the wet runway crosswind limitation. Because of the wind shift, the crew asked to land on Runway 4 Right, which was approved. They then decided that they could see the airport well enough to land visually, which would be faster than a full instrument approach. Strong gusts and wind shifts indicate the airport was under the influence of the storm. For those of us flying light aircraft, the game is over—diversion is the only option.]

11:43:59 APR American 1420, you can monitor 118.7, Runway 4 Right, cleared to land. The wind right now, 330 at 21.

11:44:05 RDO-2 18.7, we'll monitor, American 1420, thanks. Cleared to land Runway 4.

11:44:13 CAM-2 If you look at....

11:44:14 CAM-1 Those red lights out there. Where, where's that in relation to...?

11:44:18 CAM-2 There's another, there's two runways here. There's three runways.

11:44:19 CAM-1 Yeah, I know. See? We're losing it. I don't think we can maintain visual. *[The captain concedes that he can't see the airport.]*

11:44:22 CAM-2 ** yeah.

11:44:28 RDO-2 Approach, American 1420.

11:44:29 APR American 1420, yes sir.

11:44:30 RDO-2 There's a cloud between us and the airport. We just lost the field and I'm on this vector here. I have basically the last vector you gave us. We're on kind of a dog leg it looks like.

11:44:39 APR American 1420, can you fly heading 220? I'll take you out for the ILS. [There is some additional discussion about the visual approach.]

11:45:15 CAM-1 *I hate droning around visual at night in weather without...having some clue where I am.*

11:45:23 CAM-2 Yeah, but the longer we go out here the....

11:45:24 CAM-1 Yeah, I know. [The captain acknowledges that he was unsure of the flight's location, and the FO is keenly aware that the storms are very close to the airport and may compromise their arrival if there is any delay, such as accepting an ILS approach.]

11:45:29 CAM-2 See how we're going right into this crap.

11:45:31 CAM-1 Right.

11:45:47 RDO-2 Approach, American 1420, I know you're doing your best, sir. We're getting pretty close to this storm. We'll keep it tight if we have to. [During the next minute, approach control vectors the aircraft to just outside the final approach fix and then advises that the runway visual range for Runway 4 Right has dropped to just over one-half mile.]

11:46:39 APR American 1420 is three miles from the marker. Turn right heading 020. Maintain 2,300 'til established on the localizer. Cleared ILS Runway 4 Right approach.

11:46:47 RDO-2 ** 'til established, American 1420, cleared 4 Left approach.

11:46:52 CAM-1 *Aw, we're goin' right into this.*

11:46:52 APR American 1420, right now we have heavy rain on the airport. The current weather on the ATIS is not correct. I don't have new weather for ya, but the visibility is less than a mile. Runway 4 Right RVR is 3,000.

11:47:04 CAM-1 Three thousand.

11:47:04 RDO-2 Roger that, 3,000, American 1420. This is 4 Right, correct?

11:47:08 APR American 1420, that's correct, sir. Runway 4 Right, cleared to land. The wind 350 at 30, gusts 45.

11:47:10 CAM-1 Can we land? [The captain voiced some uncertainty and probably wanted the FO to check the crosswind chart or Runway Visual Range (RVR.)]

11:47:16 RDO-2 030 at 45, American 1420. *[The FO misinterpreted the winds, thinking they were aligned with the runway when they were actually 50 degrees off runway alignment. This exceeded the crosswind limitation. The crew then discussed the visibility minimums.]*

11:47:19 CAM-2 ** zero forecast right down the runway.

11:47:22 CAM-1 3,000 RVR. We can't land on that. *[The captain voices more uncertainty. Actually, the minimum RVR for Runway 4R is 2,400.]*

11:47:24 CAM-2 3,000, if you look at uh....

11:47:27 CAM-1 What do we need?

11:47:28 CAM-2 No, it's 2,400 RVR.

11:47:29 CAM-1 OK, fine.

11:47:30 CAM-2 Yeah, we're doing fine.

11:47:31 CAM-1 All right.

11:47:34 CAM-1 Fifteen. *[Captain calls for flaps.]*

11:47:44 CAM-1 Landing gear down.

11:47:49 CAM-1 And lights ** please.

11:47:53 APR Wind shear alert, center field wind, 350 at 32, gusts 45. North boundary wind 310 at 29. Northeast boundary wind 320 at 32. *[These reported winds exceed the wet crosswind limitation of 20 knots.]*

11:48:04 CAM-2 Flaps twenty-eight?

11:48:11 CAM-1 Add twenty.

11:48:13 CAM-2 Right.

11:48:13 CAM-1 Add 20 knots. *[The captain is carrying additional speed to prepare for a possible wind shear encounter.]*

11:48:15 CAM-2 OK.

11:48:13 APR American 1420, the Runway 4 Right RVR now is 1,600. *[This is below landing minimums, but the regulations allow the approach to continue if the flight is established inside the final approach fix. They can't land if the visibility stays down but in a dynamic situation the weather can improve as quickly as it can deteriorate.]*

11:48:18 CAM-2 Aw, #.

11:48:19 CAM-1 Well, we're established on the final.

11:48:21 CAM-2 We're established, we're inbound, right.

11:48:25 RDO-2 OK, American 1420, we're established inbound.

11:48:27 APR American 1420, roger, Runway 4 Right, cleared to land, and the wind, 340 at 31. North wind, boundary wind is 300 at 26, northeast boundary wind 320 at 25, and the 4 Right RVR is 1,600.

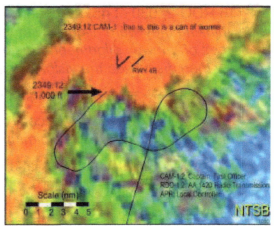

Flight path, NTSB

11:48:42 RDO-2 American 1420, thanks.

11:48:48 CAM-2 Keep the speed.

11:48:51 CAM-2 Thousand feet.

11:48:55 CAM-1 I don't see anything. Lookin' for 460. *[Referring to the msl altitude that is the decision height for landing.]*

11:49:01 CAM-2 It's there.

11:49:03 CAM-2 Want 40 flaps?

11:49:05 CAM-1 Oh yeah, thought I called it.

11:49:06 CAM-2 Forty now. 1,000 feet. 20, 40, 40, land. *[Probably referring to extra speed margin and flap indication on each wing.]*

11:49:11 APR Wind is 330 at 28.

11:49:13 CAM-1 *This is this is a can of worms.*

11:49:25 CAM-1 I'm gonna stay above it a little. *[Probably referring to glideslope.]*

11:49:25 CAM-2 There's the runway off to your right, got it?

11:49:27 CAM-1 No.

11:49:28 CAM-2 I got the right runway in sight.

11:49:31 CAM-2 You're right on course. Stay where you're at.

11:49:32 CAM-1 I got it, I got it.

11:49:33 APR Wind 330 at 25.

11:49:38.6 CAM-? Wipers.

11:49:42.3 CAM [Sound similar to windshield wiper motion.]

11:49:47.3 CAM-2 Five hundred feet.

11:49:54.6 CAM-1 Plus twenty. *[Referring to extra 20 knots of speed.]*

11:49:54 APR Wind 320 at 23.

11:49:57.5 CAM-? Aw, #, we're off course.

11:50:01.4 CAM-2 We're way off. *[Referring to the localizer.]*

11:50:02.5 CAM-1 I can't see it.

11:50:05.4 CAM-2 Got it? [Probably referring to the airport.]

11:50:06.1 CAM-1 Yeah, I got it. *[There are some altitude call outs just prior to landing.]*

11:50:21.2 CAM [Sound of two thuds similar to aircraft touching down on runway concurrent with unidentified squeaking sound.]

11:50:23.2 CAM-2 We're down.

11:50:25.4 CAM-2 We're sliding.

11:50:27.1 CAM-1 #...#.

11:50:32.9 CAM-? On the brakes.

11:50:34.2 CAM-? Oh, #....

11:50:34.6 CAM [Sound similar to increase in engine rpm.]

11:50:36.2 CAM-? Other one, other one, other one.

11:50:42.0 CAM-? Aw, #.

11:50:42.7 CAM-? ##.

11:50:44.9 CAM [Sound of impact.]

11:50:45.4 CAM-? ##.

11:50:48.0 CAM [Sound of several impacts.]

11:50:49.1 End of recording.

The Crash - The driver of a tractor-trailer passing the airport reported torrential rain when a jetliner crossed in front of his truck at an estimated altitude "of no more than 10 to 20 feet" above the top of his truck. The truck in front of him was blown from the center lane into the left lane just prior to the aircraft's passing overhead. The driver also reported intermittent golf ball-size hail and "almost continuous lightning."

At 11:51 p.m. AA1420 overran the end of Runway 4R and collided with the approach light stanchion. The estimated speed at impact was 83 knots. The captain and 10 passengers died; the remaining 134 passengers and crewmembers sustained various injuries.

According to the NTSB Operations Report, the FO gave his recollection of the last few seconds on the approach, the touchdown, and the roll-out. The approach was stabilized until about 400 feet agl, when the aircraft drifted to the right of course, and the FO estimated that they were displaced "about a runway width" to the right. "We were going right of course, and I thought it's getting more and more difficult to handle. So, I thought it would be safer to go around." The FO said that he thought he said "Go around" at that time. He looked at the captain to see if he had heard him, but the captain was intent on flying and was doing a "good job," except "his azimuth was off." The FO thought to himself that "The captain knows what he is doing, and he was flying the airplane."

The Crew - There was a significant difference in experience level between the pilots. The captain was a check airman on the MD-80 and the chief pilot at the Chicago base. He had total flying time of 10,234 hours with MD-80 pilot-in-command time of 5,518 hours. In contrast, the FO had total flying time of 4,292 hours with second-in-command MD-80 time of 182 hours. This may have accounted for his previous comment about the captain's being on top of things. The FO was a relatively new hire at American and may have been reluctant about voicing concerns to a senior captain, although other pilots described the captain as "Someone who does the right thing when no one's looking" and not intimidating to fly with.

Spoiler Factor - The Flight Data Recorder showed a momentary ground spoiler activation and almost immediate retraction. But there was no indication that the ground spoilers had been armed. Their normal deployment upon touchdown would have helped immensely in dumping lift and allowing the aircraft to stop within the confines of the runway.

According to the NTSB report, "The actual demonstrated landing distance was 2,830 feet, so the Part 121 minimum dry and wet runway lengths were 4,715 and 5,425 feet, respectively. Runway 4R/22L at Little Rock National Airport is 7,200 feet in length, so there is about an 1,800-foot margin between the required minimum wet runway length and the end of the runway."

NTSB conducted several tests to ascertain the effect of spoilers on the aircraft stopping distance. There was no conclusive proof that the spoilers malfunctioned, but it is also clear that they did not deploy.

The captain elected to use manual brakes for landing in place of autobrakes. The touchdown was on the centerline and "not that far down the runway." At touchdown, the main gear was to the right of the runway centerline and the nose of the airplane was pointed left. The FO said the touchdown was "sort of flat, sideways, and it was violent."

The captain selected reverse thrust immediately and the FO said that "…he really honked on it," with engine pressure ratios (EPRs, or power output) of 1.6 to 1.8. The AA DC-9 (MD-80) operating manual stated in part: ""The application of reverse thrust tends to blank out the rudder. The effectiveness of the rudder starts decreasing with the application of reverse thrust and at 90 knots, at 1.6 EPR (in reverse) it is almost completely ineffective.

"One of the worst situations occurs when there is a crosswind and sufficient water to produce total tire hydroplaning,.." the manual continued. "Reverse thrust tends to disrupt airflow across the rudder and increases the tendency of the airplane to drift downwind, especially if a crab or yaw is present. As reverse thrust increases above 1.3 EPR, rudder effectiveness decreases until it provides no control at about 1.6 EPR. Use aggressive manual braking or maximum autobrakes and auto spoilers. Apply reverse thrust as soon as possible after nosewheel touchdown. Do not exceed 1.3 EPR reverse thrust on the slippery portions of the runway, except in an emergency. When reversing, be alert for yaw from asymmetric thrust. If directional control is lost, bring engines out of reverse until control is regained. Do not come out of reverse at a high RPM [revolutions per minute]. Sudden transition of reversers before engines spool down will cause a forward acceleration."

The FO did not remember if the spoilers extended after touchdown. According to the flight data recorder, the spoilers did not deploy and there was no indication that they had been armed. Immediately after touchdown, the FO stated that they had no control of the airplane, and it did not feel like they had ground contact. The FO recalled that the airplane was skidding "right off the bat" in a straight line sideways to the right.

"We then started to drift to the left across the runway," and he described the sensation as "Like a roller coaster." The aircraft went to the left but came back toward the center of the runway, and it felt like they had it under control. The main concern was the speed and the hydroplaning. At one point, the FO said the airplane was "Fishtailing and it felt like we might ground loop."

The captain brought the engines out of reverse and "it looked like he was either going to do a go-around or to regain directional control." They "kind of drifted" on the runway and seemed under control but "going fast." He said the captain then went back into reverse thrust, but it wasn't working. They slid sideways down the runway and the main gear slipped off the left side while the nose gear remained on the runway.

As the slide continued, the crew saw the alternating red-and-white centerline lights that mark the last 3,000 feet of the runway. Nearing the end of the runway, at about 80 knots, the captain said, "Brakes," and the FO got on the brakes with him.

The aircraft was carrying an extra 20 knots and touched down at about 150 knots as a precaution against the reported wind shear. This solved one problem and created another. With the water on the runway there was little or no ability to stop or steer. The wind was near or beyond the crosswind limits of the aircraft under these conditions. The spoilers did not deploy, which would have helped slow the aircraft, and the design of the MD-80 caused rudder blanking when the engines went above 1.3 EPR in reverse - which further compromised steering ability.

NTSB Probable Cause - *"...the flight crew's failure to discontinue the approach when severe thunderstorms and their associated hazards to flight operations had moved into the airport area and the crew's failure to ensure that the spoilers had extended after touchdown. Contributing to the accident were the flight crew's (1) impaired performance resulting from fatigue and the situational stress associated with the intent to land under the circumstances, (2) continuation of the approach to a landing when the company's maximum crosswind component was exceeded, and (3) use of reverse thrust greater than 1.3 engine pressure ratio after landing."*

Commentary - Every airline captain and many general aviation pilots have been in similar circumstances before. As a flight gets close to the airport it's psychologically much tougher to make the divert decision. The NTSB's full report is required reading for anyone considering a professional piloting career, corporate or airline, because you will be faced with similar circumstances. Reading the transcript of the CVR, the crew knows they are getting into trouble and yet, they continue. There's an easy answer that psychologists will spend hours describing. Remember continuation bias for pilots and hindsight bias for all we second-guessers.

The pressure is caused by destination focus, professional pride, prior experience to conditions that may have looked similar and the desensitizing effects of fatigue. It's extremely powerful and not to be underestimated. *I would never make such a mistake, ever (much.)* Anyone who's flown in anything other than benign weather has pushed beyond their capabilities at some point and it was only luck that saved us.

Highlight shows tire tracks off end of runway, NTSB

A few years prior to this crash MIT Lincoln Laboratory conducted a study of thunderstorm penetrations in the terminal area: "The study also found that, farther from the airport, pilots nearly always deviated around intense storms and penetrated weaker storms and that, closer to the airport, pilots mostly penetrated the storms regardless of their intensity. Several flight-related variables were tested to determine whether they were correlated to a pilot's decision to penetrate thunderstorms. *The tests determined that pilots were more likely to penetrate convective weather when they were following another aircraft, behind schedule by more than 15 minutes, or flying after dark.*"

The evidence is irrefutable: Destination focus can be deadly. Diversion to another airport or a short period of holding to see if conditions might improve are options that we always have. Had either of the crew acted on what they intellectually knew was a rapidly deteriorating situation this would have turned into just another long evening caused by thunderstorms with an exhausted crew and bunch of equally exhausted and irritated passengers. Destination is seductive.

Always Another Dawn - (Pilot superstar misreads the weather)

Scott Crossfield, pioneer in the jet age, offers his last lesson

Test pilot, aeronautical pioneer and engineer, scientist, general aviation pilot, AOPA member, family man, nice guy. A. Scott Crossfield's biography is impressive and although I never had the privilege of knowing him personally, I read his book, Always Another Dawn, which detailed his early life up through the design and flights of the X-15 rocket plane. As a research pilot for NACA, predecessor to NASA, he made aeronautical history on November 20, 1953, when he reached Mach 2, or more than 1,320 miles per hour, in the D-558-II Skyrocket.

Scott Crossfield and X-15, NASA

His aircraft was dropped from a B-29 at 32,000 feet and climbed to 72,000 feet before diving into the history books at more than twice the speed of sound. By any measure, he was a highly accomplished pilot and a superb individual. A. Scott Crossfield died in a thunderstorm near Ludville, Georgia, on April 19, 2006, at the controls of his Cessna 210. He was 84 years old.

A reaction by a number of pilots was, "If it can happen to someone like Crossfield, it can certainly happen to me." Perhaps that's the beginning of a nagging self-doubt that will lead some away from the cockpit. For others, they shrug it off as bad luck—the wrong place at the wrong time. Despite Crossfield's book title, this dawn would be his last, but maybe there's another way to look at it. Like the thousands of crashes that preceded it and, unfortunately, those that will follow, there is a definite chain, an explanation, and a way to avoid ending life on such a tragic note.

Morning Thunder -The flight departed Prattville, Alabama, around 10 a.m. on a planned IFR route over the VORs in Rome, Georgia; Snowbird, Tennessee; and in Virginia to Roanoke, Montebello, and Casanova, to the airport in Manassas, Virginia. The requested cruise altitude was 11,000 feet. Prior to departure, Crossfield obtained several weather briefings. They described a large area of convective activity across the route and a forecast of considerable thunderstorm activity later in the day. Crossfield discussed the weather with an acquaintance, according to the NTSB, mentioning that he "might need to work his way around some weather but that it didn't appear to be anything serious."

The Cessna 210 climbed to 11,000 feet and everything proceeded routinely. ATC was well aware of the weather north of Atlanta with airline flights deviating west to avoid it while some departures were delayed. Crossfield's flight was handed off to various enroute sectors as he proceeded northeastward, but there was no discussion of thunderstorms.

He checked on to the final frequency at 11:01 a.m. and was provided the local altimeter setting but no mention of weather.

At 11:09 Crossfield transmitted, "Atlanta, this is Seven-Niner-X-Ray. I'd like to deviate south for weather." The controller replied, "Six-Five-Seven-Niner-X-Ray, roger, we'll show you deviating south for weather and your mode C indicates one-one-thousand five hundred." Crossfield did not respond. At 11:10, about 30 seconds after a turn was started, radar contact was lost at 5,500 feet.

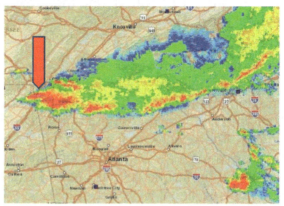

Arrow shows approx. location of crash, NTSB

Before the Cessna entered the weather, the controller's radarscope showed a band of moderate to extreme precipitation along the projected flight path, indicating a supercell thunderstorm. The aircraft's final descent rate exceeded 12,000 feet per minute; the debris distribution was consistent with a low-altitude, in-flight breakup; and impact occurred in near vertical descent.

The wreckage was located in remote mountainous terrain 3.3 nautical miles northwest of Ludville, Georgia.

The Airplane - The 1960 Cessna 210A had externally braced wings, and Crossfield had owned it since 1989. The airframe and engine had accumulated 4,987 hours total time, with nearly 1,260 hours on the engine since its last major overhaul. An annual inspection was completed the month before the accident. The NTSB determined that the aircraft was operating normally at the time of the accident.

Avionics included a BF Goodrich WX 950 Stormscope, a Garmin GNC 250 GPS/Com, and a Garmin GNS 430 GPS/Com/Nav. The GNC 250 was limited to VFR flight only. The Garmin GNS 430 system was approved for IFR domestic flight, including enroute and non-precision approaches. Crossfield subscribed to SiriusXM Aviation Weather for satellite datalink weather, but the required antenna had not been installed. No pre-impact mechanical malfunctions were noted.

The Pilot - Crossfield held a commercial certificate with airplane single-engine land, multiengine land, and instrument ratings. His third-class medical certificate was issued in December 2004. He reported a total flight experience in excess of 9,000 hours. The logbook showed he had flown 95.5 hours during the previous 12 months, 28.5 hours during the prior six months, and 23.1 hours during the previous 30 days. During the prior year, all the flights were in the accident airplane and his last flight review was completed on August 27, 2004.

Total actual instrument flight time was 423.1 hours with an additional 106.0 hours using a view-limiting device. Like many of us, Crossfield had not spent much time in the clouds recently, with 5.4 hours of instrument flight time and two instrument approaches recorded during the previous 12 months. There were no instrument flight time or approaches shown during the six months before the accident and no instrument instruction or instrument proficiency check within the previous 12 months. By regulation 61.57, it was illegal to fly under IFR.

The Weather - The accident occurred in an area of severe thunderstorms, identified as a mesoscale convective system with intense to extreme precipitation rates. The forecast predicted instrument to marginal VFR conditions during the morning hours, with isolated severe thunderstorms and moderate rain expected after 10 a.m. with cloud tops reaching 43,000 feet. Crossfield obtained five computer weather briefings on April 18 and 19. The latest was recorded on the morning of the accident at about 6:56 a.m., which included warnings of convective weather later in the day. At 8:56 a.m., the National Weather Service (NWS) warned of a cluster of severe thunderstorms moving into northern Georgia. The area was expected to further destabilize by midday because of surface heating, enhancing the threat of organized severe thunderstorms. It was not a good place to fly.

A convective Sigmet was issued at about 10:55 a.m. for severe embedded thunderstorms over northern Georgia, portions of Alabama, Tennessee, and North Carolina. The system was moving southward at 35 knots with cloud tops reaching 45,000 feet. However, this advisory was never given to Crossfield.

Failure to Communicate - Unlike years ago, virtually all ATC radar depicts precipitation in varying degrees. It has limitations but is far superior to the old days where lines and "Hs" gave a vague indication of heavy rain. At enroute centers such as Atlanta, a system known as Weather and Radar Processor (WARP) supplies controller displays with an overlay showing areas of moderate, heavy, and extreme intensity precipitation. In the worst case, the displayed precipitation can be as much as 10 to 12 minutes old in areas covered by only a single Nexrad radar.

However, single-site coverage is unusual, especially in parts of the country that experience frequent convective activity. Most enroute controllers view a composite picture of data received from several Nexrad sites. Because the composite display is refreshed every time any of the overlapping sites supply new information, what the controller sees is typically four to six minutes old and sometimes less. Controllers can set altitude filters that show precipitation up to 24,000 feet; 24,000 to 33,000 feet; 33,000 to 60,000 feet; or all levels. This is changing as the technology improves.

The display does not show turbulence or thunderstorms, although heavy precipitation is roughly correlated with thunderstorm location if there's a tight gradient. It takes some skill to interpret where convective weather might be.

The NTSB confirmed that Crossfield was not provided any severe weather advisories nor was he advised of the radar-depicted weather displayed on the controller's scope—both were required notifications. The controller acknowledged that weather was "all over" his sector in varying intensities and that northbound airline departures out of Atlanta were encountering thunderstorms, "picking their way through holes in the weather."

The controller knew about the requirement to issue weather advisories depicted on their scopes, but he did not issue a warning because he felt that the weather display was unreliable. This controller felt "That pilots have a better idea of where adverse weather is, and that he expects them to inform him on what actions they need to take to avoid it."

The Air Traffic Control Handbook 7110.65 Chapter 2 describes controller responsibilities regarding weather. Priority one is separating traffic. But the directives further state that controllers should use good judgment and first perform the action that is most critical from a safety standpoint. Weather avoidance, particularly thunderstorms, is considered a high priority.

NTSB Probable Cause -: *"….the pilot's failure to obtain updated enroute weather information, which resulted in his continued instrument flight into a widespread area of severe convective activity, and the air traffic controller's failure to provide adverse weather avoidance assistance, as required by Federal Aviation Administration directives, both of which led to the airplane's encounter with a severe thunderstorm and subsequent loss of control."*

Commentary - Crossfield knew he would be flying into an area of widespread severe thunderstorms based on his pre-departure briefing. He did not ask for any updates from either ATC or flight service. In today's environment with datalink and electronic flight bags it couldn't be much easier. But the equipment was not completely installed. The Stormscope, if working, would have had an active display. The best way to circumnavigate would have been to deviate northwest to get behind the front as most of the airliners were doing and then turn toward Manassas. Direct flight this day just wasn't possible, but it's easy to second guess.

Don't assume that ATC knows the flight conditions you're facing. If you can see the weather, advise ATC what you're doing. Ask for weather avoidance and reconfirm with every sector when storms are nearby. The controller in this case did not perform to standard and, just as with pilots, there will be skill differences. The vast majority will do everything to help. If a controller is unable to help for any reason, including workload, modify your plan and escape. It may mean landing short or deviating extensively.

If you're not getting the help you need, declare an emergency. Don't worry about FAA's hammer - Mother Nature carries a much bigger one! There will be times when some airspace is just not flyable. Being equipped with Stormscope, datalink, or even weather radar doesn't change the atmospheric conditions. It just shows what may be an impossible situation. The airlines even with all their tools are periodically skunked, but GA has much more flexibility than the airlines. Go earlier, go later, divert, cancel.

A. Scott Crossfield flew some of the most exotic and dangerous high-performance aircraft ever built under the most perilous of circumstances. In considering his tremendous service to our nation, perhaps he has performed one final and noble sacrifice. His celebrity, skill, and experience overlaid onto this tragedy may draw enough attention from the entire aviation community that pilots, controllers, meteorologists, and system designers will learn—and many lives will be saved.

Working a Hole - (C172 CFI learns that holes aren't always there)

A Cessna 172 with a CFI and student get knocked around in a Florida thunderstorm thinking that ATC is providing separation.

> *"First law of holes: When in one, stop digging."*
>
> Denis Healey, British Labour Party Politician

This incident could easily have been fatal crash but resulted in a happy ending. It highlights a common misconception that just because you are talking with ATC about thunderstorms doesn't mean the controllers have the ability, or the time, to help you stay clear. Speak up and be clear on what you need.

On March 28, 2005, a Cessna 172S carrying a flight instructor and an instrument student departed Kendall-Tamiami Executive Airport in Miami on an IFR flight plan, heading to Sarasota/Bradenton International Airport at approximately 8:45 a.m. Eastern Standard Time. A cold front extended into central and southern Florida and across the Gulf of Mexico. Areas of

Approx. flight path – black arrow, NWS archive

scattered embedded thunderstorms and rain showers were present in the frontal zone, with mostly visual conditions elsewhere.

At 8:59 a.m., the flight first contacted Palm Beach Approach Control at 5,000 feet on a northerly heading. At 9:05, the controller asked if the flight was planning to fly to Vero Beach for practice approaches before continuing to Sarasota. The pilot responded, "Negative. If we could get Sarasota direct and you could help us out with the weather, that would be greatly appreciated."

The controller replied, "I don't see any way of getting to Sarasota without going through the weather."

The pilot responded, "Uh, roger that. Just, uh, on the cells." That response clarified nothing.

The controller stated, "We're not displaying any right now. It's outside our airspace," and asked the pilot if he wanted direct Sarasota but understand that they may not be able to provide it.

The pilot replied, "Affirmative," and asked for a higher altitude. Although this may have sounded "sort of" like a deal had been made with ATC to keep the Cessna out of thunderstorms, that was not what happened. The Cessna was equipped with a panel-mounted datalink receiver capable of displaying Nexrad radar mosaics. This likely played a part in the decision to tackle the line of weather. The standard disclaimer with this equipment is that it's not to be used for tactical weather avoidance. It displays precipitation and typically updates every few minutes—though it can take considerably longer under certain conditions—similar to the Center controller's display.

The Cessna was assigned 6,000 feet and given a vector of 290 degrees toward Sarasota. At 9:13, the pilot again requested a higher altitude and was cleared to 8,000 feet. Flying toward the front, it was evident the pilot was attempting to maintain visual contact above the lower clouds to avoid the buildups.

The "Hole"

At 9:14, Miami Center asked Palm Beach to vector a different aircraft on a 300-degree heading toward a hole in the weather northwest of Palm Beach International, near Pahokee, Florida. The Palm Beach controller advised that pilot about a line of weather extending about 30 miles.

At 9:15, the Palm Beach controller advised the Cessna, "Apparently at your one o'clock and about 35 miles [there] is a hole the Center's working. We've got to keep you westbound for now; expect a turn that way shortly." The Cessna pilot acknowledged. Most would believe that ATC was providing weather advisories.

At 9:19, there was additional coordination between Miami and Palm Beach concerning other aircraft being vectored for weather. At 9:22, Palm Beach instructed the Cessna to fly 290 degrees, and at 9:25, the flight was handed off to Miami Center.

The Center controller acknowledged and issued the Fort Myers altimeter setting. At 9:28, the controller transmitted, "November-Five-Eight-Nine, *after deviations permitting*, proceed direct Sarasota and advise." The pilot acknowledged direct Sarasota, and the controller asked the pilot to advise on deviations. This was the controller's way of saying that the Cessna pilot was largely on his own to stay clear of the thunderstorms.

At 9:34:12, the pilot asked, "Center, Five-Eight-Nine, how's the weather look in front of us?" That's usually an indication that there's uncertainty in the cockpit.

The controller responded, "Five-Eight-Nine, I'm showing a line of precipitation that extends five miles along your route of flight; then you should be clear until you get about 25 southeast of Sarasota." The controller did not provide any intensity information. Things were about to get interesting.

At 9:37:23, the Cessna pilot transmitted, "Center, I got an emergency."

The controller responded, "Aircraft calling the Center, state your emergency and intentions."

At 9:37:40, the pilot continued, "Five-Eight-Nine... I'm in some serious weather. Can you help us out?"

The controller responded, "Five-Eight-Nine, maintain 8,000 feet. If feasible, turn right heading 090, weather permitting. Five-Eight-Nine, your present position, ah, heading 090 or right 100 degrees heading should get you clear of the weather soonest."

At 9:38:04, the Cessna transmitted, "I can't see a thing. We're showing 1,500 feet." The controller advised that there was no other traffic in the vicinity. At 9:38:31, the controller asked if the flight was clear of the clouds, and the pilot replied that it was not. The controller recommended an easterly heading to clear the precipitation.

The pilot responded, "I'm trying, I'm trying."

At 9:39:11, the controller reported that he had lost radar contact. About 40 seconds later, radar contact was reestablished, and the controller verified the aircraft's position as three miles northeast of Pahokee.

The controller advised that the present heading, or slightly to the right, would allow the Cessna to exit the weather in about three miles. At 9:42:23, the controller again advised that radar contact was lost and asked for an altitude report.

The pilot transmitted, "We're still in trouble, we're still in...." The controller asked the pilot to say again, but the reply was unreadable.

At 9:42:55, the CFI in the Cessna stated, "OK, we are... got Okeechobee here."

The controller asked for an altitude report, and the pilot replied, "Ah, 500." The controller concurred and reported the Cessna nearing the edge of the weather in a mile or less.

At 9:43:25, the controller told the CFI that the minimum safe altitude in his vicinity was 1,300 feet and instructed him to climb and maintain 8,000 feet.

The pilot refused and requested, "We need to go back to the airport."

The controller asked for the nature of the emergency and the preferred destination.

The pilot replied, "OK, uh, we're coming out of fine precipitation—we've got it covered, uh, we're coming out of precipitation. We got the land down below us, so we're OK now."

The controller acknowledged and again asked the pilot for his intentions.

The pilot then requested direct to the nearest airport.

At 9:44:35, the controller handed the Cessna off to Palm Beach Approach for vectors to the nearest airport.

Getting It on the Ground

At 9:45, the CFI again requested to "land at the nearest airport." The controller asked if the aircraft was in visual conditions. The pilot replied, "That's affirmative, that's affirmative; uh, if we could get to Palm Beach, I'd greatly appreciate it." The Palm Beach controller confirmed the 172's location, 10 miles northeast of Pahokee.

The nearest airport was North Palm Beach County General Aviation Airport. The controller continued, "I can get you to Palm Beach, or I can get you to North County. Your closest airport right now is North County."

The pilot responded, "OK, uh, we're OK for now. If we could get to Palm Beach, that would be great." The flight was instructed to fly 100 degrees for vectors to the right downwind for Runway 27R at Palm Beach International Airport.

At 9:47, the controller asked if the Cessna was experiencing any problems, and the CFI replied, "Roger, we're OK right now." The controller then advised that the Palm Beach winds were 220 degrees at 22 knots. The pilot acknowledged and then asked whether there were any thunderstorms in their path. He was now taking a more active role in staying clear of the weather. The controller replied, "Negative, sir."

At 9:51, the controller again queried the flight's status. The pilot responded, "Oh, we're doing fine. We're back to normal operation." The controller said, "OK, great. The wind at Palm Beach is 210 at 23, so you're going to have a really nice crosswind there." The pilot acknowledged.

At 9:52, the approach controller advised Palm Beach Tower that the Cessna, which had been having "a lot of problems," was about 12 miles northwest of the airport and requested that the Cessna be given the long runway for landing (27R).

At 9:54, the pilot advised the controller that he wished to declare an emergency due to structural damage to the aircraft and asked for fire equipment to be standing by at the airport. The controller then asked if the aircraft was still flying all right, and the pilot reported that it was, but he wanted to go directly to the airport. The controller then requested that Palm Beach Tower stop departures. She reported that the wind was 220 degrees at 22 knots, that 27R remained the best runway, and that the flight should enter a right downwind.

The pilot replied, "OK, uh, just stay with us, and we'll try to make it."

At 9:56, the pilot asked, *"Is there any way I can get a direct landing in?"* *The controller responded, "November-Five-Eight-Nine, proceed direct to the airport, sir; you can land on any runway you need to, sir. The winds are currently 220 at 22."*

The pilot stated, *"Uh, roger that; 220 at 22? I'm going to try to make Runway 27 Right." The controller then advised the pilot that he was directly between North County Airport and Palm Beach International and said, "North County is off your left if that will help any."* The pilot elected to continue to Palm Beach.

From 9:56 to 10:02, the approach controller continued to provide wind updates to the pilot and relayed clearance for the aircraft to land on any runway at Palm Beach. The aircraft landed safely on Runway 27R at 10:02.

From the Cockpit

The CFI later recounted his story. On the assigned heading, the CFI said the flight encountered rain and turbulence, so he elected to make a 180-degree turn and exit the precipitation. While turning, the airplane encountered a downdraft, with an immediate loss of about 2,000 feet, followed by increased turbulence.

During the turbulence, the door hinge pins fractured, and the passenger-side window blew off, striking the right-side horizontal stabilizer. That's when they declared an emergency.

The instrument student's view was similar. He remembered that Miami Center had said they had five miles of rain to pass through. There was no mention of heavy rain or cells.

"After a few bumps," the student said, the instructor took control and began a 180-degree left turn to exit the showers. At about 90 to 120 degrees of turn in a 30-degree bank, they encountered strong updrafts and downdrafts, and the airplane went into a 90-degree bank. The right door was blown off the hinges, pushed back about eight inches, and the window blew out.

They continued flying with the attitude indicator tumbled, as the instructor finally regained control at 6,000 feet, at which time the attitude indicator returned to showing normal wings-level flight. Apparently, they encountered more strong turbulence because the Cessna flew out of the storm at a mere 500 feet MSL. Fortunately, most of Florida is close to sea level.

The Weather

The National Weather Service Radar Summary Chart for 1519Z (10:19 a.m. EST) depicted an area of very light to light intensity echoes over southeast Florida and portions of the Everglades, with another area of strong to very strong echoes extending west-southwestward from the East Coast across central Florida to the north of Lake Okeechobee, into the Gulf of Mexico.

Several intense to extreme intensity echoes were embedded within that area. Another area of echoes extended from the general vicinity of the accident site and eastern Lake Okeechobee northeastward off the coast, where a solid line of heavy to extreme intensity echoes was located. Echo tops in the vicinity of the encounter were identified at 39,000 feet, with cell movement to the northeast.

NTSB Probable Cause

"…The flight crew's inadvertent encounter with turbulence in clouds while being vectored by ATC while operating in instrument meteorological conditions. A factor in the accident was the ATC personnel's (ARTCC) inadequate weather avoidance assistance."

Commentary

About three minutes elapsed from when the pilot last asked about the weather until he declared an emergency. During that time, the aircraft moved about five miles while the storms were moving at roughly right angles to the flight path. The "hole," such as it was, either filled in or moved about four miles.

This is threading the needle and should only be attempted with good onboard radar in a heavy aircraft. Most experienced pilots would describe the situation as a "sucker hole," especially given the intensity of the storm/front as evidenced by the cloud tops. They needed to be much farther away from the weather.

The pilots likely couldn't see the buildups and believed that ATC wouldn't let them blunder into a cell. Miami Center is equipped with the latest digital weather display capability, designed to show precipitation levels now described as "moderate," "heavy," or "extreme." The old terminology, which referred to Level 2 through 6 on the National Weather Service VIP (video integrator processor) scale, is now outdated.

The Center radar displays are not suitable for tactical maneuvering in rapidly developing situations. A strategic look by the Center controller would have shown that the weather was far less intense to the north and that a deviation of about 50 miles would have avoided all the heavy weather. That was the best solution other than landing short of the front. This information would have been the same as what was displayed on the aircraft's datalink receiver.

The NTSB also felt that controller training on radar equipment and light-aircraft capabilities had been somewhat lacking. I will second-guess the CFI slightly, though he obviously did a good job getting the aircraft on the ground. Once an aircraft has encountered severe turbulence and visible damage, it's best to land as quickly and gently as possible. The instructor gave the impression that, after getting things under control, he was doing fine. He said, "We're back to normal operation." However, structural damage—no matter how slight—can be a big deal.

With the right door apparently hanging askew and the window missing, North County Airport might have been a more conservative option, assuming the runway and wind conditions were appropriate for the aircraft. Shedding pieces of the aircraft while searching for the perfect airport could suggest that an off-airport landing might be the better option. It's hard to know without being in the cockpit.

How well did the pilots understand what was being displayed and how close they could safely maneuver to a powerful and rapidly moving system? According to the CFI, the datalink did not display any heavy or extreme precipitation. Independent verification might show otherwise.

Mid-Level Mayhem: A Piper Jetprop Encounters Weather in the Flight Levels

The Shortest Distance May Not Be a Straight Line

When getting bumped by building cumulus clouds at 10,000 feet, it's natural to think that another 10,000 feet could make a flight easier and safer. Many times, it does. However, aircraft flying in the lower flight levels (typically FL180 to FL290) are subject to some of the worst thunderstorm hazards imaginable.

FL250 is simply not high enough to clear the biggest storms, and it's a long way down to the cloud bases, where rain shafts may be seen—and sometimes avoided. At higher altitudes, there's no quick bailout to a nearby airport, an option that is sometimes available to lower-altitude flights. At FL250, the storm tops may be as far above as the ground is below.

The Flight - In June 2002, the pilot of a Piper Malibu Jetprop contacted the Macon, Georgia, Flight Service Station (FSS) at 4:52 p.m. Eastern Daylight Time and requested a briefing for a flight from Rowan County Airport in Salisbury, North Carolina, to Raleigh-Durham International Airport in Raleigh-Durham, North Carolina, and then on to Florida's Marco Island Airport.

Jetprop – great performer but only gets to mid-level on a severe storm, AOPA

The forecast for the Carolinas, Georgia, northern Florida, and coastal waters predicted thunderstorm activity with tops as high as FL450 (45,000 feet). The specialist stated there were "looming thunderstorms" in the area. It was a typical summer evening, with thunderstorms liable to crop up almost anywhere. It was unknown if the aircraft was equipped with datalink.

On the IFR flight plan from Raleigh-Durham to Marco Island, the routing was over Myrtle Beach, South Carolina, and Orlando, with a proposed departure time of 6 p.m. The planned altitude was FL260, with a true airspeed of 250 knots and an estimated time en route of two hours and 45 minutes. The aircraft had four hours of fuel on board.

The flight departed Raleigh-Durham at 6:28 p.m. EDT, and all was routine until 8:02 p.m., when the pilot requested permission to leave the frequency to check the weather. The Gainesville, Florida, FSS Flight Watch specialist advised the pilot of "cells east of St. Augustine; they continue to move east at around two-zero knots" and also noted that convective Sigmet 05E was in effect for southern Florida. The specialist suggested a more westerly routing "toward the Tampa-St. Pete area and then southward" to avoid an area of thunderstorms. At this point, the airplane was about 60 miles east-northeast of Jacksonville.

At 8:06 p.m., the pilot reported back on frequency to the Jacksonville controller, who approved deviations as needed. The pilot replied that no deviations were necessary at that time.

At 8:27:36 p.m., the pilot requested to deviate west of course to fly through "a little hole." The PA-46 was just east of Daytona Beach, Florida. When asked how far he needed to go, the pilot said, "About 10 or 12 miles." ATC then asked if the flight could fly a heading of 170 degrees. The pilot responded that he could not and that the aircraft was "blocked in on the east side." The controller acknowledged, approved the deviation, and instructed the pilot to proceed direct to Marco Island when able. The flight then turned about 20 degrees to the right.

At 8:29:44, the flight was handed off to the next sector. The pilot

A westerly deviation would have avoided the wx – Red arrow shows approx. location, NWS archive

acknowledged and stated, "OK...(unintelligible) and a little hole here." Radar indicated the Malibu was in the vicinity of some heavy weather returns. The controller stated that he observed the weather area and was aware of other pilots deviating away from it on the west side

At 8:29:49 p.m., the ground track plot showed that the pilot had observed a three- to five-mile gap between two thunderstorm clusters and attempted to fly through an area of light radar echoes between the two large areas of heavier echoes. At 8:29:53 p.m., the pilot made an unintelligible transmission to the controller.

At 8:33:36 p.m., the aircraft departed level flight and began an uncontrolled descent from FL260. Radar showed that one cluster of thunderstorms had moved east-northeast, but other Level 3 and 4 thunderstorms were present in the vicinity. (When this accident occurred, precipitation was described in terms of Video Integrator Processor (VIP) levels. Level 3 roughly equates to today's terminology of heavy precipitation, and Level 4 is roughly equivalent to extreme precipitation.)

Witnesses observed the airplane emerge from the clouds about 300 feet above the ground in a spiral, with the right wing missing. NTSB data showed that the PA-46 descended rapidly in a left turn, reaching a maximum descent rate of 20,700 feet per minute. According to ballistic trajectory estimates, the radar pod and wing panel separated at around 26,000 feet just prior to the airplane's rapid descent. The private pilot and two passengers were fatally injured. The right wing and horizontal stabilizer had separated from the aircraft, and the main wing spar was broken inside the cabin area.

Pilot

The pilot held a private certificate with an instrument rating issued almost 10 years earlier. The last dated entry in his logbook was on April 10—six years prior. FAA records indicated that he completed a 16-hour Piper PA-46T Malibu ground and flight recurrent course less than a month before the accident. His flight time was estimated at 2,800 total hours, with 380 hours in the Malibu.

The Aircraft

The 1988 Malibu had been modified, replacing its Continental piston engine with a Pratt & Whitney PT-6 turboprop engine that had 900 hours in service since being overhauled. This modification provided the airplane with additional speed and altitude capability. The aircraft was equipped with onboard radar. The total time on the airframe was 2,814 hours, and the most recent annual inspection had been completed three months before the accident. The NTSB found no fault with the aircraft.

NTSB Probable Cause

"...The pilot's inadequate weather evaluation and his failure to maintain control of the airplane after entering an area of thunderstorms, resulting in an in-flight separation of the right wing and the right horizontal stabilizer and impact with the ground during an uncontrolled descent."

Commentary

The Malibu flight instructor who last trained the pilot had also provided his initial PA-46-310 Malibu training in 1996. The pilot was upgrading from a

Mooney to the Malibu—his first pressurized aircraft. This accident occurred in 2002.

The CFI had noticed during both initial and recurrent training that the pilot "pushed himself dangerously close when making weather decisions in this class of airplane." He seemed to "lack a healthy respect" for the destructive forces of thunderstorms and took "delight" in seeing how close he could come to pushing the envelope. The CFI had cautioned him just two weeks prior to the accident that his decision-making was deficient and that he needed to exercise "greater care" when flying in and around "adverse weather systems."

It's instructive to examine the weather conditions on that June evening. The nearest weather reporting station was about 12 miles southwest of the accident site. The reported weather at 8:19 p.m. was:

- Winds: 180 degrees at 9 knots
- Visibility: 7 miles
- Sky condition: Scattered at 2,000 feet, broken at 5,500 feet, and overcast at 7,000 feet
- Temperature: 24°C
- Dew point: 23°C
- Altimeter setting: 29.85 inches of mercury

At the local level, conditions seemed relatively benign. However, a broader view showed a prefrontal trough extending northeast to southwest across northern Florida, ahead of a cold front located over southern Georgia and Alabama. A broad southwesterly flow over the state provided ample heat and moisture, as indicated by the narrow temperature-dewpoint spread.

More significantly, no fewer than six convective SIGMETs were in effect either just before or just after the crash. All indicated a large area of thunderstorms moving from the west at 10 to 25 knots, with tops reaching FL450. During late afternoon and early evening, widespread thunderstorms are common in summertime Florida and many other parts of the country. Isolated air-mass storms are one thing—widespread convection is something else entirely.

This pilot was aware of the weather but chose to disregard the Flight Watch specialist's recommendation to deviate to the western part of the state and avoid the entire storm system. Taking that advice would have extended his flight by perhaps 30 minutes and might have necessitated a fuel stop.

The Malibu/Mirage series has been somewhat susceptible to thunderstorm-related accidents in the past, but not due to any fault of the aircraft itself. The FAA conducted a special certification review after a number of in-flight breakups, nearly all of which were associated with convective weather. Poor pilot decision-making was repeatedly identified as the probable cause. Onboard radar, datalink, and ATC recommendations can help identify where the worst weather is—but ultimately, it's up to pilots to avoid it.

Convection at Altitude

According to the AOPA Air Safety Institute's accident database, the past 25 years have shown a significant decrease in PA-46 accidents related to convective weather encounters. In the late 1980s and early 1990s, such incidents occurred almost annually. However, over a 20-year period, only nine accidents have shown clear indications of thunderstorm encounters. About half occurred at flight levels 220 to 240, while the rest were in the mid- to low teens.

Pilot experience varied widely—from highly experienced aviators to those with little experience in high-performance, high-altitude operations. In addition to the required high-altitude physiology training for flying above FL250, pilots should also receive training in high-altitude meteorology, negotiating with ATC, and using airborne weather radar. Some newly qualified "flight level" pilots may not fully understand or appreciate the risks of the high-altitude environment.

Along with formal training, practical experience with a mentor pilot is highly recommended. Airlines and most corporate flight departments require years and hundreds—if not thousands—of hours of experience before their pilots are entrusted with flying near thunderstorms. That level of caution and respect for severe weather is critical for all pilots operating at high altitudes.

Deadly Surprise: Southern Airways 242 Makes History

This landmark accident changed the way the FAA and the National Weather Service provided information to pilots.

> **"We've lost both engines, and, uh, I can't, uh, tell you the implications of this—we, uh, only got two engines."**
>
> — *Southern Airways Flight 242*

A common sight in the South in 1977, Wikipedia

Seasoned pilots know that Thor can be unpredictable, but sometimes even they get caught off guard. This situation seemed so similar to previous ones that the pilots relied on a strategy that had always worked before. This time, it didn't.

Heavy Weather in the Southland

April 4, 1977, was a typical spring day. A strong cold front extended from a low-pressure system near the Great Lakes and was forecast to move across Arkansas, Louisiana, and Kentucky. The area forecast, issued by the National Weather Service (NWS), called for marginal VFR conditions, with visibilities of 3 to 5 miles in haze and 1,000- to 2,000-foot overcasts, with occasional IFR conditions. Cloud layers extended up to 19,000 feet, and scattered thunderstorms with tops reaching 35,000 feet were predicted. A few severe storms were expected to top out at 55,000 feet.

A Southern Airways DC-9 departed Atlanta around 2 p.m. EST for a short flight to Muscle Shoals, Alabama. Flight 242 left Muscle Shoals at 3:21 p.m. and landed at Huntsville, Alabama, at 3:44 p.m. on the return trip to Atlanta.

During the turnaround in Muscle Shoals, the crew received tornado watches and several SIGMETs for embedded thunderstorms along the flight's planned route. Southern's dispatchers provided terminal sequence reports (METARs in today's terminology), which indicated cloudy, warm, and windy conditions at Atlanta's Hartsfield Airport. The sky was broken at 2,700 feet, broken at 5,000 feet, and overcast at 25,000 feet. Visibility was reported at 5 miles, with winds out of the southwest at 31 knots, gusting to 47 knots. However, no mention was made of the squall line that had developed and was visible on the National Weather Service radar at Athens, Georgia, at 3:34 p.m. Four-tenths of the area was covered, and echoes were moving east-northeast at 55 knots. The most intense precipitation returns were near Rome, Georgia, with tops reaching 46,000 feet. The crew had flown through this area earlier with nothing more than some rain and light turbulence.

The crew did not leave the cockpit in Huntsville, and it was the first officer's turn to fly back to Atlanta. At 3:54 p.m., Flight 242 departed

Huntsville with clearance to 17,000 feet. The flight was routed directly to Rome, Georgia, and the cockpit conversation quickly turned to weather.

The captain remarked, *"Well, the radar is full of it; take your pick."*

At 3:56 p.m., the controller informed Flight 242 that his radar scope showed heavy precipitation about five miles ahead of the flight. The crew asked whether it appeared worse than what they were already experiencing. The controller, using weather attenuation devices, replied that it might be heavier. The crew acknowledged the report, and the flight was handed off to Memphis Center.

At 3:58 p.m., the captain said, *"If it doesn't get any heavier than this, we'll be all right."*

The first officer replied, *"Yeah, this is good."*

Memphis Center then advised Flight 242 that a SIGMET had been issued for the area and recommended monitoring VOR broadcasts before switching to Atlanta Center.

At 3:59 p.m., the captain commented, "Here we go… hold 'em, cowboy." It was a prophetic statement.

After checking in with Atlanta Center, the captain suggested the first officer reduce speed due to turbulence. Between 3:59 p.m. and 4:02 p.m., the center discussed weather conditions with other air carrier flights in the area. A TWA flight requested deviations, while an Eastern Airlines flight reported, "It was not too comfortable, but we didn't get into anything we would consider the least bit hazardous." Timing and location are everything.

Transcript

All times are in Eastern Daylight Time (p.m.). The following transcript is lengthy, but it provides valuable insight into how events unfolded.

- Capt = Flight 242's captain
- FO = First officer
- S242 = Radio transmissions from Flight 242
- AC = Atlanta Center
- AA = Atlanta Approach
- CAM = Information from the airplane's cockpit area microphone

4:03:48

Capt: "Looks heavy; nothing's going through that."

4:03:54

Capt: "See that?"

4:03:56

FO: "That's a hole, isn't it?"

4:03:57

Capt: "It's not showing a hole; see it?"

4:04:05

CAM: *(Sound of rain.)*

4:04:08

FO: "Do you want to go around that right now?"

4:04:19

Capt: "Hand fly at about two eighty-five knots."

FO: "Two eight five."

4:04:30

CAM: *(Sound of hail and rain.)*

4:04:53

S242: "Southern Two-Forty-Two, we're slowing it up here a little bit."

4:04:53

AC: "Two-Forty-Two, roger."

4:05:53

FO: "Which way do we go, cross here or go out—I don't know how we get through there, Bill."

Capt: "I know. You're just gonna have to go out…"

FO: "Yeah, right across that band."

4:06:01

Capt: "All clear left approximately right now; I think we can cut across there now."

4:06:12

FO: "All right, here we go."

4:06:25

FO: "We're picking up some ice, Bill."

4:06:29

Capt: "We are above 10 degrees." *(Discussing outside air temperature.)*

FO: "Right at 10."

Capt: "Yeah."

4:06:30

AC (to TWA 584): "I show the weather up northwest of that position north of Rome, just on the edge of it—I tell you what, maintain one five thousand."

4:06:38

TWA 584: "Maintain one five thousand; we paint pretty good weather one or two o'clock."

4:06:41

FO: "He's got to be right through that hole about now."

4:06:42

AC: "Southern Two-Forty-Two, descend and maintain one four thousand at this time."

4:06:46

Capt: "Who's that?"

4:06:48

FO: "TWA."

AC: "Southern Two-Forty-Two, descend and maintain one four thousand."

4:06:53 S242: "Two-Forty-Two down to fourteen."

4:06:55 AC: "Affirmative."

CAM: (Heavy hail or rain sound starts and continues until power interruption.)

4:07:00 AC: "Southern Two-Forty-Two, Atlanta altimeter two-niner-five-six. Cross 40 miles northwest of Atlanta at two-five-zero knots."

CAM: (Sound similar to electrical disturbance.)

4:07:57 CAM: (Power interruption for 36 seconds.)

4:08:33 CAM: (Power restored.)

CAM: (Sound of rain continues for 40 seconds.)

4:08:34 AC: "Southern Two-Forty-Two, Atlanta."

4:08:37 FO: "Got it, got it back, Bill; got it back, got it back." [Probably referring to the engines.]

4:08:42 S242: "Uh, Two-Forty-Two, stand by."

4:08:46 AC: "Say again."

4:08:48 S242: "Stand by."

4:08:49 AC: "Roger. Maintain one-five thousand if you understand me; maintain one-five thousand, Southern Two-Forty-Two."

4:08:55 S242: "We're trying to get it up there."

4:08:57 AC: "Roger."

4:09:15 S242: "Okay, uh, Two-Forty-Two, uh, we just got our windshield busted. We'll try to get it back up to fifteen—we're at fourteen."

4:09:24 FO: "Fifteen thousand."

4:09:25 AC: "Southern Two-Forty-Two, you say you're at fourteen now?"

4:09:27 S242: "Yeah, uh, couldn't help it."

4:09:30 AC: "That's OK. Uh, are you squawking five-six-two-three?"

4:09:36 FO: "Left engine won't spool."

4:09:37 S242: "Our left engine just cut out."

4:09:42 AC: "Southern Two-Forty-Two, roger. And, uh, you lost your transponder. Squawk five-six-two-three."

4:09:43 FO: "I am squawking five-six-two-three. Tell him I'm level fourteen."

4:09:49 S242: "Five-six-two-three, we're squawking."

4:09:53 AC: "Say, you lost an engine and, uh, busted a windshield?"

4:09:56 S242: "Yes, sir."

4:09:59 Capt: "Autopilot's off."

FO: "I've got it; I'll hand-fly it."

4:10:00 AC: "Southern Two-Forty-Two, you can descend and maintain one-three thousand now. That'll get you down a little lower."

4:10:04 FO: "My (deleted), the other engine's going, too. (deleted)"

4:10:05 S242: "Got the other engine going, too." [Pilot is not clearly communicating that the second engine has failed.]

4:10:08 AC: "Southern Two-Forty-Two, say again."

4:10:10 S242: "Stand by—we lost both engines."

4:10:14 FO: "All right, Bill, get us a vector to a clear area."

4:10:16 S242: "Get us a vector to a clear area, Atlanta."

4:10:20 AC: "Uh, continue present southeastern-bound heading. TWA's off to your left about 14 miles at fourteen thousand and says he's in the clear."

4:10:25 S242: "OK."

4:10:27 S242: "Want us to turn left?"

4:10:30 AC: "Southern Two-Forty-Two, contact approach control on one-two-six point nine. They'll try to get you straight into Dobbins."

4:10:35 S242: "One-two—"

4:10:36 FO: "I'm familiar with Dobbins; tell them to give me a vector to Dobbins if they're clear."

4:10:38 S242: "Give me, uh, a vector to Dobbins if they're clear."

4:10:41 AC: "Southern Two-Forty-Two, one twenty-six point nine. They'll give you a vector to Dobbins."

4:10:45 S242: "Twenty-six nine, OK."

4:10:50 FO: "Ignition override—it's gotta work by (deleted)."

4:10:56 CAM: (Power interruption for 2 minutes, 4 seconds.)

4:13:00 CAM: (Power restored.)

4:13:03 Capt: "There we go."

FO: "Get us a vector to Dobbins."

4:13:04 S242: "Uh, Atlanta, do you read Southern Two-Forty-Two?"

4:13:08 AC: "Southern Two-Forty-Two, Atlanta approach control. Uh, go ahead."

4:13:11 S242: "Uh, we've lost both engines. How about giving us a vector to the nearest place? We're at seven thousand feet."

4:13:17 AA: "Southern Two-Forty-Two, roger. Turn right heading one-zero-zero. Will be vectors to Dobbins for a straight-in approach to Runway One-One. Altimeter two-niner-five-two. Your position is 15—correction, 20 miles west of Dobbins at this time."

4:13:18 FO: "What's Dobbins' weather, Bill? How far is it? How far is it?"

4:13:31 S242: "Okay, uh, one-forty heading and 20 miles."

4:13:35 AA: "Make a heading of one-two-zero, Southern Two-Forty-Two. Right turn to one-two-zero."

4:13:40 S242: "Okay, right turn to one-two-zero. Uh, you got us our squawk, haven't you, on emergency?"

FO: "Declare an emergency, Bill."

4:13:45 AA: "Uh, I'm not receiving it. But radar contact—your position is 20 miles west of Dobbins."

4:14:03 FO: "Get those engines (deleted)."

4:14:24 S242: "All right, listen, we've lost both engines, and, uh, I can't tell you the implications of this. We, uh, only got two engines. How far is Dobbins now?"

4:14:34 AA: "Southern Two-Forty-Two, uh, 19 miles."

4:14:40 S242: "OK, we're out of, uh, fifty-eight hundred. 200 knots."

4:14:44 FO: "What's our speed? Let's see, what's our weight, Bill? Get me a bug speed."

4:14:45 AA: "Southern Two-Forty-Two, do you have one engine running now?"

4:14:47 FO: "No."

4:14:48 S242: "Negative, no engines."

4:14:50 AA: "Roger."

4:14:59 Capt: "One twenty-six." [Probably referring to the final approach speed.]

FO: "One twenty-six."

4:15:04 Capt: "Just don't stall this thing out."

FO: "No, I won't."

Capt: "Get your wing flaps."

CAM: (Sound of lever movement.)

4:15:11 FO: "Got it, got hydraulics, so we got...."

Capt: "We got hydraulics."

FO: "What runway? What's the heading on the runway?"

S242: "What's the runway heading?"

AA: "Stand by."

S242: "And how long is it?"

AA: "Stand by."

Capt: "Like we are, I'm picking out a clear field."

FO: "Bill, you've got to find me a highway."

Capt: "Let's get the next clear open field."

FO: "No (deleted)."

Capt: "See a highway over—no cars."

FO: "Right there, is that straight?"

Capt: "No."

AA: "Southern Two-Forty-Two, the runway configuration—"

FO: "We'll have to take it."

AA: "—at Cartersville is three-six-zero, running north and south. The elevation is 756 feet. I'm trying to get the length now—it's 3,200 feet long."

CAM: [Beep on gear horn.]

CAM: [Gear horn steady for four seconds.]

S242: "Uh, we're putting it on the highway, we're down to nothing."

FO: "Flaps."

Capt: "They're at fifty."

FO: "Oh (deleted), Bill, I hope we can do it."

FO: "I've got it, I got it."

FO: "I'm going to land right over that guy."

Capt: "There's a car ahead."

FO: "I got it, Bill. I've got it now. I got it."

Capt: "OK."

Capt: "Don't stall it."

FO: "I gotta bug."

FO: "We're going to do it right here."

FO: "I got it."

CAM: [Sound of breakup.]

End of tape.

The Crash

Flight 242 crashed on a road and slid for roughly 1,830 feet before colliding with a convenience store/gas station. Sixty-two people on board the aircraft died, including the cockpit crew. Eight people on the ground also perished. Remarkably, 21 passengers survived, along with the flight attendants.

According to one of the surviving passengers, who held a commercial pilot certificate, the flight was routine until the aircraft encountered severe turbulence, followed by very heavy precipitation, a lightning strike on the left wingtip, and hail. The hail increased in size and intensity, then the right engine quit, and the left engine quit shortly thereafter. Coinciding with the turbulence, power was reduced on the engines. The passenger estimated that the turbulence lasted one to two minutes, the heavy rain continued for 45 seconds, and the hail for about the same amount of time. The engines quit just after the hail stopped. Shortly after that, the auxiliary power unit (APU) was started. The two flight attendants had similar recollections.

The Crew

The captain had been flying for Southern Airways since 1960 and had over 19,000 hours of flight time, with about 3,200 hours in the DC-9. However, he had only recently been upgraded to captain a few months prior to the accident.

The first officer had been with Southern for four years, with 3,800 hours of total flight time and just over 200 hours in the DC-9.

Aftermath

According to the post-accident analysis by the NWS, the storm system was one of the worst on record, spawning 20 tornadoes and 30 severe thunderstorms. It was clearly not business as usual. According to the radar plot shown above...

Flight 242 went through the line very close to the largest thunderstorm cell.

Never before had an air carrier aircraft had both engines snuffed out by a

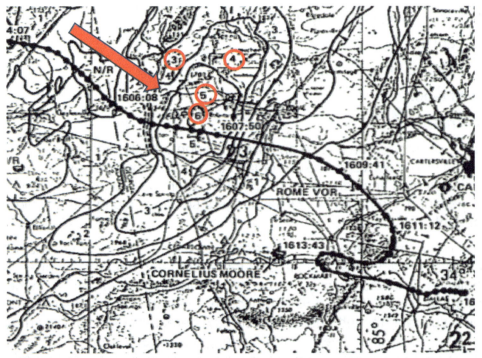

Flight path is black line, red arrow shows tight gradient. The VIP numbers 3,4,5,6 correlate to Moderate, heavy, very heavy, extreme – The flight flew directly into the worst part of the storm!, NTSB

thunderstorm. The probability was considered so remote that the aircraft flight manual had no guidance on what to do if a DC-9 became a glider. The maximum gliding distance from 14,000 feet was estimated at 34 miles. The crew managed to coax about 32.5 miles out of the aircraft, a remarkable performance for an untrained maneuver.

The board surmised that shortly after the loss of both engines, the flight may have entered visual conditions. The pilots, busy attempting to restart both engines and the APU to restore electrical power, chose to stay visual. This decision may account for the 180-degree turn westbound.

The crew had no information on the storms west of Rome. Initially, the captain decided—based on his airborne weather radar returns—that the storms were too severe to penetrate. However, for some reason or due to a misinterpretation of the radar display, he changed his mind.

In the transcript, the first officer appears to take more command of the situation than the captain. He is hand-flying the aircraft and instructing the captain that they need vectors to an open area.

NTSB Probable Cause:

The total and unique loss of thrust from both engines occurred while the

Engine Damage, NTSB

aircraft was penetrating an area of severe thunderstorms. The loss of thrust was caused by the ingestion of massive amounts of water and hail, which, in combination with thrust lever movement, induced severe stalling and damage to the engine compressors.

Major contributing factors included the failure of the company's dispatching system to provide the flight crew with up-to-date severe weather information regarding the aircraft's intended route of flight, the captain's reliance on airborne weather radar for penetration of thunderstorm areas, and limitations in the FAA's ATC system, which precluded the timely dissemination of real-time hazardous weather information to the flight crew.

Board member Francis McAdams dissented, stating that the majority belief "…was merely a statement of what happened rather than an explanation of why." His determination of probable cause is, in my view, more accurate. He referenced the crew's conversation about what was showing on the airborne radar just prior to penetration: "This accident involves the captain's critical decision to penetrate rather than avoid a known area of severe weather…. It is obvious that the captain flew a route, or directed the FO to fly a route, into an area which the aircraft should not have entered. Southern Airways, and all air carriers, prohibit flying into convective storms because these types of storms are known to be serious hazards. The primary hazard relates to forces in these storms which can destroy an aircraft structurally; however, other hazards exist which are not well defined…. The loss of thrust…might have been unusual…but should not be considered entirely unexpected, given the multiple hazards associated with flight into severe convective storms."

Commentary

This accident reminds us that onboard weather radar can sometimes lead to dangerous situations. The newer onboard radars are much better, but they still have limitations. When used in conjunction with Nexrad, there's really no reason to get caught unprepared. The NTSB accident report notes: "The system could display targets at three range selections—30 miles, 80 miles, and 180 miles. The system was designed to display weather in two modes: normal and contour. In the normal mode, precipitation is displayed as luminescent areas on the dark background of the indicator. In the contour mode, the areas of heavy precipitation are electronically eliminated to produce a dark hole (contour hole) surrounded by luminescent areas of lighter precipitation. Southern Airways and the manufacturer's operating manual state that contour holes should be avoided by at least 10 miles. Additionally, any weather displayed beyond a range of 75 miles indicates areas of significant rainfall, regardless of the presence or absence of contour holes, and should be avoided."

The captain and the dispatch system could have anticipated that with storms moving that fast and growing that tall, a 20- to 30-mile deviation to the south would have been lifesaving. The Southern Airways captain failed to recognize the explosive growth of what had been a benign line of weather just two hours earlier and didn't fully utilize the resources available to him.

This accident led to the creation of Center Weather Service Units (CWSU), staffed by meteorologists from the National Weather Service to advise both ATC and pilots on the nature of severe weather. Today, there is renewed emphasis on weather, and the CWSU may be replaced with a different system. With Nexrad, both for ground and airborne observation, there should be fewer surprises like this, but when Mother Nature goes vertical, we need to stay far away—horizontally!

CHAPTER 5
TECHNOLOGY, MAGENTA LINES, AUTOMATION AND BLACK BOXES

"Somewhere in the heavens, there is a great invisible genie who, every so often, lets down his pants and pisses all over the pillars of science."

—Ernest K. Gann, airline pilot, author of many aviation classics

Today's aircraft are remarkably better than our early attempts. As noted in Chapter 2, regarding human failings, there's often a technical fix that's much easier to implement than changing human behavior. For example, the single biggest safety improvement in airliners was the arrival of jet engines. The introduction of tricycle landing gear greatly reduced landing accidents. These aren't examples of automation, but they demonstrate the value of technology.

Tech Improvements — Reciprocating (piston) engines are reasonably reliable, but they have many moving parts, often working in opposite directions. The basic design was developed over 100 years ago. These powerplants have received little attention from either manufacturers or the FAA, largely due to economics. Small production numbers and product liability contribute heavily to the cost — which raises the chicken-or-egg dilemma. As we'll discuss in Chapter 6, *Emergencies*, piston engine failures result in the second-highest number of GA (general aviation) fatalities annually, in addition to massive insurance and lawsuit expenses. This is an area that needs more attention. Engine redesign is expensive, but so are lawsuits. There are no easy answers.

Smokey, Oily, lots of moving parts

Smooth Steady Power

Jet engines are much simpler, with everything mostly going in the same direction. The airline safety record improved dramatically as airliners became turbine-powered, but not because pilots suddenly became miraculously better. It was the jet engine. Not only was the engine more reliable, but it also expanded the altitude and speed envelope, which helps immensely with weather avoidance. Training also improved tremendously with the advent of simulators.

In the beginning, airports didn't have runways; they were just open fields, so all takeoffs and landings could be made directly into the wind. This solved the pesky crosswind problem. While purists may scoff at anyone who doesn't fly a tailwheel aircraft, the landing gear geometry greatly favors nosewheels. They are much easier to handle on the ground and won't ground loop easily. Ham-fisted or footed pilots still botch tailwheel operations, and in the early 1990s, the FAA decided that specialized training and a CFI's endorsement would improve safety. Nosewheels have largely replaced tailwheels, except for specialized aircraft, classics, kits, and sport planes. Ask any insurance company about their loss experience with tailwheel aircraft. The typically higher premiums reflect the higher level of skill required. That doesn't mean you shouldn't learn—just respect the geometry.

Fuel mismanagement has also significantly decreased. Range optimism, where there's never a fire at the crash site (exhaustion), and the inability to get fuel into the engine, even though there's plenty available (starvation), have been greatly reduced in newer aircraft. Fuel consumption monitoring has improved with digital flow transducers. Warning lights, independent of pilot misprogramming or wishful thinking, typically advise 30 minutes before quiet ensues. Automobiles have had fuel warning lights for decades, long before light aircraft, but the aviation industry finally caught up. Should we wonder why?

What does all of this have to do with automation? It's part of a growing technological prowess, and in the case of jet engines and many turboprops, they are now highly automated, with auto-throttles, flight control systems, and temperature management controls. Auto-throttles, while implicated in a few crashes where pilots didn't understand how the system worked (a training AND design issue), greatly simplify the flying task. Likewise, fuel management has become automated on many aircraft. Simpler is better, but if the system malfunctions, which is rare, the pilot must quickly call up seldom-used skills.

> *"Every complex problem has a solution which is simple, plausible—and wrong."*
>
> — *H.L. Mencken, journalist, critic, curmudgeon*

The Technology Conundrum

There are two sides to this argument, which will irritate those with strongly held opinions. Few people get lost these days with the advent of GPS. Knowing the distance and direct bearing to every airport makes life easy—much easier than using four-course ranges, NDBs, pilotage, or even VORs.

After seeing Horizontal Situation Indicators (HSI) and Flight Directors (FD), my respect for airline and bizjet pilots diminished somewhat. Flying the six-pack and scanning multiple instruments in proper sequence, with no autopilot, kept the workload high in primary IFR trainers of yore. As an instrument instructor, I spent hundreds of hours teaching students to develop the "moving map" in their heads. Today, it's so much easier to fly through cloudy skies. The consolidation of navigation and primary flight displays made scanning and interpretation a breeze. Add in terrain and weather avoidance, and task loading goes way down, provided the data are presented intelligently.

Pilots still need to decide what to do with all that information, but it's all there in today's glass cockpits. The glass can be complex and requires effort to master. With all the eggs in just a few baskets, the ability to revert to back-ups is still a skill to develop and maintain. How long that will remain is an open question, and we Luddites who learned all the old skills will forever denigrate those who have it easier. However, the glass evolution has proven its ability to simplify location awareness. Additionally, autopilots have become extremely capable.

Sometimes, there's too much data and not enough useful information cluttering both the display and the thought processes. That's a design issue, but if that's the equipment you're flying, guess who's responsible?

Integrated flight display, Aspen Avionics

The 6-pack of yore, FAA

First generation HSI, Bendix King

Automation: Love and Hate

Simple automation is extraordinary in reducing workload while allowing the pilot to concentrate on higher-order tasks, yet it's essential to helps retain core skills when needed. A basic two-axis autopilot provides much of the benefit with a fraction of the complexity of Flight Management Computers or Systems (FMC/FMS). Single-pilot hand-flying for prolonged periods is tiring, and in large aircraft, it consumes additional fuel. At high flight levels, hand-flying becomes more challenging, as the stall/overspeed margin is narrow.

Century IV Simple Autopilot, Mitchell Flight Control Systems

The FAA requires an autopilot for Part 135 single-pilot operations. Many experienced aviators also consider the autopilot a no-go item for Part 91 instrument or high-density operations. When you're busy programming, negotiating with ATC, assessing weather, and dealing with other high-workload situations, it's nice not to have to worry about course and altitude, as the basic autopilot handles all of that. It provides an additional margin of safety.

Automation works spectacularly well most of the time, which makes it harder to stay mentally engaged. The technical term for this is "task underloading." A high comfort level creates a Titanic-like thought process. There are endless hypothetical arguments to the contrary, but millions of hours are flown annually using automation. The safety record is better with technology, despite some of the exceptions you'll read about here. Traffic

Collision Avoidance Systems (TCAS), ADS-B traffic, and Terrain Avoidance Warning Systems work most of the time, but they're not perfect. It's a work-in-progress, even though we've been at it for decades.

As automation becomes ever more capable, pilot complacency and the inability to override a malfunction without extra effort and practice is a growing problem in partially or nearly completely automated aircraft without superior reliability. However, extremely high levels of reliability are coming with the advent of advanced operating systems. Air carrier Autoland for Category II and III landings (extremely low or no ceiling or visibility) are routine today, but special training, airport and aircraft equipment, and recurrent certification are required. I'm not aware of any Category II or Category III crashes, but extraordinary measures are taken to ensure that it all works in certification, with extensive and recurrent training.

> *"You know you've achieved perfection in design, not when you have nothing more to add, but when you have nothing more to take away."*
>
> — *Antoine de Saint-Exupery, pilot, author*

The level of attention needed for these landings is limited to only a few minutes of intense concentration, not to the hours that many flights demand. A mitigating factor is that, while enroute, the aircraft is a long way from the ground, so presumably, there's time to intervene and re-corral the beast if it escapes. That didn't work in the Air France 447 crash (When Nothing Makes Sense, this chapter).

Complex and Non-Standard Designs –

Our personal computers, smartphones, and appliances can be irritating, overly complicated, inconvenient, and poorly designed, but in avionics (and automotive infotainment systems), those nuisances become potentially deadly. Despite rigorous certification standards, the FAA currently hasn't figured out if it should, or even if it can, standardize core avionics functionality. The avionics manufacturers have been largely left to their own devices, and now that we're in about the fifth GPS generation, the systems have become much more user-friendly, but they're all different, and quirks still abound.

The avionics Tower of Babel collapse, where there is no longer a standard pilot interface, creates significant learning and training challenges. Back in the avionics dark ages, one could move from one aircraft to the next by understanding VOR and ADF. They weren't that complex, and if you could fly one VOR or ADF, you could fly them all. Today, a pilot consistently flying one FMS can develop proficiency, but when switching between aircraft with different systems, it becomes sub-optimal at best and dangerous at worst. Programming logic is sometimes an oxymoron. It also complicates life tremendously for simulator manufacturers and instructors.

The FMS that controls flight and navigation functions on legacy units can be so complex that understanding programming and failure modes is a big deal. Older FMS and GPS avionics with less intuitive interfaces than newer designs will continue to trip up unsuspecting fliers for years because installed equipment has a long economic lifespan. There's no easy fix there—you'll just have to learn it.

One of the most complex GA units was the Apollo/UPS-AT/Garmin GNS 480. This very advanced,

Garmin GNS 480, Garmin

now legacy system, which Garmin acquired, was based on the Boeing 757 FMS. Once learned, it was very good, but the learning curve was steep. The airlines claimed to spend about a week teaching their pilots on the FMS, using part-task trainers as well as their big simulators. But some avionics marketing types said that the average GA pilot could learn it all in a day. Being well below average at the time, I mastered it in six months, flying about 10-15 hours per month with some serious home study. No simulator existed for it back then.

The latest generation of equipment is much more user-friendly, with alphanumeric touchscreens and mostly logical page layouts and flight planning. But newer equipment has hundreds of thousands, or even millions, of lines of software code, where bugs invariably hide and sometimes manifest themselves in ugly ways.

Touchscreens - Blessing or curse? That depends on how big the screen area is and whether turbulence is a factor. Bigger is better, both for font size and symbols, to improve readability and the ability to touch the right part of the screen when the atmosphere is bouncy. I've had multiple occasions where an ill-timed bump resulted in my switching frequencies or misprogramming and having to go back and do it again.

The other aspect is the longevity of electromechanical instruments versus electronic ones. Personal experience has shown that altimeters, airspeed indicators, and some of the old-style navcoms can soldier on for decades, while the more complex electronic versions might last several years before needing an expensive fix. It all depends, and perhaps someone has done a study to prove me wrong.

Pilot Engagement - Situational awareness, or SA, is lost when humans drop out of the loop, intentionally or otherwise. When the systems control everything, the pilot is under-loaded, and the mind begins to wander. Fatigue, complacency, complexity, inattention, and unfamiliarity compound the problem. The NTSB's term for this is "Automation Complacency."

New avionics perform in ways that were unimagined just a few decades ago. But until the "magic" is completely reliable—whenever that happens—pilots must learn at least some of the earlier disciplines. Or there needs to be a completely independent backup that allows immediate reversion to continue flight safely. We may not be too far from that. Advances in hardware and software have brought Autoland and pilot incapacitation modes into reality for higher-end aircraft. Light aircraft, where these features are really needed, will be the last to get the upgrades, as usual, because of economics.

If the automation experiences a critical meltdown, you're at the mercy of a faceless programmer who never anticipated the conditions that may lead to your personal disaster, especially if there are no suitable backups. The dire prediction of a bloodbath, where GA pilots get into IMC and are unable to program complex units, hasn't occurred. There are occasional mishaps, but not the doomsday scenario that many feared. However, NASA's Aviation Safety Reporting Program (ASRP) reports are full of vertical and horizontal wanderings, proving that completely abdicating the PIC role to electrons is a bad idea.

The equipment fails far less than humans, but it's an "aw shucks" moment when there's a bug and no viable Plan B. Several major airline accidents have shown that high levels of automation, coupled with a loss of basic flying skills, can be catastrophic. We'll explore a few of those here.

"What's it doing now? Look, it's doing it again."

Reportedly the two most oft-heard pilot comments regarding advanced avionics.

Airbus Antics –

Air carrier automation accidents have cost hundreds of lives and billions of dollars. Airbus, which builds marvelously easy-to-fly aircraft, has had several major miscues where the automation played a direct part in confusing the crew to the point that aircraft were lost.

Airbus mushing into the trees - BEA

Airbus cockpit, Airbus

One of the first major accidents involving automation occurred in 1988, as Airbus was establishing itself as a major competitor to The Boeing Company. A new Air France Airbus A320 was being demonstrated during the Habsheim Airshow (LFGB). It was the aircraft's first flight with passengers and the first public exhibition of civilian fly-by-wire technology. The aircraft was designed with a flight protection mode that would not allow a stall, and with auto-throttles programmed for approach, it was supposed to fly right to touchdown.

Using automation, the plan was to slow the aircraft to its minimum speed and disable the automatic angle-of-attack (AOA) protection system, which would normally increase power when the AOA reached 15 degrees. The first officer was to manually increase engine thrust to maintain altitude and then apply full power for a steep climbing turn as the end of the runway approached. The captain reported that he had performed this maneuver 20 times before.

A low-speed, gear-down flyover was planned at 100 feet AGL. Instead, the Airbus descended to just 30 feet and struck trees at the end of the runway. While most occupants and the crew escaped, three fatalities occurred. The official French investigation faulted the pilots for being too low, too slow, and reaching maximum AOA with the engines at flight idle and power applied too late. Both pilots were highly experienced, with more than 10,000 hours of flight time each. The captain, as the technical pilot in charge of the A320 training division, had been involved in testing the aircraft and its systems for more than a year.

Since the airport was not listed in the flight computer, a visual approach was begun only six miles from the airport. The approach was still not stabilized when the aircraft reached the runway threshold. A late runway change, due to the spectators' location, added further distraction. As they approached the end of the runway, the first officer applied takeoff power, but the aircraft did not respond as expected, remaining in a flight protection mode.

The flight data recorder (FDR) showed that the engines performed within their design specifications, going from flight idle (about 29 percent, according to the BEA) to 83 percent in about five seconds, although the crew contended that they did not respond as expected. A TV documentary later claimed that the FDR had been tampered with and suggested that a computer malfunction had unfairly implicated the pilots. Airbus, predictably, issued a detailed rebuttal, and the official report stands.

Ironically, the purpose of the low pass was to demonstrate that the aircraft was smarter than the pilots and would not allow a stall. The entire accident sequence—from midfield to impact—took only about 15 to 20 seconds. It was not an auspicious beginning for the fly-by-wire era. However, in fairness, major advances often come with missteps—that's how aviation progressed in the first place, with Icarus leading the way. When exercising caution and maintaining a firm belief in Murphy's Law (if something can go wrong, it will), the odds improve. Automation miscues are far fewer now than they were two decades ago. The current generation of "the Bus" is much improved from the early models as the bugs are gradually removed, but it remains an ongoing process.

The Max Mess - (B737 Max critical system fails twice)

Where economics attempted to override engineering. It didn't work.

Boeing doesn't come through unscathed either. The Boeing 737, the world's most popular airliner, had been in continuous production since the 1960s. Over the years, it became bigger, more fuel-efficient, and much more sophisticated. But when two nearly new Boeing 737 Max 8 aircraft crashed within months of each other—Lion Air Flight 610 (Indonesia, October 29, 2018) and Ethiopian Airlines Flight 302 (March 10, 2019)—fingers pointed in every direction: the manufacturer, the FAA, the pilots, and the maintenance techs. On November 8, 2018, the FAA issued an Emergency Airworthiness Directive. The Maneuvering Characteristics Augmentation System (MCAS) was identified as the source of the problem, triggered by bad angle-of-attack (AOA) sensor input.

AOA sensors are generally reliable, and transport-category aircraft typically have at least two. However, Boeing decided that only the left-side AOA would control MCAS, introducing a single point of failure.

AOA Sensor, Boeing

There was an optional "mis-compare" annunciation system to alert the crew if the left and right sensors disagreed, allowing a quick assessment and fix. Neither of the accident aircraft had this feature installed.

B737 Max, The Boeing Co.

B737 Max Flight Deck, The Boeing Co.

Why Was MCAS Needed?

Competing against its archrival Airbus, Boeing sought to improve the 737's fuel efficiency while avoiding significant certification expenses and minimizing transition training costs for airlines. Larger, more fuel-efficient engines were mounted farther forward on the wing because they wouldn't fit underneath. This change altered the thrust vector and affected the aircraft's handling in certain flight regimes.

No problem—just add a "simple" software patch to automatically adjust the AOA under specific conditions. The big selling point? Minimal transition training for pilots and no required simulator time. Economics remains the driver of flawed decisions.

To make the Max handle like earlier 737 models, MCAS was designed to activate only within a narrow part of the flight envelope to prevent pilots from over-pitching. It was disabled when flaps were deployed or when the autopilot was engaged. Given these limitations, Boeing decided there was little need to inform or train pilots about MCAS—an oversight that, in hindsight, proved disastrous.

In March 2019, just days after the crash of ET302, the FAA grounded the Max. Boeing subsequently shut down production for over a year. What was supposed to be a simple fix for a minor aerodynamic quirk turned into a major crisis.

A "Black Swan" Event?

The first accident was somewhat of a "Black Swan" event—unexpected but not entirely unpredictable. In both crashes, the aircraft had just taken off with a malfunctioning or damaged AOA sensor. Shortly after takeoff, MCAS, believing the aircraft was pitched too high, automatically re-trimmed to lower the nose. If the pilot activated electric trim, MCAS would deactivate—but only for a few seconds before re-engaging to lower the nose again.

In any design, once a system is disabled, it should stay disabled. In both accidents, the pilots initially used electric trim but were bewildered by what was happening—they had never been trained on MCAS.

> *"Generally, you don't see that kind of behavior in a major appliance."*
>
> — *Dr. Peter Venkman, Ghostbusters*

Boeing had introduced a serious flaw into a well-proven product. The FAA, in turn, mismanaged the certification process. Complacency affects organizations just as it affects individual pilots. Boeing, which had built more jet airliners than any other company, assumed the Max was just another 737 upgrade. The FAA, relying on Boeing's track record, followed a well-worn certification path. MCAS was seen as a minor modification—though some knowledgeable voices within the industry raised concerns that were ultimately overridden.

Emergency Airworthiness Directive

Following the Lion Air crash, in November 2018—four months before the Ethiopian Airlines crash—the FAA issued an Emergency Airworthiness Directive (AD). Boeing also sent an Operations Management Bulletin to all 737 Max operators outlining the procedure for an MCAS malfunction. However, the Ethiopian Airlines crew apparently never saw it.

That crew never reduced power and did not follow the emergency checklist, striking the ground at well over never-exceed speed. This raised many questions about pilot training, system redundancy, and, most importantly, whether the pilots even understood what was happening. After the emergency AD was issued, all Max pilots should have been on high alert and better prepared.

A Whitewashed Investigation

The official report from the Ethiopian Accident Investigation Board (EAIB) did not address human factors or organizational issues. It listed nine contributing factors but omitted any discussion of the crew's response or training. The airline's management should also have been implicated for failing to notify pilots and the training department about the AD.

Additionally, strong evidence suggested a bird strike had destroyed the left-side AOA vane, triggering MCAS. However, this detail was also absent from the final report. Flight Data Recorder analysis closely matched bird strike signatures from other crashes. Ironically, the EAIB had previously investigated an incident at Addis Ababa airport and recommended addressing the well-known bird concentration problem. None of that made it into the final report.

The NTSB filed dissenting comments, which the EAIB ignored, despite ICAO requirements to consider them. The French BEA, which downloaded the flight recorders, concurred with the NTSB's assessment and also submitted a dissenting opinion. This case underscores the importance of having an independent body conduct crash investigations, separate from government transportation departments and state-owned airlines.

"If ever there was an Icarus moment, the second Max crash illustrates it."

Author

Training for the unexpected –

The NTSB board members were invited by Boeing to fly a 737 MAX simulator to observe both the original crash scenario and the post-accident software modifications. We received a 30-minute pre-brief on what to expect and how the system worked.

Stabilizer Trim Cutout switches on rear of 737 pedestal, Boeing

Boeing's memory checklist: Set a pitch attitude of 10 degrees nose-up, reduce power to 80 percent, use electric trim to stabilize, activate the dual stabilizer trim cutouts, and then hand-fly the aircraft. The MCAS malfunction was completely controllable, even before the software fixes. Manual re-trimming was required, but pilots should have been able to manage that.

A key point: In our demonstration, there was no confusion—we had been trained and fully expected the system to behave erratically. Had the Ethiopian crew received the required training, as outlined in the emergency Airworthiness Directive (AD), the second crash likely would not have happened.

Design Challenges

Once the emergency AD and Operations Bulletin were issued to alert all MAX operators of the issue, Boeing and the FAA believed this would be sufficient while a redesign was in progress. The idea was simple: If MCAS malfunctioned, just follow the memory checklist. But the Ethiopian Airlines Flight 302 crew never received that training.

When an aircraft's flight control system begins misbehaving, reducing levels of automation quickly often helps. Ironically, in this case, turning **on** the autopilot would have disabled MCAS. The ET302 crew eventually engaged the autopilot, but by that point, the aircraft was unrecoverable.

The very concept of flight envelope protection failed catastrophically in this instance. The pilots were bewildered and did not understand what was happening—especially given MCAS's persistent nature. They also did not respond in the way that Boeing and the FAA had assumed they would. The expectation was that, within about 10 seconds, the crew would correctly diagnose the problem and take appropriate action. That assumption was far from reality.

The NTSB issued several recommendations urging the FAA to take human factors into serious consideration when designing complex aircraft systems. Pilots can quickly become overwhelmed by cascading alarms as systems or sensors fail. While automation solves some problems, it also creates others.

"New" Thinking

There are several key takeaways from this tragedy:

- Transport-category aircraft should never have single-point failures in critical systems. This principle also applies to light aircraft, within the bounds of reasonable complexity and economics.

- Unambiguous annunciation is essential to prevent overwhelming pilots with multiple alarms from highly automated systems. Tell the crew the most critical problem and its root cause—not all the derivative issues.

- Designing errors out of the system is the best way to improve safety. While training is always necessary, reducing the potential for human error minimizes the reliance on pilots to be fail-safe.

Boeing's decision not to disclose MCAS's existence to operators—and to provide only limited information to the FAA—is inexcusable. The FAA also bears responsibility for failing to ask the right questions. Anything that interacts with the flight control system is critical, and pilots should be thoroughly trained. A simple 15-minute simulator session on MCAS would have been sufficient, but that would have undermined Boeing's marketing pitch and increased costs.

The final tally for fixing the 737 MAX exceeded $20 billion—and, more tragically, 346 lives were lost.

Another 737 Safety Mishap

A few years later, while Boeing was already under scrutiny for design and quality control, another incident occurred: an improperly secured door plug blew off a newly delivered MAX 9. (Keep that seatbelt on!) Fortunately, there were no serious injuries, but the aircraft had just come out of a Boeing plant, and proper safety protocols were apparently not followed.

Management in any manufacturing organization must prioritize engineering and quality control. Financial concerns should be secondary. Accountants, CEOs, and some shareholders might disagree, but if the product doesn't perform well, neither will the balance sheet.

It's easy to criticize Boeing after these events, but it's also important to remember that thousands of Boeing airliners fly safely every single day. Aircraft manufacturing is an incredibly complex business. If we demanded similar human performance standards in other high-stakes industries, Boeing might not look quite so bad.

That said, these Boeing failures should become landmark case studies in business, aviation, and engineering schools—examples of how not to do things.

Uncommanded but Never Unexpected

Before leaving this topic, anyone who flies with an autopilot or electric trim should be well-versed in handling uncommanded trim movement—also known as "runaway trim" or an autopilot "hard-over."

Procedures may vary slightly by aircraft, but the objective remains the same: **when the system behaves erratically, shut it down.**

Trim wheel marked to show when in motion, Author

The NTSB recommended decades ago that manufacturers mark the manual trim to make it easier to see when in motion. That never happened—but you can do it yourself using electrical tape or contrasting paint from the "aviation aisle" at your local home improvement store.

Typically, there is a large red button on the pilot's yoke that, when pressed and held, will deactivate both the autopilot and the electric trim. However, the trim may re-engage once the button is released. The next step is to pull the electric trim circuit breaker, which should be collared to make it easily identifiable among the others, preventing any fumbling.

Shut off the autopilot and manually fly the aircraft (the horror!). This is a bold-print memory item for every aircraft equipped with such systems.

Additionally, do you know the minimum altitudes during climb-out and descent where the autopilot can be safely engaged or disengaged? If the system malfunctions, quick action and sufficient altitude are the only things that can save the situation.

Collared circuit breaker for quick ID, author

Note the autopilot (A/P) circuit breaker (CB) located next to the pitch trim CB. In the highly unlikely event that shutting down both the trim and autopilot doesn't resolve the issue, immediately inform ATC that you'll be temporarily out of communication and turn off the master switch.

In an all-electric aircraft, this action will transfer primary flight displays to battery power. If you're becoming task-saturated, notify ATC so they can help reduce your workload.

Software SNAFUs

Sometimes, systems that have performed perfectly go wacko when an update is made. Collins Aerospace GPS units in flight management computers (FMC), used in multiple high-end business jets and several airliners, failed and caused hundreds of flights to be canceled or rerouted. It affected transponder functions and collision avoidance systems, which are required to operate above Flight Level 280. GPS functions were degraded, forcing those who still elected to fly to revert to using VORs. A software upgrade was the problem. Sometimes, one manufacturer's change doesn't play well with another designer's subsystems, even though they have coexisted for years. How many times have you installed a software update to your computer and had some significant malfunctions? It happens regularly.

In aviation, low-probability, high-consequence events are the bane of safety engineering. Glowing promises from manufacturers, such as "There's never been a failure of a HAL 9000 series computer" (from the movie *2001: A Space Odyssey*), are to be taken with some skepticism. However, as computers and systems become more reliable and software design improves, there will be near-perfect safety.

Go Down in Levels of Automation

American Airlines learned from the Cali, Colombia, crash in 1995 (High Terrain Tangle, This chapter), and professional crews have recognized that sometimes the "magic" can increase the workload significantly. American Airlines, and others, now teach pilots to go down in levels of automation as flight conditions change from the original plan. If the flight management system (FMS) is flying properly and everything is proceeding as planned, life is good. Leave it alone, but verify. As soon as ATC changes something or the software does something unexpected, it's time to assess whether to salvage the automation or go to manual mode.

Sometimes, it takes longer to reprogram than to just fly the aircraft and get pointed in the right direction. Go to heading mode and ask for a vector if many changes are needed. Be wary of nearby high terrain or obstacles. A change close to the airport, when everything is set for arrival or departure, is a guaranteed distraction when it's least desired. There's a strong tendency to think "I'm busy but can handle this" until it devolves into "Yikes, I'm overloaded!" Ask for help. That's not weakness—it's smart.

But for every rule, there is usually an exception. The two MAX crashes probably would not have occurred if the automation had been used. That's similar to crashes in VFR into IMC sections (Weather Chapter, Vineyard Spiral, and Calabasas). If you are beginning to underperform (hard as that is to admit) and there's no known problem with the automation, use it at its most basic level, i.e., heading and altitude. In many newer autopilots, pushing the blue button will bring the aircraft to straight and level flight. Once things are stabilized, advise ATC that you need some help and calmly unwind the mess.

Less is More

A microprocessor may have the ability to deliver additional functions, but that doesn't mean another layer of complexity should be added to satisfy the gadget freaks among us or someone's "great" marketing idea to make the unit more "capable." Another ten thousand or hundred thousand lines of programming code come with massive mayhem potential. Until the failure modes are well known and corrected, simpler is generally safer. We may get to that point, but for now, go for the simplest solution.

Too many times, when trying to make something just a little easier or to "integrate it," the execution suffers. In today's automobiles, it often takes multiple pages on an MFD just to adjust the heater or air conditioning. We used to be able to just turn one knob—how's the new approach better?

Links in the Chain - Thumbnails of automation acting up.

> *"No amount of testing can prove software right, but a single test can prove software wrong."*
>
> —*Amir Ghahrai, Sr. Test Consultant, University of Southampton*

Here are some real-world examples of pilot misunderstanding and possible software slip-ups that did not result in a crash. These NASA ASRS reports are only the tip of the automation miscue iceberg. Think of automation as the flying crew member. Effective monitoring requires that instructions are understood and executed correctly. There are two possibilities when things don't work: Either we misprogrammed the box, or it's taking us electronically where no one has gone before. Either way, it's got to be addressed.

Possible Software Corruption:

A Cirrus SR22 pilot reports: "In cruise, while I was IMC on the edge of icing conditions, most functions on my PFD failed. I had no comms for 3-4 minutes, no transponder, etc. All functions had an 'X' through them on the PFD. Shortly thereafter, I lost the moving map on the MFD of the G1000. My autopilot continued to operate, and I initiated a 180-degree turn back to [a suitable airport]. My COMM came back up, and I was able to take vectors to the field. Warning sounds continued throughout all of this. I continued toward [the airport] where the ceiling was approximately 800 ft. I loaded the ILS approach, and it seemed to load as I got a glideslope. The Perspective [the Cirrus branding for the Garmin panel] seemed to be working. I still didn't have a map on the MFD. The terrain awareness system started going off as I was on the approach, telling me to pull up. I landed safely.

I used a [Macintosh computer] to do a recent update. I may have some corrupt files as a result of using a Mac to update the data cards. In looking at the cards, there were indeed two corrupt files on the cards that essentially caused the G1000 to overload and come close to locking up, which is why I lost the functions I lost. It was trying to 'fix' the corrupt files. I have been flying the Cirrus for 2 years and had a Columbia 400 with a G1000 for 5 years prior, and had never heard of the potential issue of using a Mac and the resulting corrupt files. I have two other friends who had been doing the same thing."

"We Must Watch it...Like a Hawk": A Boeing 737 crew was caught off-guard during descent. The threat was real and had been previously known. The crew did not realize that the aircraft's vertical navigation had reverted to a mode less capable than VNAV PATH.

From the Captain's Report:

"While descending on the DANDD arrival into Denver, we were told to escend via [the standard terminal arrival procedure]. We re-cruised the current altitude while setting the bottom altitude in the altitude window. Somewhere close to the DANDD intersection, the aircraft dropped out of its vertical mode, and before we realized it, we had descended below the 17,000-foot assigned altitude at the DANDD intersection to nearly 16,000 feet. At once, I kicked off the autopilot and began to climb back to 17,000 feet, which we did before crossing the DANDD intersection.

Reviewing the incident, we still don't know what happened. We had it dialed in, and the vertical mode reverted to CWS PITCH (CWS P). Since our software is not the best and we have no aural warnings of VNAV SPD or CWS P, alas, we must watch it ever more closely—like a hawk."

From the First Officer's Report:

"...Would be nice to have better software—the aircraft constantly goes out of VNAV PATH and into VNAV SPEED for no reason, and sometimes the VNAV disconnects for no reason, like it did to us today."

> *"Computers are good at following instructions, but not at reading your mind."*
>
> —Donald Knuth, Computer Scientist, Mathematician, Professor Emeritus, Stanford University

Mode Changes Are Insidious

Learning From Icarus

A Boeing 737-800 captain became distracted while searching for traffic during his approach. Both he and the first officer missed the FMA mode change indication, which resulted in an altitude deviation in a terminal environment.

From the Captain's Report: "Arrival into JFK, weather was CAVU. The captain was Pilot Flying; the first officer was Pilot Monitoring. We planned and briefed the visual Runway 13L with the RNAV (RNP) Rwy 13L approach as backup. Approach cleared us direct to ASALT, cross ASALT at 3,000 feet, cleared approach.

During the descent, we received several calls for a VFR target at our 10 to 12 o'clock position. We never acquired the traffic visually, but we had him on TCAS. Eventually, Approach advised, 'Traffic no factor, contact Tower.' On contact with Tower, we were cleared to land.

Approaching ASALT, I noticed we were approximately 500 feet below the 3,000-foot crossing altitude. Somewhere during the descent, while our attention was on the VFR traffic, the plane dropped out of VNAV PATH, and I didn't catch it. I disconnected the autopilot and returned to 3,000 feet. Once level, I reengaged VNAV and completed the approach with no further problems." From the First Officer's Report: "FMA mode changes are insidious. In clear weather, with your head out of the cockpit clearing for traffic in a high-density environment—especially at your home field on a familiar approach—it is easy to miss a mode change. This is a good reminder to keep instruments in your cross-check on those relatively few great weather days."

Does that apply in Class B and C airspace, where ATC is also watching? Conventional wisdom has it that too much time is spent heads-down in the cockpit rather than outside during VFR. Are the times changing with the hardware?

The Ol' Trusty iPad Wasn't So Trusty: This was reported by an air carrier captain regarding his electronic flight bag (EFB):

"With Jeppesen FliteDeck-Pro, High IFR chart selected, if the EGTT Flight Information Region (FIR) line is pressed (selected/highlighted) and the communication box is selected, the application freezes. Control is only regained if the application is completely closed and then reopened. It will not reopen on the last page used. I have verified this glitch on several iPads."

ASRS Synopsis:

An air carrier captain reported that the Electronic Flight Bag (EFB) freezes while using the Jeppesen FliteDeck-Pro application, and functionality is only restored after restarting the EFB.

This could be more than a small nuisance during an approach, especially in a single-pilot aircraft.

Liquid-Free Zone: Do we really need to explain this?

The European Aviation Safety Agency (EASA) issued a formal advisory following Airbus's request not to spill liquids into the electronics. It leads to "inconsistent outputs" in the flight control systems that may result in engine shutdown.

There have been at least two diversions after pilots engaged in some "moisture testing" of the hardware—by setting a drink on the control pedestal, which spilled, affected the software, and ultimately affected the hardware—the engines. This is inconvenient when over the ocean.

So, liquid shorts out electronics—who knew? So far, only coffee and tea have been tested, but based on preliminary results, it doesn't look promising for other drinks.

A modest solution: Provide cup holders (and use them), located well away from any critical equipment. Ernie Gann's *Genie Theory* regarding processed liquids isn't so fanciful after all.

Looking Ahead – Approaching Perfection in Controllability

Disclaimer: I consulted with one of the companies (Skyryse) discussed here.

Now, let's discuss the other side of the equation.

What if aircraft handling could be made much easier in all flight regimes? What if it didn't require constant practice to maintain proficiency? What if the system design reduced the chance of failure to one in a billion? Ten years ago, that would have been a stretch—but let's all be from Missouri, as in *"Show Me."* Today may be quite different.

Skyryse's approach was to use a basic helicopter as a demo platform. Rotor-wing aerodynamics are daunting, and helicopters are more challenging to fly. However, this operating concept can work on any aircraft. It goes well beyond even the advanced autopilots of today.

It is described as an aircraft "operating system" (*Skyryse OS*). There are two methods of control that far exceed current advanced autopilots: operation via touchscreen or a single side-stick controller.

Briefly, one can select *hover*, and the machine will hover. Select *takeoff* and program the direction and altitude on the touchscreen,

Robinson R-66, Robinson

Skyryse simplified flight deck, Skyryse

or use the side-stick controller, which combines four controls into one. Envelope protection prevents the aircraft from being badly mishandled.

Autorotation, which is necessary in the event of an engine failure, is one of the more complex skills for helicopter pilots. Recently, the system demonstrated the world's first automated autorotation, where it set up the glide while the pilot selected the point of touchdown.

To dispel my skepticism, Skyryse demonstrated their flying prototype. With just 30 minutes of simulator training, I was doing a tolerable job of flying the helicopter. With about 20 hours of training (an arbitrary estimate), one could likely manage the aircraft very well and safely. As with all general aviation training, completing the FAA certification is just the beginning—pilots must continue learning and developing proficiency to stay safe. However, with this operating system, pilots can advance much further in a fraction of the time.

With triple redundancy—using different hardware and software for each system—this technology has great potential to simplify flight. And simpler is safer.

Skyryse's goal was not to create an autonomous vehicle but to take an existing aircraft and make it much easier to fly. The system can be applied to any aircraft, opening up flying to more people and, most importantly, making aviation much safer and more forgiving. It is reinventing how aircraft and humans interact.

Some may argue that this is "dumbing down" aviation, but simplification is a good thing. As noted in the opening paragraphs, advancements like jet engines, tricycle landing gear, basic autopilots, auto-throttles, TCAS/traffic information, weather-in-the-cockpit, terrain avoidance, and GPS navigation could all be considered "dumbing down." Yet, each of these innovations has drastically reduced workload and, most importantly, increased safety. The numbers prove it.

Urban Air Mobility

Urban Air Mobility - One of many

Now, onto the hotbed of urban air mobility vehicle companies, which are simultaneously developing electrically powered aircraft that are pilotless—at least onboard. More than 30 different startups are currently vying for a potentially lucrative market, provided the economic, technical, and certification challenges, as well as, most importantly, the safety concerns, can be addressed.

There are as many designs as there are designers—much like the early attempts at aircraft. A marketplace shakeout will likely occur, but designers and dreamers are how we got into the air in the first place. A common theme among these concepts is an onboard fly-by-wire system with the ability for a ground controller to take command. Will it work? Eventually.

> *"I'm not saying we purposely introduced bugs or anything, but this is kind of a natural result of any complexities of software... that you can't fully test it."*
>
> —Will Wright, Game Designer, Co-founder of Electronic Arts

Aircraft automation remains in an advanced transitional phase, even though it has been around for over 90 years. The lessons of Cali, Air France 447, Asiana, and the Boeing 737 MAX are written in blood and dollars—for pilots, airlines, avionics designers, manufacturers, and regulators alike.

The yin and yang of automation versus pilot proficiency will continue to be debated. Automation used to be secondary. Today, it flies so well most of the time that manual flight skills seem to have become an afterthought, except in recreational flying.

However, the crew or pilot should be able to disable—or at least manage—the automation and fly the aircraft until we move beyond the transition phase. Fly-by-wire is now standard on airliners, business jets, and military fighters. It may eventually come to all but the most basic aircraft, improving safety overall. However, software bugs or component failures buried deep within systems may take years to uncover and often seem to appear when least expected.

"If a major amount of training, button pushing, or rebooting is required to operate the magic, then the magician needs to get smarter."

Author

The genie lives!

Poor training, a software upgrade, a mental lapse, or overconfidence in complex design or software may allow the great invisible genie to make an unwelcome appearance at the most inopportune time and place. Until we can securely bottle the genie, at least some level of piloting ability will still be required.

For the purists—just as there are beautiful antique automobiles with manual transmissions and radio/heater controls that require a bit more skill—there will always be low- or no-automation aircraft. These provide an opportunity to connect with the machine and showcase true piloting skills.

When Nothing Makes Sense - (Air France 447 drops in from FL380)

One of the world's most advanced airliners crashes when the crew is overwhelmed with complexity and forgets basic airmanship

Airbus 330, Air France

- *An Airbus A330 encounters extreme temperature and moisture conditions at cruise altitude, causing all three pitot tubes to freeze.*

- *Automated systems disconnect, and the crew becomes confused by cascading failure warnings.*

- *Unable to interpret the situation, the crew pulls the aircraft into a stall, descending from 38,000 feet before crashing into the South Atlantic Ocean.*

- *The report describes chaos between the two first officers as they struggle to make sense of the instrument readings while desperately summoning the captain.*

"Flying for the airlines is not supposed to be an adventure. From takeoff to landing, the autopilots handle the controls. This is routine, in a Boeing as much as an Airbus. And they make better work of it than any pilot can. You're not supposed to be the blue-eyed hero here. Your job is to make decisions, to stay awake, and to know which buttons to push and when. Your job is to manage the systems."

—Bernard Ziegler, Former Senior VP of Engineering at Airbus

Mr. Ziegler, a strong proponent of fly-by-wire technology, retired long before the granddaddy of all automation accidents. The loss of an Airbus A330 due to a minor but critical hardware malfunction—one the pilots were unable to recover from—was a stark contradiction to his belief that automation does a better job of flying than humans, most of the time.

The loss of one of the world's most advanced aircraft over the South Atlantic, in cruise flight, is akin to the sinking of the Titanic. This wasn't supposed to happen. Has the state of aeronautical advancement truly progressed beyond such mishaps—or does the Boeing 737 MAX controversy suggest otherwise?

This Airbus accident and its subsequent investigation qualify as historic, marking a tragedy in which 228 lives were lost. The eventual recovery of the main wreckage and flight data recorders—nearly two years after the crash, from ocean depths of almost 13,000 feet—is a remarkable achievement in itself.

The Flight—Air France 447

Air France 447, an Airbus A330-203, departed from Rio de Janeiro, Brazil, on May 31, 2009, at approximately 7:30 p.m. for its 12-hour flight to Paris. Due to the lengthy journey, an extra first officer (FO) was aboard to allow for duty rotation.

The aircraft leveled off at FL350, cruising at Mach 0.82, with autopilot two and auto-thrust engaged. The weather was calm, and everything appeared normal. However, as the A330 entered the Intertropical Convergence Zone (ITCZ), conditions became more challenging.

The ITCZ, located near the equator, is characterized by constant atmospheric lifting and convective activity due to intense solar heating. Pilots and mariners are well acquainted with the convergence of trade winds between the Northern and Southern Hemispheres. That evening, severe storms were present, prompting several flights to divert around the area. However, AF447 continued on its course, penetrating the turbulent zone.

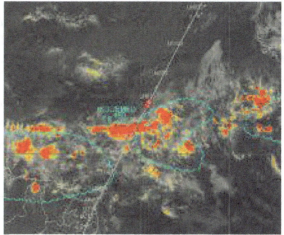

Flight Path Through ITCZ , BEA Report

The Flight

Air France 447, an Airbus A330-203, departed from Rio de Janeiro, Brazil, on May 31, 2009, at about 7:30 p.m. for a 12-hour flight to Paris. Due to the extended trip, an extra first officer (FO) was aboard to ensure proper duty rotation.

The flight leveled off at FL350, cruising at Mach 0.82, with autopilot two and auto-thrust engaged. The weather was calm, and the situation appeared normal. However, as the A330 entered the Intertropical Convergence Zone (ITCZ), conditions became more complex.

The ITCZ, located near the equator, is characterized by constant atmospheric lifting and convective activity due to intense solar heating. Pilots and mariners are familiar with the convergence of trade winds between the Northern and Southern Hemispheres. That evening, severe storms were present, and several flights diverted around the area that AF447 continued to penetrate.

The Incident

About three hours after takeoff, the onboard radar detected precipitation in the zone. The sun had set, and cloud tops were well above the operational ceiling of the A330. The crew discussed the possibility of climbing for a smoother ride, but the aircraft was still too heavy with fuel, and the outside air temperature was too warm.

The captain called for the relief FO and listened as the two FOs briefed. The FO in the right seat, the least experienced of the three crewmembers, was designated as the pilot flying (PF). Six minutes after the captain left the cockpit, the crew made a 12-degree heading change, presumably to avoid precipitation. During this period, despite the pitot heat being on, all three pitot systems iced up. High atmospheric moisture content, a common occurrence in the ITCZ, is thought to be the cause.

At this point, the automatic systems, deprived of key airspeed inputs, began to disconnect and issue failure warnings.

The Bureau d'Enquêtes et d'Analyses (BEA), the French equivalent of the NTSB, reported: "The autopilot, then the auto-thrust, disconnected, and the PF said, 'I have the controls.' The aeroplane began to roll to the right, and the PF made a nose-up and left input. The stall warning triggered briefly twice in a row. Displayed airspeed on the left primary flight display (PFD) dropped from about 275 knots to 60 knots."

The report describes chaos between the two FOs as they struggled to make sense of the instrument readings while desperately summoning the captain. The times are noted in UTC/GMT:

"At 2 hours, 10 minutes, and 16 seconds, the pilot not flying (PNF) said, 'We've lost the speeds,' then, 'Alternate law protections.' The PF made rapid roll control inputs, more or less from stop to stop. He increased the aircraft's pitch attitude up to 11 degrees in 10 seconds. The PNF said that the aircraft was climbing and asked the PF several times to descend. The latter then made several nose-down inputs, resulting in a reduction in pitch attitude and vertical speed (rate of climb). The flight was then at about 37,000 feet and continued to climb.

"At 2 hours, 10 minutes, and 51 seconds, the stall warning triggered again in a continuous manner. The thrust levers were positioned in the takeoff power detent, and the PF made nose-up inputs. The recorded angle of attack, around 6 degrees at the triggering of the stall warning, continued to increase.

"The PF continued to make nose-up inputs. The aircraft's altitude reached its maximum of about 38,000 feet, with a pitch attitude and angle of attack of 16 degrees.

"At 2 hours, 11 minutes, and 37 seconds, the PNF said, 'Controls to the left,' took over priority without any callout, and continued to handle the aircraft. The PF almost immediately took back priority without any callout and continued piloting.

"Around 2 hours, 11 minutes, and 42 seconds, the captain reentered the cockpit. During the following seconds, all of the recorded speeds became invalid, and the stall warning stopped after having sounded continuously for 54 seconds. The altitude was then about 35,000 feet, the angle of attack exceeded 30 degrees, and the vertical speed was about 10,000 feet per minute (emphasis added)."

The aircraft's pitch attitude did not exceed 15 degrees, and the engines' N1s were close to 100 percent. In the description below, the highest pitch attitude reached 16.2 degrees. The aircraft was subject to roll oscillations to the right that sometimes reached 40 degrees. The PF made an input on the sidestick to the left stop and nose-up, which lasted about 30 seconds.

"At 2 hours, 12 minutes, and 2 seconds, the PF said, 'I have no more displays,' and the PNF said, 'We have no valid indications.' At that moment, the thrust levers were in the idle detent. About 15 seconds later, the PF made pitch-down inputs. In the following moments, the angle of attack decreased, the speeds became valid again, and the stall warning triggered again."

"At 2 hours, 13 minutes, and 32 seconds, the PF said, 'We're going to arrive at level one hundred.' About 15 seconds later, simultaneous inputs by both pilots on the sidesticks were recorded, and the PF said, 'Go ahead, you have the controls.'

"The angle of attack, when it was valid, always remained above 35 degrees (emphasis added). From 2 hours, 14 minutes, and 17 seconds, the ground proximity warning system (GPWS) sink rate and then pull-up warnings sounded. The recordings stopped at 2 hours, 14 minutes, and 28 seconds. The last recorded values were a vertical speed of minus-10,912 feet per minute, a groundspeed of 107 knots, a pitch attitude of 16.2 degrees nose-up, a roll angle of 5.3 degrees left, and a magnetic heading of 270 degrees."

The Crew

The 58-year-old captain had nearly 11,000 hours of total flight time, more than 6,200 hours as a captain, and more than 1,700 hours in the A330. The PNF first officer was 37 years old, with more than 6,500 hours of total flight time, and was the most experienced on the A330, with more than 4,500 hours. The PF FO was 32 years old, with less than 3,000 hours of total flight time and slightly more than 800 hours in the A330.

All three crewmembers had received A320 training in unreliable indicated airspeed indications and piloting in alternate law stall conditions. The PF FO had received simulator training in the A320, described as "preventive

Figure 2 Position of the Pitot probes on the Airbus A330

Figure 3 Pitot probe (with protection caps)

A nearly fail-safe system, BEA

recognition and countermeasures to approach to stall." The similarity between Airbus models was reportedly handled by type differences training in subsequent models covered in simulation and ground training.

The Aircraft

The accident A330-203 entered service in April 2005 and had accumulated about 19,000 flight hours and approximately 2,600 cycles (which roughly equate to flights). As previously noted, Airbus's design philosophy was to make the aircraft as automated and simple as possible.

This normal simplicity can lead to some rather complex outcomes when things don't work as planned, especially with multiple faults. This is compounded by the fact that, most of the time, the systems work as designed—so crews get little practice in handling complex abnormal operations and even less in actual conditions, especially those involving high-altitude stall recovery. Excellent simulation and thorough practice are essential, but both are expensive, which sometimes creates dissonance in the accounting department.

In a greatly simplified description of an extremely sophisticated system, the A330 has four operational modes as a fly-by-wire aircraft: normal law, alternate law 1, alternate law 2, and direct law. In normal law, there is complete envelope protection, which essentially prevents the pilot from putting the aircraft into an unflyable configuration—including stalls. In the alternate laws, the protections progressively diminish, and in direct law, the protections are lost. The sidesticks control the various control surfaces directly, just as in light GA aircraft. In alternate or direct law, the angle-of-attack protections are no longer available, and a stall warning is triggered when the greatest of the valid angle-of-attack values exceeds a certain threshold.

The event that started the cascading errors was the simultaneous icing of all three pitot systems. Compounding the situation was the fact that the tubes would freeze and thaw periodically, thus providing confusing airspeed indications. This fault had been identified 13 times before on A330/340 aircraft. The certifying authorities knew this, but apparently, it didn't happen often enough to trigger an immediate airworthiness directive because, in all cases, the aircraft never departed the flight envelope. The other aircraft reverted to alternate law when the automatic systems disconnected, but all the crews were able to maintain altitude within 1,000 feet.

The BEA report noted that the indications on the A330 at FL350 were a drop in indicated airspeed from Mach 0.8 to about 0.3, and the true airspeed indication changed from 461 knots to 182 knots. The indicated altitude dropped about 300 feet, and the displayed windspeed changed from a 30-knot headwind to a 249-knot tailwind.

The BEA said, *"The obstruction of the pitot probes by ice crystals during cruise was a phenomenon that was known but misunderstood by the aviation community at the time of the accident."*

"After initial reactions that depend upon basic airmanship, it was expected that it would be rapidly diagnosed by pilots and managed where necessary by precautionary measures on the pitch attitude and the thrust, as indicated in the associated procedure. The occurrence of the failure in the context of flight in cruise completely surprised the pilots of flight AF 447."

"The apparent difficulties with aeroplane handling at high altitude in turbulence led to excessive handling inputs in roll and a sharp nose-up input by the PF. The destabilization that resulted from the climbing flight path and the evolution in the pitch attitude and vertical speed was added to the erroneous airspeed indications and Electronic Centralized Aircraft Monitor (ECAM) warning messages, which did not help with the diagnosis. The crew, progressively becoming destructured, likely never understood that it was faced with a 'simple' loss of three sources of airspeed information."

"Until we get truly fail-safe designs, aircraft will have to be flown as they are—not as we wish or believe them to be."

— *Author*

Commentary

The hindsight view is clear: The pitot heat system should have been addressed as soon as the problem surfaced. A critical triple-redundant system that fails cannot be ignored. This is a rare case where the hardware failed and then tripped up the avionics.

The captain chose to leave the flight deck as the aircraft was entering the ITCZ. This area has a well-known severe weather reputation, and that was the reality on the night of the crash. Weather complacency occasionally catches airline captains, just as it does GA pilots.

Loss of control in GA aircraft is a leading fatality producer, but it also happens to seasoned crews who have forgotten the basics. With engines operating normally and level flight attitude, nothing is likely to go too far wrong unless the pilot misunderstands the situation.

As light GA aircraft begin to emulate airline cockpits, training must address complexity. System designers and certification authorities might consider that simpler, more intuitive, and more robust systems are better than adding new features with the potential to confuse.

Stall avoidance education begins in primary training but sometimes is not carried through into advanced flight—especially at high altitude, where the margins are very thin between over-speed and stalling.

The BEA noted, "Following the autopilot disconnection, the PF very quickly applied nose-up sidestick inputs. The PF's inputs may be classified as abrupt and excessive. The excessive amplitude of these inputs made them unsuitable and incompatible with the recommended aeroplane handling practices for high altitude flight."

Delicate handling is required at high altitudes, and a flight-level stall recovery will be completely different from what is typically trained at low altitude. Complex systems and seldom-practiced events can leave us unprepared when the basics of flight are ignored.

Too Low, Too Slow - (Asiana 214 Crew Fails to Understand Systems)

"Keep thy airspeed up, lest the earth come from below and smite thee."

— *William Kershner, Military pilot, author, CFI, test pilot*

On July 6, 2013, a Boeing 777-200ER, operating as Asiana 214, struck a sea wall during a visual approach to San Francisco International Airport. The 777 series had been in service for decades with an excellent operational record. Losing one on a routine visual approach in day VFR,

370

no-wind conditions was unthinkable. Unlike many other air carrier crashes, there was no system failure, no dark night, and no inclement weather. This crash could have been included in Chapter 3, Takeoff and Landings, but there was a misunderstanding of the automation.

The Flight

Asiana 214 was vectored for a straight-in visual approach to Runway 28L and intercepted the final approach course about 14 nautical miles from the threshold, slightly above the desired three-degree glide path. The flight was cleared for the approach and requested to maintain 180 knots to the final approach fix.

The Approach

The approach started poorly, with the aircraft 400 feet above the desired three-degree glide path at the 5 nautical mile point. To increase the descent rate, the pilot flying (PF) incorrectly selected an autopilot flight level change speed (FLCH SPD), which resulted in a climb because the aircraft was below the altitude selected. The pilot disconnected the autopilot and moved the thrust levers to idle, which directed the autothrottle to the HOLD mode, which does not control airspeed. Apparently, he thought the autothrottles were still active. The pilot then pitched the nose down and increased the descent rate. Neither the pilot flying, the pilot monitoring, nor the observer noted the change in autothrottle mode to HOLD, which is essentially the same as "OFF." This mode might have been better labeled, perhaps with a warning tone.

As the airplane descended through 500 feet above airport elevation, the runway precision approach path indicator (PAPI) showed that the airplane was slightly above the desired glide path. Asiana's procedures required the approach to be stabilized at that point. The airspeed, which was rapidly decreasing, had just reached the proper reference (approach) speed of 137 knots. However, with the thrust levers in idle, the descent rate was about 1,200 feet per minute, nearly double the descent rate of 700 fpm needed to maintain the glide path. A go-around should have been initiated. The Boeing descended well below the glide path, and the airspeed decayed to 112 knots as the pilot increased back pressure to decrease the descent rate. At 200 feet, the crew noticed the low airspeed and low flight path conditions but delayed a go-around until below 100 feet. There was not enough energy remaining to allow a go-around or stop the descent.

The cockpit voice recording is remarkable in that the aircraft was in a high sink, low-speed condition very close to the ground with a PAPI showing all red, yet neither of the two other pilots challenged the pilot flying.

From the NTSB Cockpit Voice Recorder – Times are Pacific Daylight Time. Sound sources are identified, Cockpit Area Mic is from flight crew audio panel. ? = unintelligible, * = expletive.

11:27:17.5 Instructor: Landing checklist complete, cleared to land.

11:27:19.8 Cockpit Area Mic: On glide path, sir.

11:27:21.2 Pilot: Check.

11:27:32.3 Electronic Voice: Two hundred (referring to altitude).

11:27:34.8 Instructor: It's low.

11:27:34.8 Pilot: Yeah.

11:27:36.0 Cockpit Area Mic: ?*

11:27:41.6 Electronic Voice: One hundred.

11:27:42.8 Instructor: Speed.

At 11:27:43, the pilot increases the power.

11:27:44.0 Cockpit Area Mic: ? speed * *

11:27:45.8 Electronic Voice: Fifty.

11:27:46.4 Sound similar to stick shaker lasting for approximately 2.24 seconds.

11:27:46.6 Electronic Voice: Forty.

11:27:47.3 Electronic Voice: Thirty.

11:27:47.8 Instructor: Oh #, go around!

11:27:48.6 Electronic Voice: Twenty.

11:27:49.5 Cockpit Area Mic: Go around.

11:27:49.6 Electronic Voice: Ten.

11:27:50.3 Cockpit Area Mic: ? oh.

11:27:50.3 Cockpit Area Mic: Sound similar to impact.

11:27:55.4 Cockpit Area Mic: Ah, what's happening over there?

The main landing gear and the aft fuselage struck the seawall. The tail separated, and the airplane slid along the runway. It then lifted partially into the air and spun about 330 degrees before coming to a stop. There were three fatalities, including two passengers who were not wearing seatbelts. Fire broke out in the right engine, but, remarkably, of the 307 people on board, 99 percent of the occupants survived.

The Captain (In Training, Left Seat) – According to the NTSB

"The captain, age 45, held an airline transport pilot certificate with type ratings in the Airbus A320 and the Boeing 737, 747-400, and 777. His records showed no accidents, incidents, violations, or company disciplinary actions. His total flight time was 9,684 hours, including 3,729 hours as PIC. He had 33 hours of 777 flight time and 24 hours of 777 simulator time. The captain was hired by Asiana Airlines in 1994 as a cadet pilot with no previous flight experience and received ab initio flight training in Florida from 1994 to 1996. He began First Officer training on the 737 in 1996, upgrading to 737 captain in 2005. He transitioned to A320 captain in 2007 and began transition training to 777 captain in March 2013. He completed ground training, full flight simulator training, and a simulator check. His 777-simulator proficiency check was completed on May 18, 2013, and his line-oriented flight training check on May 30, 2013. Training records indicated that during the simulator stage of his 777-transition training, the PF performed six visual approaches, two without an ILS glideslope, receiving a grade of "good" each time. In addition, the PF performed one circling approach and one sidestep maneuver in the simulator."

Instructor – Right Seat

"The instructor, age 49, held an airline transport pilot certificate with type ratings in the Boeing 757/767 and 777. There were no accidents, incidents, violations, or company disciplinary actions in his record. His total flight time was 12,307 hours, with 3,208 hours in the 777. He was hired by Asiana in 1996 and had transitioned to captain on the 777 in 2008 after flying as a captain on the B767. The instructor had been trained in visual approaches into SFO and had performed well in simulator training every year since 2008. This was his first flight as a supervising trainee captain."

Some of the NTSB findings are summarized here:

1. "The flight crew's mismanagement of the airplane's vertical profile during the initial approach led to a period of increased workload that reduced the pilot monitoring's (Instructor) awareness of the pilot flying's actions around the time of the unintended deactivation of automatic airspeed control."

373

2. "At about 200 feet, one or more flight crew members became aware of the low airspeed and low path conditions, but the flight crew did not initiate a go-around until the airplane was below 100 feet, at which point the airplane did not have the performance capability to accomplish a go-around."

3. "Insufficient flight crew monitoring of airspeed indications during the approach likely resulted from expectancy, increased workload, fatigue, and reliance on automation."

4. "As a result of complexities in the Boeing 777 AFCS and inadequacies in related training and documentation, the pilot flying had an inaccurate understanding of how the autopilot flight director

PAPI - all red, airspeed 18 knots below ref, full flaps, and power at idle,
NTSB

system and autothrottle interacted to control airspeed, which led to his inadvertent deactivation of automatic airspeed control."

5. "A review of the design of the 777 automatic flight control system, with special attention given to the issues identified in this accident investigation and those identified by the Federal Aviation Administration and European Aviation Safety Agency during the 787-certification program, could yield insights on how to improve the intuitiveness of the 777 and 787 flight crew interfaces, as well as those incorporated into future designs."could yield insights about how to improve the intuitiveness of the 777 and 787 flight crew interfaces as well as those incorporated into future designs."

A/T Mode	A/T Annunciation	A/T Mode Description
Thrust reference	THR REF	Thrust set to the reference thrust limit displayed on EICAS.
Speed	SPD	Thrust applied to maintain target airspeed set using the MCP or FMC.
Thrust	THR	Thrust applied to maintain the climb/descent rate required by AFDS pitch mode.
Idle	IDLE	Occurs when A/T controls the thrust levers to the aft stop.
Hold	HOLD	Occurs when A/T removes power from the servo motors. In this mode, A/T will not move the thrust levers.
No mode		A/T is armed but not engaged. This is the only state where the A/T automatic engagement function is potentially active.[41]

Autothrottle modes and annunciation. What is the purpose of "HOLD? Would "OFF" make better sense? NTSB

In any design, less is more, and too many choices tend to confuse. By attempting to make things just a little easier, the designers introduced confusion.

"I don't need that much organization in my life."

—Jimmy Buffett, Singer, Songwriter

NTSB: "By encouraging flight crews to manually fly the airplane before the last 1,000 feet of the approach, Asiana Airlines would improve its pilots' abilities to cope with maneuvering changes commonly experienced at major airports and would allow them to be more proficient in establishing stabilized approaches under demanding conditions. In this accident, the pilot flying may have better used pitch trim, recognized that the airspeed was decaying, and taken the appropriate corrective action by adding power."

NTSB Probable Cause: *"The flight crew's mismanagement of the airplane's descent during the visual approach, the PF's unintended deactivation of automatic airspeed control, the flight crew's inadequate monitoring of airspeed, and the flight crew's delayed execution of a go-around after they became aware that the airplane was below acceptable glidepath and airspeed tolerances.*

Contributing to the accident were: (1) the complexities of the autothrottle and autopilot flight director systems that were inadequately described in Boeing's documentation and Asiana's pilot training, which increased the likelihood of mode error; (2) the flight crew's nonstandard communication and coordination regarding the use of the autothrottle and autopilot flight director systems; (3) the PF's inadequate training on the planning and execution of visual approaches; (4) the PM/instructor pilot's inadequate supervision of the PF; and (5) flight crew fatigue, which likely degraded their performance."

Commentary: The accident chain here is extensive and obvious in hindsight, as they often are. Fatigue played its part after a long flight across the Pacific. An experienced captain, but new to the B777, a new instructor pilot, a lack of full understanding, and over-reliance on automation led to a most unusual accident.

An important link in the chain was ATC's request to maintain speed until the final approach fix. The "slam-dunk" or high-speed approach request, while not unusual, always increases pilot workload. These ATC techniques can work against the stabilized approach. It's up to the pilot to determine whether he or she can comply.

To test the difficulty of the approach, the NTSB had test pilots from both Boeing and the FAA, who were rated and current in the 777, fly the accident profile in a simulator. The results are telling. The NTSB reported: "When starting at the accident profile conditions (above glidepath), the pilots experienced difficulty in achieving stabilized approaches. On 4 of the 10 accident profile runs, they were unable to achieve a stabilized approach." That's not a very good record. Of the 6 successful runs, "...the pilots used descent rates in excess of 1,000 fpm below 1,000 feet AGL..." — i.e., completely unstabilized.

> ***"Change the plan if it looks like you'll arrive 10 seconds behind the aircraft."***
>
> *—Author*

The natural desire is to cooperate to avoid being thought incapable. Proficiency in routine procedures is expected, but if you're sinking, speak up. It may be inconvenient, but you'll get the benefit of the doubt, and honest aviators will admit to having been there before. Crashing, however, is always considered bad form.

Crossing the final fix on altitude is essential. In any clean airframe, going down and slowing down means playing catch-up all the way down final. The two Asiana pilots had collectively logged more than 20,000 flight hours and had probably salvaged this situation before. It had always worked.

Fatigue and complacency are insidious. On long trips with little to do for hours on end, the most challenging part, the landing, lurks. With thousands of flights in air carrier aircraft, both this pilot and instructor were perhaps overly comfortable in their roles. Both knew angle-of-attack and power settings to fly a visual approach, and yet, a fundamental error was made.

Complex systems should be designed to be operated by tired and distracted humans. That's a tall order because we are infinitely variable. Would some level of automation standardization help? Airbus and Boeing autothrottles operate differently. The accident report shows the many ways the automated flight control systems function on different aircraft and the many ways different modes can be activated or deactivated within the same system.

Each additional feature increases the opportunity for a software or human programming problem. The benefits versus the complexity are too often ignored in the pursuit of a "competitive advantage." It also helps if there is one best way to activate a system, as opposed to having multiple ways to reach a desired mode. We should be able to access that single activation method from whatever submenu or page the system is operating in. Simpler is safer.

Crew resource management requires intervention before a condition becomes unrecoverable. One can be respectful and still save the day.

CHAPTER 6
EMERGENCIES AND OTHER SURPRISES

"Emergencies are funny things. You don't have them until you're having them."

Author

Precious Cargo

The pros often pass the quiet hours aloft talking through and simulating emergency thought processes. Without practice, those critical abilities won't be there. Like flood insurance or fire extinguishers, the practice is worthless until needed—and then worth every penny.

Too often, we just react instead of thinking through the problem. It's tough to stay calm and collected in a startling event when there are only a few seconds to get it right. A pure gut reaction is frequently sub-optimal. Usually, there's time to work it out. Quick-response emergencies, such as engine failure on takeoff or fires, are rare, but that's when the planned reaction and the bold-printed memory checklist items are needed the most. Let's look at the more common problems first.

Go Slow—When asked what the first reaction to an emergency should be, an old airline captain replied that he'd wind the clock on the instrument panel (back when wind-up clocks existed). Then, he'd get out the checklist so as not to miss anything, if it's not a memory item.

Situations where it may take, arbitrarily, 20 to 30 minutes for something to become troublesome are defined as "abnormal conditions." Although a dead alternator with a dying battery in instrument conditions needs serious attention, nothing is immediately threatening, provided it's caught quickly. Malfunctions such as comm, nav, or flight instrumentation can—and should—be anticipated with backup systems. Large aircraft have multiple backups, and for light aircraft, eliminating critical single-point failures, wherever possible, is key. Will your phone suffice in case the iPad overheats or dies, leaving you chartless? Do you have a fire extinguisher? A CO detector? A handheld transceiver? A personal ePIRB? All are useless until needed.

When something unexpected crops up, and it's not easily fixed but may create a possible problem, inform ATC. If you're not in contact with ATC, now might be a good time to join the conversation. But, as always, flying takes priority. If there's ANY uncertainty, **DECLARE AN EMERGENCY**. You'll get the help you need. Ask for a precision approach and radar vectors when in IMC. If it's not immediately critical, "Pan, Pan" means, "I can handle this but may need some assistance at some point." "Minimum Fuel" is another clue to ATC that the flight is okay for now, but a little delay could make it a bigger issue. After landing, assess whether you put yourself into that situation and resolve to carry more fuel next time.

Transfer workload to the ground where possible. Don't keep heading to the original destination unless it's the closest place. You'd be amazed at the number of pilots, both VFR and IFR, who ask for help and then figure things will somehow get better or at least not get worse. Sometimes, that turns out badly.

Take time to set up correctly. A quick intercept to the final approach course invites getting behind, with a possible loss of control. Unless the aircraft is in immediate danger of crashing or on fire, the nearest airport or runway may not be the best. Look for long runways and better weather, if possible. A failing engine may provide power for as long as needed, or it could quit right now. Accepting a downwind or crosswind runway could be a great idea, but if you wait too long, that option may close.

"There's always time to do it right and seldom time to do it over."

—*Time-honored management mantra (often ignored)*

Landing Gear—Landing gear malfunctions require a slow, steady approach. It can be taken to extremes, however (see Flameout this chapter). Twice in my flying career (so far, anyway), the gear would not extend. Trying to fix it in the pattern or during an IFR approach is a bad idea. Move away from the airport and climb. IFR? Miss the approach, advise ATC, and ask for a delaying vector or holding pattern. Use the autopilot. In some airplanes, the backup gear extension system may not allow a second chance. A single charge of nitrogen in a "blow-down" type backup system can be wasted if the gear switch isn't in the right position. In both of my cases, following the checklist worked perfectly.

Not long ago, the NTSB released a report on a single-engine aircraft with a gear malfunction. This airplane had an electrical failure, and the backup procedure is to slow down and release hydraulic pressure so the gear will free fall. The pilot didn't know to do that. Some basic systems knowledge and consulting the checklist might have saved this aircraft from the scrap heap.

Slow down to reduce the air pressure on the gear, which makes it easier for both you and the backup system to function. Troubleshoot but forget the heroic efforts. Let the insurance company deal with it if the emergency extension doesn't work.

Don't try this at home or the airport either!

The picture here involved an impromptu race down the runway with a car precariously positioned under a ruptured retractable gear and a lame main gear. The pilot called a buddy with a car that had a sunroof, who raced down the runway while the passenger acted as the emergency extension mechanism. It ended happily when the gear was miraculously pulled into place, and the aircraft landed successfully. Lady Luck smiled on the gladiators, but the risk was real.

The conservative approach to landing gear malfunctions invariably leads to no injuries and relatively minor damage to the aircraft, although many older machines may be uneconomical to repair. Unpracticed formations to "check" the gear can have unintended consequences. It takes only a small gust of wind or an unanticipated response from either driver or pilot to result in fire, destruction of both machines, or the loss of all participants— you get the picture (see *Down and Locked*, this chapter). Tower flybys tell you nothing. The gear may be almost down, but they can't tell if it's locked.

Electrical—If the alternator stops feeding juice to the battery, there's usually time (maybe 20 to 30 minutes—you are keeping up with routine battery maintenance, right?) to find a runway if you're in IMC. Electronic flight instruments often have an emergency battery built in that typically recharges during routine flight. When was it last replaced?

What's the bailout option if there's widespread low IMC? Spot the problem when it first occurs. Shed the load by shutting down non-critical equipment. List the high power-draw items on the emergency checklist. Once the lights start to dim or the radio fades, the situation becomes more complex. When ATC figures out that you're incommunicado, they'll clear the airspace. Do what's needed to get on the ground safely.

Pretty hard to see, Author

Better annunciation is helpful, not the mini needles that were so popular years ago. Most newer aircraft identify the problem immediately. Consider a retrofit for better monitoring if you're doing much night or instrument flying.

Lost communication in instruments is unnerving, perhaps, but not critical. Methodical checking of mic jacks, switch settings, headsets, etc., will often solve the problem. A hand-held radio plugged into the aircraft's external antenna is a great idea. A simple coax plug connected to one of the external antennas works wonders compared to the "rubber-duck" antennas, which work okay on the ground but are often just decorative aloft. Keep a copy of the lost comm procedures with your checklists.

Vacuum Failure – Dry vacuum pumps that powered most attitude and heading indicators long ago were a source of periodic loss-of-control accidents. Years ago, IFR aircraft were often equipped with only a single pump, and the pilot's partial-panel skill was all that stood between a safe landing and disaster. There are no statistics on how many pilots successfully dealt with it, but a failure to successfully address a pump failure always made headlines.

Single point Failure, Author

The pumps failed with some regularity, and so did the seldom-used pilot backup skills. The fix was simple: add a second pump or change the instrument system. Today, these crashes have largely been eliminated by redundant systems or using electrically powered gyros. If you're still flying in IMC or at night with old technology and a single pump, a backup system is highly recommended. Our partial panel skills are about the same as they've always been—reasonably good right after training, and not very good a few months down the line.

Hangar Story – When the VAC annunciator illuminated on the Piper Arrow I was flying from Chicago Midway to Wichita, Kansas, in solid and widespread IMC, it was unsettling. The vacuum gauge confirmed the pump's demise as the attitude indicator slowly rolled over. Thank goodness the annunciator was there. A 400-foot overcast with 1.5 miles visibility made this more than an academic exercise. Vectors to the Peoria, Illinois ILS were the best option since there was no nearby VFR.

Typical of the time, the autopilot was useless because it derived guidance from the failed instruments. You picked a fine time to leave me, Lucille. The sole vacuum pump failed at 25 hours because the previous installer did not blow out the vacuum lines when the earlier pump failed. That was contrary to the installation instructions. Contamination from the old pump destroyed this one at the most inopportune time. Before the next IFR flight, a backup system was installed, and the lines were cleaned.

"Put all your eggs in one basket and then watch that basket."

– Mark Twain, Writer, Humorist

Maintain, Maintain, Maintain – Good maintenance is essential, but you knew that. Unfortunately, not everyone adheres to best practices, and crashes result. New pilots get very little instruction on the mechanical innards of aircraft. Some people just want to fly and not be bothered with the details. But these are not airborne automobiles and demand more respect.

The FAA issued an emergency AD against the Beech V-tail Bonanza back in the 1980s for not quite meeting spin/spiral recovery certification standards. Beech offered a free fix on airworthy aircraft brought back to the factory, a notable and welcomed response. A Beech employee noted that many of the aircraft arrived in such poor repair that the factory refused to work on them until they were returned to airworthy condition. Most of them had logbooks attesting to their pristine condition at the last annual.

To shed that amateur standing, learn about the hardware. Flight instructors should encourage students and newly rated pilots to spend quality time in the shop, if possible. Mike Busch's book, appropriately named Engines, is heartily recommended to provide insight into what goes on under the cowling. As an owner, participate in at least a few owner-assisted annuals to really learn your machine. Some type clubs offer maintenance clinics where experts look over your aircraft and make recommendations. Highly recommended.

If you can't measure it, you cannot manage it – Recording and analyzing is now prevalent in many activities. Professional sports reaped the benefits of recording and analyzing plays. Watched the Olympics recently? They measure performance to 1/100 of a second. The National Football League started reviewing plays around 1976, but not without controversy. However, it upgraded the officiating tremendously.

The value of engine monitors can hardly be overstated. Back in the 70s, 80s, and early 90s, most light aircraft with piston engines had minimal instrumentation. One cylinder was it—maybe. Those instruments might tell you, "Yup—you're about to (or have already) become a glider." It's a lot better today on new aircraft, and there are relatively inexpensive upgrades for legacy machines.

One of the reasons airlines excel in crash prevention is that they measure and monitor everything on board the aircraft, including the pilots. Flight Data Recorders (FDRs) and Cockpit Voice Recorders (CVRs) are built to withstand being burned, bashed, or dunked. (The media often refer to them as "black boxes," but they're actually orange—go figure.) While these devices are invaluable after a tragedy, their main value lies in preventing crashes in the first place by routinely downloading data.

The NTSB has been advocating for the benefits of Flight Data Monitoring (FDM) for years. FDM is a simplified and much less expensive version of the FDR for general aviation (GA). It has been recommended for all Part 135 operators to help them manage their risks.

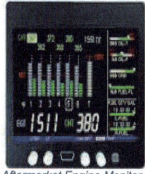

Aftermarket Engine Monitor, JPI

But it's not enough to simply have the gear on board. The real payoff comes from capturing the data and getting it analyzed. Under Flight Operations Quality Assurance (FOQA), computer algorithms measure almost everything the pilots and aircraft do and compare it to what should have been done during that phase of flight. Datalink to the ground in real time is now technically feasible, and some large or advanced aircraft do just that. Deviations outside the normal range are flagged, analyzed, and categorized by hardware or software. Corrective action, if necessary, is then taken before anything bad happens.

I installed an engine monitor on my old aircraft, which tracks critical temperatures, fuel flow, alternator output, and records it in two-second intervals. Every few months, I upload the data to my maintenance monitor who compares it to the rest of the Bonanza fleet equipped with the same engine. and receive a report card not only for my engine but also for how it also evaluates how I'm managing the powerplant, not just the hardware.

The right image shows the readout with a cracked exhaust stack. Savvy Aviation (I subscribe to their analysis service), a maintenance management and monitoring organization for GA, immediately sent the customer the following: "Check cylinder 1 for an exhaust leak. EGT 1 diverges from the others under power, when the volume of exhaust gases is greatest." Obviously, this is critical due to the potential for an engine fire and/or possible carbon monoxide incapacitation.

This is just one example of the many things that can go wrong. Without regular internal monitoring, many engine problems would be missed during a preflight inspection. Aircraft with hard-to-remove cowls exacerbate the problem. Looking through an oil fill/dipstick access door is worthless. Now, we're depending on luck to avoid an extremely serious situation.

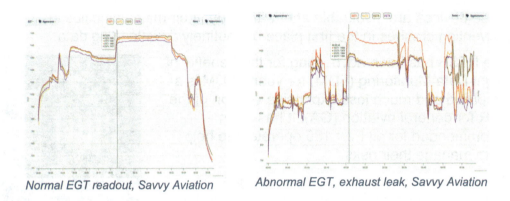

Normal EGT readout, Savvy Aviation   *Abnormal EGT, exhaust leak, Savvy Aviation*

Monitoring the pilot – Several owner groups of turboprops and jets are encouraging their members to participate in FOQA-like programs to identify pilots who may be having difficulty flying within proper tolerance on approaches. The purpose is NOT to punish but to help. Improper procedure can certainly be improved with training but will not improve by simply ignoring it.

Rarely does either the pilot or aircraft fail instantaneously, without some warning. If a crash occurs, product liability should be addressed early on. If something legitimately and improperly broke, the manufacturer should pay up and fix it for the rest of the fleet. Deferred maintenance is on the owner, and a bad repair belongs to the shop. If, in 60 to 80 percent of crashes, the pilot was a contributing factor, it's not up to the rest of us to pay for those mistakes. This approach should significantly reduce product liability costs for manufacturers and shops. The savings should be shared with customers who are equipped and participate in regular analysis. Likewise, aircraft insurance should reflect these savings.

Don't like the hardware spying on you? Don't plan on flying most newer aircraft or becoming a professional pilot. Consider that it may save your life. Since the early 1990s, most cars and trucks have had event data recorders that track speed and braking. Manufacturers use this data to defend themselves from bogus liability claims, and it works.

Forced landing - Change for the parking meter? Never mind, there's an app for that.

It's going to be a bad day when the fan stops

Pilots like to tell passengers that if the engine quits in a single-engine aircraft, they'll glide down like a bird and find a soft place to land. That's whistling past the graveyard. Sometimes it works out, but circumstances (or luck) play a significant role in happy outcomes. NTSB statistics show that there are a significant number of fatal accidents annually after an engine stoppage for whatever reason. In more than 30% of the cases, the cause is undetermined.

If the engine quits over a Kansas wheat field on a VFR day, great. But if it goes out over downtown Dallas, the West Virginia mountains, or Lake Michigan, the crash dynamics are not favorable. Darkness and clouds dramatically increase the odds of a bad outcome.

> *"If the engine stops for any reason, you are due to tumble, and that's all there is to it."*
>
> — *Clyde Cessna, founder of Cessna Aircraft Company*

There are some things that can reduce potentially bad results, and we'll discuss them here. However, preventing the problem is far better than playing catch-up after combustion has ceased. Many so-called "failures" are simply stoppages—there's no fuel on board, or the pilot hasn't switched tanks. Avoid the week's Darwin Award by ensuring there is adequate fuel in an appropriate tank.

Mud daubers and other insects are attracted to fuel tank vents, especially on little-used aircraft parked outside. This is difficult to catch on a preflight inspection, and there's usually just enough fuel in the lines, etc., to put the aircraft into a bad position. Cover the vents and remember to remove them before flight. Confer with your maintenance provider on checking the system if the aircraft hasn't been flown for a while.

Fuel Contamination – We've all heard that reciprocating engines don't like water. If the aircraft has been parked outside and it rained, a thorough sumping job is essential. While the FAA talks about condensation if the tanks are less than full, a far greater likelihood is water intrusion through worn fuel cap seals. They must be replaced regularly. The awful truth is that there is often just enough uncontaminated fuel in the fuel lines to get through the runup and allow the aircraft to become airborne. The water arrives at the worst possible time. Occasionally, a fuel dispenser is at fault—trust but verify.

Hangar Story – *I had been visiting a friend in Vermont, and the aircraft was parked on the ramp. It had rained heavily the night before, and when I sumped the left wing, there was no dividing line between the water and any light blue avgas. I must be seeing something, but it didn't smell like gas—never had that much water before, just a few bubbles. I discarded the sample and sumped again twice more before getting to something that would burn—nearly a pint. The fuel cap seals hadn't been replaced in three years on a company aircraft.*

Jet Fuel in Piston Aircraft – Jet fuel in a piston aircraft is always a disaster. Advise the line person that you want 100LL or avgas (unleaded, if available), and none of that nasty, greasy, smelly stuff they sell for turbines. It always gets a laugh, but now they're thinking about it. If you can't be present to observe fueling, which is a great idea, always check the receipt. They might misfuel the aircraft, but they'll *never* foul up the billing.

Diesel Exhaust Fluid (DEF) for airport fuel trucks will also bring a multiengine jet down, and there's no way for the pilot to check in advance if the fuel coming out of the truck is compromised. The NTSB published a Safety Alert on this after several incidents. Murphy is out there.

In-Cockpit Video – Another area that is very helpful on a preventive basis is in-cockpit video. Again, the real payback is in prevention versus after the crash. At an NTSB board meeting regarding a tour helicopter that crashed in Hawaii, cameras mounted on other tour aircraft flying in the same area showed the lousy weather that the VFR pilot deliberately entered. There was no camera on the crash aircraft. The videos were being sold to tourists as souvenirs, but they could have easily been used by the Directors of Flight Operations to monitor how well their pilots were adhering to company safety policy and VFR regulations—which they weren't. Video was recommended for Part 135 operations.

Engine Stoppage at Altitude – Higher cruise altitudes provide critical time when an engine gets indigestion, a much larger radius of action, and better communication options. It also eliminates those pesky tall tower entanglements. Double the cruising altitude above ground, and you quadruple the area available for forced landings. Instead of cruising at

3,000 feet, consider 6,000 feet. Obviously, there are other considerations such as ride, oxygen, headwinds, clouds, icing, etc., but all other things being equal, the higher altitude buys time and a much wider area to park the beast when things get quiet.

Turning downwind will extend the gliding distance, BUT landing into the wind reduces the impact forces when a crash will be off airport. If the best field is downwind, try to fly a traffic pattern, provided there's altitude. It's much easier to judge the glide path to touchdown from a downwind or base leg because you can "play the turns." Straight in is tough, especially with a strong wind since the ability to reduce descent with power is gone. Chances are good we'll wind up short.

A total engine failure at 6,000 feet AGL means about five minutes or so before ground contact in most light aircraft. Remember to breathe (it's important) and turn toward the nearest airport or suitable landing area. There's time to troubleshoot the problem; there's not much we can check, but check it all. Just keep flying and look for something soft and cheap to hit.

Besides the GUMP before the normal landing, there's another GGUMMPP(S) memory item that might be useful in a stoppage situation. Maintain control. Refer to printed checklists when there's time. Memory items need to be reviewed regularly. After memory items are accomplished, get out the checklist. Keep flying.

Glide – Trim for best glide attitude. Head to the nearest airport or open area.

Gas – Switch fuel tanks and try the auxiliary fuel pump (if so equipped).

Undercarriage – Up for now, and decide whether you want it down for landing.

Mixture – Adjust mixture as required. If the engine is able to run at all with the fuel pump on HIGH, the mixture may need to be leaned for smoothness.

Magnetos – Try right, left, and both. Will it run better on any setting?

Pump – If equipped with an electric fuel pump and you overlooked it earlier, try it now.

Prop – If no start and it's controllable, feather it (high pitch – aft), and the glide distance will improve significantly.

Switch Tanks – Maybe there's a blockage.

Often, the engine doesn't completely fail. Don't give up altitude until an airport or open area is within reach. Advise ATC that you'll stay high for as long as possible. If the vibration gets worse, smoke appears, or oil is streaming back over the windshield, shut down and feather the prop (if it's controllable) while you still have that ability. Squawk 7700 and switch to 121.5, but only if you're not already talking with ATC. Let's get help on the way and have ATC put in the pizza order so it's waiting after we make that successful forced landing. Don't forget to activate your personal locator beacon. Don't have one? Get one.

Hangar Story – *While taking type-specific training for engine-out practice in my aircraft, the instructor said to maintain 2,000 feet AGL until short final. Huh? Lower full flaps at 120 knots (Vfe), put the gear down, and maintain 120 until you're sure the ground will unpleasantly intervene. Then flare. With some serious swallowing and considerable body English, I landed just beyond the thousand-foot marker and easily made the mid-field turnoff on a 4,000-foot runway. This will vary by aircraft and wind conditions, but the space shuttle couldn't have done it better. My aircraft and Wiley Coyote's anvils have a lot in common.*

There are times when quick reaction matters – Takeoff. The engine starts running roughly and losing power. At 60 knots, a typical liftoff speed for light aircraft, the length of a football field is passing every three seconds. Climbing at 80 knots, two football fields are being covered in less than five seconds. Engines are reliable, and we take them for granted, but a solid plan is additive to any luck we might enjoy.

Aviate, Navigate, Communicate – You've heard all this before. Fly first and talk later. ATC will be agreeable to just about anything, so don't waste mental bandwidth on chatter. We need to land now, and there's nothing ATC can do in the short term to help you fly any better. Need a vector? Ask for it.

On takeoff, how much runway is needed to stop on pavement or overrun if the engine quits just as we're becoming airborne? The manufacturers don't include this critical bit of information for single-engine aircraft. They should. The accelerate-stop distance for single-engine aircraft can be roughly estimated by taking the ground roll takeoff distance, adding an arbitrary 500 feet for reaction time (about 3-4 seconds), and then adding the ground roll for landing. For example, if the takeoff ground roll is 1,100 feet, the engine partially fails at lift-off + 500 feet (reaction time, several seconds) + 800 feet (the landing roll), it equals about 2,400 feet. Your distance will vary, but it's a starting point.

2,400' Accelerate-Stop Distance Example

1,100' Ground Run	500' reaction	800' Ground Roll

Coincidentally, it works out pretty closely to the recommended 50 percent safety margin added for a normal takeoff to clear the 50-foot obstacle. The guesstimated surprise reaction time is about 3-4 seconds, but your reaction may vary. If all power is lost, the decision is made—the aircraft will stop. But total instant engine failure is rare, so let's assume it's still producing some power, and we're already over the trees. Should we try to limp around the pattern? How high are we? Are there alternative landing sites nearby? Is the problem getting worse, and what are the options if it quits completely? There's plenty to think about, and a planned reaction gets us to a solution much faster.

Some things to consider prior to every takeoff:

- **Use all the runway** – The pavement behind us is worthless.

  - **Safety pilot rule of thumb** – About 90 percent of flying speed is needed by the midpoint of the runway. If not, reject the takeoff and sort out why on the taxiway or at the maintenance shop. If normally airborne at 70 knots, reject the takeoff if less than 63 knots at the midpoint. Obviously, this doesn't apply to a 10,000-foot runway. Still, if it's sluggish on acceleration, stop and figure out why.

  - **What's the minimum altitude** where a turn back to the airport might be considered? Hold that thought—we'll get back to it.

In a twin, more options require more proficiency:

As the saying goes, the second engine will take you directly to the scene of the crash. Generally, if the gear hasn't started up when a problem occurs—reject. If the airspeed is well below blue line (Vyse)—reject. At high-density altitudes and weight, the normally aspirated piston twin performs as a single with power divided into two packages—reject. Loss of either engine before reaching maneuvering altitude usually means an off-airport landing. But not always—and there's the conundrum. It takes practice and near-perfect execution.

In single or twin, think angle of attack:

The full-power climb attitude is perfect to stall when the (an) engine takes the day off. Holding the climb attitude and hoping things will somehow get better is a bad idea—controlled flight will cease in a few seconds. The aircraft will pitch over and assume a glide attitude if you let it. Depending on the aerodynamics, it could be close to optimal glide speed—or not. Trim for best glide (or blue line in a twin, corrected for weight). Holding a lot of back pressure, the natural reaction, means a stall is imminent. This maneuver is best practiced in a simulator or training device, provided it approximates the control feel. It can also be done at altitude, but be careful about just chopping power in training. That can mess up crankshaft counterweights, which may give the engine severe indigestion later in life.

No runway ahead? Look for any open area, but if none, just land under control:

That's the best you can do. The belts and shoulder harnesses should be cinched up tight before taking the runway.

Impossible turn or merely improbable?

The improbable turn to get back to the airport is impossible below a certain altitude. For trainers and light sport aircraft, that altitude might be about pattern altitude, but it's significantly higher in heavier machines. It's been studied and debated for decades, and the best we so-called experts can mumble is that "It depends." On crosswind leg, things start to look a little better. Partial power is a gift, but how long will it last?

There's anecdotal evidence and crashes to prove it—that even practicing the improbable turn after takeoff may be as deadly as actual events. If you decide to do this, choose the training environment and instructor very carefully. Practice initially at least several thousand feet AGL. In a takeoff scenario with the fuel pump on high, the engine at idle will probably flood. Don't create a real emergency while training for one. Agree on when to quit if there's any doubt about a successful outcome.

Forty-five degrees is considered the optimal bank angle:

It provides the fastest rate of turn with only a 20 percent increase in stall speed. The nose will want to drop in the turn—let it. But the sight picture will be alarming this close to the ground, and the consequences of anything less than perfect technique will be bad. Perhaps it's better to forget the runway and concentrate on any open space straight ahead or 30 to 40 degrees on either side of the runway heading.

There's no time to fumble with a checklist in a takeoff emergency:

Memory items are just that. I've added a final written reminder into my before-takeoff checklist – **PPS - PP:** Pitch to glide, fuel pump on, switch tanks. If no start, feather the prop (if controllable) and turn off the fuel pump.

If there's a strong crosswind and all else being equal, turn into it—to stay closer to the runway. If getting back to the airport is not going to work, that headwind becomes a big help in slowing the touchdown speed in the crash. A strong headwind on takeoff helps initial climb but becomes a big enemy on a turn back. The tailwind increases touchdown speed and impact forces. The rule of thumb says that if the tailwind component is more than 10 knots, stay the course and land into the wind. Factor all this into pre-takeoff planning.

Did someone say takeoffs were simple?

Much of this has not been scientifically tested in the broad variety of aircraft we fly. It's just a best guess, and as with takeoff/landing distances, the average pilot will likely not match all the test parameters, from pilot skill to wind conditions. There are no statistics on successful returns right after takeoff because they don't hit anyone's database. Many factors go into a successful return—total power loss or partial? At what altitude? Aircraft glide characteristics? Wind conditions? Crosswind runway? Pilot training and proficiency? I've probably left out something. It's much better to pay close attention to engine condition, behavior, and maintenance. But if it happens—have a plan. How's that for straddling the fence?

Twins: Think fast and accurately

FAR Part 25 has extensive guidance regarding takeoff distance for transport category aircraft (those with more than one engine). Balanced field length provides the option to stop on the remaining runway or continue takeoff after an engine has failed. However, for light twins, there are no guarantees—zero, zip, nada. For piston aircraft, stopping typically requires less distance than continuing, and there's no guarantee that continuing is even possible.

Piston twins lose 50 percent of power and 80 to 90 percent of climb capability after an engine stops. This is not to rekindle the single-versus-twin argument, since there are no records showing how many pilots have successfully managed a single-engine emergency in a twin. There's twice the maintenance with a twin and twice the chance of an engine-out emergency. Regular practice is essential. Many precautions are needed to realistically practice in-aircraft engine-outs with any degree of safety, and quality simulation does a better job. The airlines and most corporate operators abandoned in-aircraft engine-out practice decades ago.

When doing twin demo flights for Cessna, we were always careful to have about 5,000 feet AGL and half fuel on board before simulating an engine shutdown. If things started to become directionally complicated, the bailout

391

was to reduce power on the operating engine, trade altitude for airspeed, and recover. With motion visual simulators, it was a revelation to see how the same aircraft that was relatively docile in the demos became a beast in high-risk situations. A simulation of coming out of Denver at gross weight and summer density altitudes showed that, in most cases, the performance just wasn't there. The twin was a single with twice the opportunity for an engine failure.

Hangar Story

At FlightSafety (FSI), a prospect would come in for a demo on one of our visual-motion piston twin simulators. I was always glad to oblige, and of course, they would be expecting an engine failure. Beforehand, I'd make a wager that if they were able to land successfully, even off-airport but under control, I'd buy lunch. But if they crashed, they might want to seriously consider taking the course. In 11 years, I bought four lunches.

In-flight fires don't "just happen"

They often indicate deferred or improper maintenance. Engine compartments, exhaust systems, turbos, clamps, and fuel/oil lines must be carefully inspected on a regular basis and replaced when needed. For example, the clamps used on aircraft exhaust systems are specifically designed and typically expensive. Often they are time-limited and must be replaced at prescribed intervals. Some do-it-yourself types, and sadly some mechanics, have used hardware store materials with disastrous results. Engines and certain systems are flight-critical.

Aftermath of an engine fire, NTSB

Critical time for cabin fires

Carrying a fire extinguisher buys critically needed time for cabin fires. Halon derivatives are recommended because they create much less mess than chemical extinguishers, though they tend to cost more.

> *"Electricity is really just organized lightning*
>
> — *George Carlin, comedian*

With an electrical fire (See "Something's Burning - this chapter"), advise ATC and shut off the master switch. Troubleshoot, but be careful about resetting circuit breakers. The first indication of overheating is usually the smell of burning insulation. Do not assume the problem will somehow resolve itself. Smoke indicates a much bigger problem.

Engine fires are extremely rare and extremely dangerous. Confirm the fire and shut down immediately. Look for an off-airport site unless you're already in the pattern. In too many cases, the nearest airport isn't near enough. A forced landing on your terms may not be great, but it will be much better than any other alternative.

Cabin fires are also extremely serious. Today, the concern is lithium-ion (LI) batteries that power our iPads, cell phones, and laptops. Talk about "burner" phones. According to the FAA, between 2006 and 2023, there were 481 incidents involving LI batteries. These occurred on air carrier or cargo aircraft that are well equipped to deal with the challenge. There are no statistics for general aviation.

General aviation is less prepared. These electronic devices catch fire frequently enough that airlines have containment bags in the cabin and fire extinguishers in the cargo holds. However, most extinguishers will not put out an LI battery fire. My untested solution is a set of BBQ gloves and a large pair of pliers. Open the door, discard the iPad, and land nearby to close the door. I suspect the iPad will not be worth going back to look for. This method has never been tested, so my usual worthless guarantee applies. When buying replacement batteries for your devices, do not go with the low bidder. Get the best quality available.

We're going to Crash—Now What?

"If you're faced with a forced landing, fly the thing as far into the crash as possible"

R.A. (Bob) Hoover – One of the greatest test and air show pilots

Bob's advice is pure physics. Detailed crash survival is beyond the scope of this discussion, but age, health, and physical fitness all play a role, according to medical literature. The typical sudden-stop crash impact lasts less than one-quarter of a second, and the G force is directly tied to speed. A high-speed stop can generate over 100 Gs, which is not survivable as internal organ attachment points fail.

In the most general terms, humans can tolerate about 40 Gs from front to back, and about half that in the other two axes. The table below shows that as the time to stop decreases by half, the G force doubles. Slowing down before impact is critical, and flaps help. Crashing downwind is not helpful either, because the groundspeed is additive in that case. A strong headwind will help a lot if the touchdown location is favorable. The G numbers have been rounded for easier readability, and this is an approximation.

Speed - KTS	1 sec/Gs	.5 sec/Gs	.25 sec/Gs	.12 sec/Gs	.06 sec/Gs
60	3.1	6	12	24	48
70	3.7	7	15	30	60
80	4.2	8	17	34	68
Survivability	Great!	OK	Not-so-good	Very bad!	Game Over!

Fuel Burns – The fuel system needs to remain intact and not spray fuel over a hot engine. The U.S. Army began installing flexible fuel bladder systems decades ago in its helicopters. Affectionately known as "Robbie Tanks" after Dr. Harry Robertson, who invented them, these tanks have saved countless military helicopter crews from burning. The NTSB asked the FAA to require Robbie Tanks or equivalent systems in new helicopters. It would be nice to see them in GA airplanes as well. Yes, it adds to the cost and a bit to the weight, but it will greatly reduce the chance of fire.

A really hard landing, NASA

Fly into any crash scenario as opposed to lawn-darting into the ground after a stall or going inverted. Please remain seated. If the aircraft is a "classic" and doesn't have shoulder harnesses—please get them installed ASAP *before any other upgrades.*

"The angle of arrival determines survival"

Unknown

NTSB still investigates too many fatal crashes where the occupants would have survived with little or no injury if they had been properly belted in. Don't like that idea? Then wear helmets. The hard truth is that there is an element of luck in where the flight comes to a stop. The airframe parachute has proven its worth if deployed appropriately.

Spreading the impact both longitudinally and vertically makes the crash survivable and greatly reduces injury. Fly between solid objects where possible, and take solace in the fact that, with a little luck, this will make for a great hangar story. Just keep flying until all the parts stop moving. As noted, all you need is a second or so to keep the forces survivable.

Closer to touchdown, run GGUFMMMP again or use the checklist. Just get 'er done and do NOT stall. Here, we undo everything that was checked earlier—note the added "F." This is all done on a time-available basis, and it may vary based on the aircraft:

Glide – Vitally important. Not too much or too little speed. Feather or move a controllable pitch prop to low RPM as it significantly reduces drag.

Gas – Fuel selector "Off" – reduce fire chances.

Undercarriage – Depends on where you're landing. Water – leave it up; most everywhere else, probably down once landing is assured. That's a judgment call.

Flaps – Reduce speed to make the crash more survivable – after landing is assured.

Mixture – Idle cutoff.

Mags – Off.

Master – Off when done with anything electrical – gear/flaps to reduce stall speed, etc.

Pump – Off if it's still on.

Here are a few other items that will help with a successful outcome:

Seat belts tight.

ELT – Activate if possible, and do the same with your Personal Locator Beacon (PLB).

Transponder to 7700 – if not on an ATC-assigned code.

Front cabin doors – Unlatch and wedge something into them to keep them from jamming on impact. On many aircraft, after opening the door, moving the door handle to the "lock" position will extend the locking pin and keep the door from closing. Gull-wing doors are problematic if the aircraft goes inverted, so work on making a controllable touchdown.

If all this seems self-evident, it is. But every year, there are hundreds of crashes of all types with no planned reaction and a worse outcome than if a few things had been done right. Practicing and thinking about the unthinkable can make it better. More positively, only 25% of GA crashes are fatal, but it has to be the right kind of crash.

Don't worry about saving the aircraft. Yes, it's your pride and joy, or it's a rental. Sacrificing the hardware is far better than sacrificing yourself or your passengers. Let's turn the traitor into beer cans or composite decking.

Can we just quit stalling? The NTSB put Loss of Control (LOC) on its Top-Ten list some years ago. A stall in VMC and a spiral in IMC are two completely different animals with similar outcomes, and the risk management for each is different. Let's start with VFR stalls.

Only about two percent of stall-related fatal accidents occur on training flights, according to the FAA. The accepted training procedure is to climb to a safe altitude, clear the area, and gently ease up to the edge of the flight envelope. Everybody is waiting, watching, and ready to recover either at the first indication or right after the break. What's missing is the surprise and distraction that makes real-world stalls so pervasive and potentially deadly. The pitch attitudes in practice are often well above the horizon, leading to the erroneous concept that if the nose is pointed low, all is well. The stall attitude with full flaps should change that thinking a bit—likewise, so will an accelerated stall (but not with flaps extended).

> *"Death is just nature's way of telling you to watch your airspeed."*
>
> *– Anonymous*

In the Automation Chapter, we discussed the crash of Air France 447, an Airbus A330, that plunged more than 30,000 feet into the South Atlantic Ocean. The automatic envelope protection system and the bewildered crew all ceased to function.

Colgan 3407, a Bombardier Q400 on autopilot, leveled at the final approach fix altitude near Buffalo, New York. In altitude-hold mode, the autopilot performed as designed and commanded the aircraft into an impending stall. It then handed the aircraft off to the surprised captain, who finished the job (Chapter 2, Colgan Calamity).

The crew mis-programmed or misunderstood the auto-throttles on an Asiana Airlines Boeing 777 and failed to recognize the stall (Chapter 5, Too Low.)

For GA, stalls usually happen in the pattern. Stalls can be prevented if pilots visualize that the aircraft must be pointed close to where it's actually going, not where the pilot thinks it's going. For reasons known only to bureaucracy and some obstinate airframe manufacturers, we still don't require angle-of-attack (AOA) indicators. These simple devices measure what the wing is actually doing.

Angle of Attack Indicator, FAA

The Luddite argument is that if the inattentive pilot doesn't pay attention to airspeed and warning horns, why would they pay any more attention to an AOA? I'm going to saw myself off a tree limb here, but if pilots were trained from the very beginning to fly the wing rather than the airspeed indicator, they might not experience as many low-altitude stalls. The military almost exclusively flies with AOA.

"Get rid at the outset of the idea that the airplane is only an air-going sort of automobile. It isn't. It may sound like one and smell like one, and it may have been interior decorated to look like one; but the difference is—it goes on wings."

Wolfgang Langewiesche, Author of Stick and Rudder

Wolfgang Langewiesche's classic book *Stick and Rudder*, written in the 1940s, is still the definitive work. Perhaps not too far off, we may see full envelope protection that's 99.9999 percent reliable on new aircraft. Legacy aircraft will have to be flown as they always have until retrofit protection is installed, if ever.

FAA: Over the past decade, more than 70 percent of all GA stall-related accidents occurred in the traffic pattern. It's a high-workload environment coupled with multiple distractions.

Takeoff: If the runway is shorter than what's needed to clear an obstacle, no amount of nose-up is going to change that. A longer runway is needed, or a lighter load.

Base-to-final turn: We've already discussed the tailwind on the base leg that leads to an overshoot and the strong desire to wrack it around to the runway centerline (Chapter 3, *Takeoffs and Landings*). Our attention should be on alignment with the runway and bank angle. If you cheat with rudder to slew the nose around, all the ingredients for a cross-controlled stall and spin are baked in. That may have been mentioned before in the literature.

A slower aircraft ahead: He just missed the first turnoff and now he's blocking your runway. Maybe slowing down and increasing the S-turns can salvage this. Really? The other pilot's speed or position is irrelevant to your minimum safe control speed. The physics won't work, regardless of the miscreant's lack of airmanship. Fly the pattern and final approach on your terms, not the other guy's. Go around early, and don't depend on someone else's airmanship. Leave more space next time.

Go Around: Power (smoothly), right rudder, pitch slightly up, flaps, and gear in that order. Practice with a CFI until it's second nature. Do this as often as necessary to develop and retain the mental and muscle memory.

Engine failure: Just keep flying.

Training: In training, how should stalls be taught? One traditional view is to always take it to the break and perhaps beyond. Provoke the aircraft to see how it behaves. The other side advocates that as soon as the aircraft gives an indication of a stall—be it a horn, warning light, buffeting, or mushiness—recover. This is what you should do in the traffic pattern or really, any time. Train like you fly and fly like you train, in other words. This doesn't preclude initial and periodic demos with some practice of full stalls to demonstrate what happens without intervention. But the idea in normal flight is not to let it ever get close to the break.

The FAA's thinking has been all over the airspeed and controllability spectrum: First it was slow flight, then minimum controllable airspeed (MCS), and then back again. There really isn't a good reason to operate at MCS other than to show you can—and that proves exactly what? The purist will argue that it shows mastery. The converse view is that, except in the academic sense, in the real world, you recognize that the AOA is getting way too high long before that. Take your pick.

After a number of airline crashes, when the FAA was researching recovery from thunderstorm windshear in the 1980s, it was thought that flying the aircraft out on stick shaker and max power would work. But if the wing isn't flying before power is added, unless there's a really good thrust-to-weight ratio, power will introduce more rapidly changing vectors into an unstable situation. Remember that backside of the power curve thing?

Current guidance is to move away from minimum altitude loss after a stall and instead, regain control first. Having flown with students who had forgotten about the rudder despite encouragement from the right seat, the concept of regaining control before adding maximum power strikes me as a great idea. Ramming the throttle home before control is regained often results in a powered spin cycle—great for washing machines, but not for airplanes. Power is destabilizing when the wings aren't flying. A few military aircraft may be able to power out of a stall, but if the aircraft is mushing downward, point it that way, get the wing working, and then bring in power.

For jets flying the high flight levels in very thin air, there is little space between Mach overspeed and stall—the "coffin corner." It's not unusual to trade 5,000 feet of altitude or more. It requires a pitch down of about eight degrees to recover from a high-altitude stall in heavier jets. So what? At altitude, you've got altitude to lose, so use it. Best done in a simulator.

Spins: Spin requirements for private pilot certificates were dropped well over half a century ago, but that debate continues. The statistics support the FAA's decision since more people crashed in spin training than in actual spins, but there will always be anecdotal evidence and strong opinions to the contrary. Rich Stowell's book *Stall/Spin Awareness* is recommended for a thorough review of the topic.

Spin training for my initial CFI certificate was perfunctory at best: one lesson, a demo, and a practice recovery once in each direction. There were occasional real-life opportunities provided by students, usually in a departure stall where they didn't quite understand the coordination bit. (I thought I'd explained and demoed well, but perhaps not.) There was a very thorough debrief. Next lesson, we'll do it again—and this time, remember your feet and relax back pressure to get the nose down (no need for negative Gs, thank you).

Spin training from a well-qualified instructor in an aerobatic aircraft is a great idea, and if not done sooner, consider post-certification training. A non-aerobatic but spin-approved aircraft is also fine, provided it is well-maintained. Be mindful of the center of gravity. Some aircraft have a dual personality. A Cessna Skyhawk is only approved for spins when it's in the utility category. Put someone or something in the back seat or baggage area, and a spin may become unrecoverable because it has now shifted into the normal category. Piper Cherokees and Warriors, despite their benign stall behavior, are not approved for spins at any time. After just one turn, no one knows if recovery is possible—just don't attempt it.

A little upside-down time doesn't have to be tremendously gut-wrenching. A good aerobatic instructor will instill confidence and skill. Experimentation with non-approved aircraft, despite some of the idiocy depicted on the Internet, makes you a Darwin Award candidate.

Spirals: Some mistakenly refer to spins and spirals in the same breath, but they're two completely different animals. In a stall, the angle of attack

(AOA) is high, and the aircraft is not going where the nose is pointed. In a spiral, the aircraft is pointed exactly where it's going—somewhat forward and mostly down. We'll be on the ground shortly, folks. Quick and correct action is essential in both cases. If control is lost in the clouds or at night, after some meandering, the aircraft will typically wind up in a spiral. A non-survivable crash typically ensues.

Students and CFIs practice garden-variety spirals in training under the guise of "unusual attitudes." But this training doesn't serve very well when confronted with the real thing, not unlike stall training. In training, we don't want to go above maneuvering speed to avoid over-stressing the aircraft during recovery. In a real spiral, the airspeed will be much higher, and the low-speed recovery technique typically taught has a critical flaw. We are sometimes trained to think that if the nose is low, it must be raised by pulling on the control yoke. In a well-developed spiral, PUSH is the proper response.

Here's why: Assume the aircraft is trimmed for level flight at 120 knots. Push the nose over, hold it, and let the speed build to 10 to 20 knots above cruise. Stay comfortably below maneuvering speed for the demo. Keep the wings level and let go of the yoke. The aircraft will pitch up, go through a series of oscillations, and return to trim airspeed—this is positive stability in technical terms. Have plenty of altitude and make sure there's no traffic below.

A well-developed spiral, FAA

This is for discussion purposes only. Allow a wing to drop and watch what happens. With asymmetrical lift the aircraft goes into a tightening turn. Speed builds quickly because the flight was at cruise power and gravity helps to pull the aircraft down. With a few seconds of inattention or confusion after the spiral has developed, the airspeed will accelerate into never-never land—say, 160 knots or more in a trainer.

Recovery After the Spiral Has Developed:

1. Power to idle.

2. Roll back to level flight and PUSH smoothly on the controls. The aircraft is going to seek trim airspeed (120 knots in this example) very quickly and will pop up like a cork if you do nothing. Above maneuvering speed (weight dependent), the risk of over-stressing the airframe is almost guaranteed if mishandled. G-forces may be significant, and you'll feel the aircraft start to pitch up aggressively, so balance is essential in response. Full forward elevator certainly isn't needed—balance and finesse are key when above V_a. Likewise, not hitting the ground is highly desirable.

3. Return to level flight, and when the speed drops back into the green arc, smoothly bring power back up to cruise. Let's hope there's enough altitude to recover. With a retractable, put the landing gear down as soon as possible during recovery—the gear doors may depart, but so what? Don't blow the doors off in training—the insurance company will be upset.

If not instrument-rated and encountering IMC, TURN ON THE AUTOPILOT, if so equipped. In a retractable, put the gear down right away before the spiral develops. It will prevent the rapid speed buildup when pitch control is lost. Many new autopilots have an "Aw Shucks" mode that will return the aircraft to level.

As with stalls, spiral prevention is FAR better than cure. VFR pilots in IMC often develop a bad case of vertigo, and the odds of recovery are not good. I've had one really well-developed case of vertigo after a series of prolonged spins (13 turns) in a Pitts with a well-qualified instructor in VMC. Once the internal gyros have tumbled, it will take a massive amount of willpower to level the wings, even in VMC. It took about ten seconds to semi-reorient myself and overcome the strong inclination to roll back into the turn. It was several minutes before the sensory illusion completely dissipated, even with a full horizon outside the windshield.

> **"Risk can be reduced but never completely eliminated."**
>
> *(Somebody smart must have said this; if not, I'll take credit.)*

Walking, bicycles, e-scooters, and pogo sticks are included. There are many ways to parse statistics, and as one knowledgeable observer noted, "If you torture the data enough, it will confess to anything." The old saying about the most dangerous part of the trip being the drive to the airport is true for the airlines, but not for light GA. On a fatality-per-vehicle-mile basis, light GA is roughly on par with motorcycles, with one major difference, as near as we can equate them. In about half of motorcycle crashes, it's caused by another driver. In light GA, crashes are about 80 percent pilot-induced—where the cause can be determined.

Down and Locked: (A Simple Gear Malfunction Leads to a Major Crash)

A minor malfunction turns into a major crash and the death of a U.S. senator.

Pilots sometimes turn a minor mishap into a major accident. Breathless media reporting notwithstanding, a landing-gear malfunction on a retractable-gear aircraft should be no more than a belly slide, some skin and prop damage, and an engine teardown. Not to worry—the airport management will have the runway cleared in a very short time. Heroic attempts to preserve the aircraft are sometimes successful, but they can also have horrendous results. This one did, with flaming wreckage falling into a schoolyard, the death of U.S. Senator John Heinz of Pennsylvania, and four pilots.

On April 4, 1991, a Piper Aerostar departed from the Williamsport-Lycoming County Airport in Williamsport, Pennsylvania, at 10:22 a.m. Eastern Standard Time for an IFR flight to Philadelphia International Airport. The

Piper Aerostar, AOPA

flight was operated as a charter, with a captain, copilot, and one passenger (Heinz) on board. The weather was VFR, and the flight was cleared for an ILS approach to Runway 17 shortly after noon.

While on the approach, the captain reported that the nose landing-gear light had failed to illuminate. This could be caused by anything from a burned-out bulb to a hydraulic leak or an electrical problem.

Help is on the way: Just as the Aerostar began its approach, a Bell 412 helicopter departed from its company helipad at Philadelphia for a local VFR flight with two seasoned pilots aboard. The pilots heard the communications regarding the unsafe nose-gear indication. The Aerostar was told to maintain 1,500 feet to allow the helicopter to pass underneath. As the Bell passed under the Aerostar, one of the Bell pilots said, "Looks like the gear is down."

The Aerostar pilot acknowledged the helicopter's transmission and replied, "I can tell it's down, but I don't know if it's locked...." The pilot may have been referring to a reflection of the nose landing gear visible on the propeller spinner from the cockpit. The tower acknowledged the transmission, advised that the helicopter was no longer a factor, and cleared the Aerostar to land on Runway 17.

The controller then declared an emergency, and the tower supervisor alerted the airport's rescue and firefighting units. Runway 17 arrivals were terminated, and a clear communications channel was established. So far, everything was proceeding toward an uneventful outcome. At 12:03:35, the controller offered the Aerostar the option of making a low pass by the tower to observe the position of the nose gear, which the pilot accepted. The tower observed that the gear appeared to be down.

Tower flybys do no harm, but they don't help much either. The tower will not be able to make a positive determination one way or the other. The pilot's options remain exactly the same as before: treat the gear as unlocked and land the aircraft with as much weight shifted as far aft as possible if it's the nose gear. If it's the main gear, the procedure is to hold

Bell 412, Bell Helicopter

the affected wing aloft as long as possible and anticipate that there will be a sudden turn to the failed side. Consider that landing with the gear completely up is likely to result in less damage and easier directional control.

Formation – The controller asked the Aerostar to enter the downwind leg for Runway 17 and advised that the Bell helicopter was in position to take a further look at the nose gear if the Aerostar pilot approved. The controller provided vectors until the pilots saw each other and joined up in formation.

The aircraft were now on an extended downwind leg for Runway 17 at about 1,100 feet. At 12:08:52, the Bell first officer told the Aerostar pilot, "We're going to come up behind you on your left side, so just hold your heading." The Aerostar pilot responded that antennas were ahead and he might need to change heading by 15 degrees to the left.

At 12:09:30, the helicopter pilot said, "Aerostar, we're gonna pass around your right side now; take a look at everything as we go by." The Aerostar acknowledged. Around 12:10:00, the Aerostar pilot again stated that the gear did not appear to be down and locked. At 12:10:16, the Bell pilot stated, "Everything looks good." The Aerostar pilot replied, "OK, appreciate that. We'll start in." The last transmission was abruptly terminated by a loud noise. Shortly thereafter, the controller noticed a smoke plume to the north of the airport. Subsequent radio attempts to contact either aircraft were unsuccessful.

Weather – Ceiling and visibility were unlimited. The wind was reported from several stations as anywhere from 8 to 10 knots with occasional gusts to 15. There were no gusts reported at Philadelphia at the time of the collision.

Formation flight at any time is challenging, and even more so when the aircraft are not similar. The helicopter was generally below and behind the Aerostar, so the fixed-wing pilot needed to maintain heading and altitude while the helicopter maintained adequate distance. Witnesses differed on exactly what happened—some saw the helicopter climb, while others said the airplane veered right. It was agreed that the rotor blade hit a wing panel, rendering both aircraft uncontrollable. There were three fatalities on the Aerostar and two on the Bell. On the ground, there were two fatalities, one severe burn injury, and four minor injuries, with more than $6 million in property damage alone, not counting lawsuits. That's a terrible price to pay for what should have been a minor incident at worst.

The Crews – The Aerostar pilot held an airline transport pilot (ATP) certificate with a multiengine rating and a flight instructor certificate with airplane and instrument ratings. He had more than 1,500 hours of single-engine flight time and more than 400 hours in twins. A Part 135 pilot-in-command check in the Aerostar had been administered by the FAA just 10 days earlier on March 26, 1991. The inspector recalled they had covered emergency gear-extension procedures in the oral part of the check. At the time of the accident, the pilot had 72 hours as captain and 42 hours as first officer in the Aerostar.

The accident flight was his second revenue flight as captain of the PA-60. On April 1, 1991, his first revenue flight in the PA-60 had not gone well, either, from a mechanical perspective. The flight was aborted because of a surging engine. The charter passenger, an experienced PA-60 pilot himself, had more than 300 hours in type. He noted that the captain had difficulty starting the engines and needed instruction on proper starting techniques. The takeoff roll was described as "pretty erratic" because the captain overcontrolled the electric/hydraulic nosewheel steering. At altitude, the right engine began to "surge about 200 rpm," which the passenger believed to be a problem with the fuel controller. The captain did not respond to the emergency, and the passenger suggested an immediate return to the airport. Subsequent inspection found a defective fuel controller.

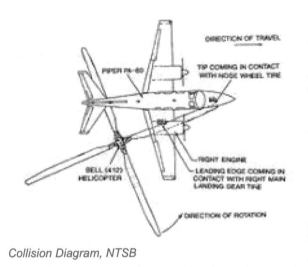

Collision Diagram, NTSB

The Aerostar copilot was not required by regulation, although Senator Heinz insisted on having two crewmembers when he flew. The copilot had about 1,300 hours of single-engine experience and 200 hours in multiengine equipment. He had been on duty the night before from 9 p.m. until 6 a.m. and reportedly slept from about 6:30 a.m. to 9 a.m. While the copilot's flight and duty time did not exceed the limitations for unscheduled one- and two-pilot crews, the NTSB believed it was ill-advised for him to fly, having been on duty the entire previous night.

Both helicopter pilots held ATPs for multiengine rotorcraft, and both had over 8,000 hours of flight time, with several thousand hours in company helicopters. None of the pilots had any record of formation flight instruction. The chief pilot for the flight department that owned the helicopters noted that company policy for "in-flight" inspection of another aircraft was not to get closer than 300 feet, the length of a football field. That's good advice.

By the Book

Oddly enough, the FAA-approved flight manual for the Piper PA-60 did not contain emergency gear-extension procedures in Section 3, the emergency procedures section. However, in the systems description, there is information about what to do if hydraulic pressure is lost. Essentially, the landing gear will free-fall to the down and locked position using gravity and springs when the gear handle is placed in the down position. In addition, the landing-gear warning horn will sound if the throttles are retarded to idle and the gear is not locked. This is a good method to check a gear anomaly, and most aircraft with retractable gear employ a similar method for checking gear position.

The training/check pilot for the charter company said he did not instruct the Aerostar captain on the operation of the landing-gear warning horn but had explained the push-to-test function of the gear indicator lights—not exactly a thorough checkout.

Without a cockpit voice recorder, it is impossible to determine whether the captain took any action to isolate the problem to the indicator light or verify that the nose gear was locked in the down position. He did not mention any tests during tower communications.

The NTSB examined another Aerostar and concluded that while it was possible to see if the gear was down from a safe distance, it was impossible to verify by visual inspection at any distance whether it was locked. This is true of any aircraft, whether viewed by the tower or another aircraft.

NTSB Probable Cause

"... [was] the poor judgment by the captain of the Aerostar to permit the in-flight inspection after he had determined, to the best of his ability, that the nose landing gear was fully extended, the poor judgment of the captain of the helicopter to conduct the inspection, and the failure of the helicopter crew to maintain safe separation. Contributing to the accident were the incomplete training and checkout that the captain received from the charter operator and insufficient oversight by the principal operations inspector for the FAA flight standards district office."

Commentary

This scenario started well enough. Advising the tower of a potential problem is always a good idea. Emergency equipment can be standing by, and technical help may be available to troubleshoot.

The NTSB believed that the Aerostar pilot's inexperience was a significant factor. A more experienced pilot likely would have accomplished any emergency procedures at a safe altitude and landed the airplane, accepting the possibility of minor damage during the landing roll. Incidents like this happen with some regularity and do not become major accidents.

The in-flight inspection should have been declined since there was nothing to be gained. Would you let a complete stranger take responsibility for the safety of your aircraft in such a critical maneuver? None of the pilots had any experience in formation flight, especially with such dissimilar aircraft.

A review of the literature on the aerodynamic interaction between fixed- and rotary-wing aircraft in close proximity noted two potentially hazardous aerodynamic concerns: (1) turbulence-induced blade stall and settling experienced by rotary-wing aircraft while flying in the turbulent area behind and below a fixed-wing aircraft, and (2) opposing pitch changes experienced by both aircraft when one flies close behind and below another.

The textbook *Aerodynamics for Naval Aviators* specifically refers to the case of one aircraft inspecting the landing gear of another. The lower aircraft may experience a nose-up pitching moment and the higher aircraft, a nose-down pitching moment. The author states that the opposing pitch-moment changes can be large and must be anticipated, or a collision may result. Engineers at Bell Helicopter confirmed that the Bell 412 would experience such a nose-up pitch change.

If the gear malfunctions, get away from the ground, out of the traffic pattern, and up to a safe altitude. Then, methodically run the emergency gear-extension procedure. This is one of those "wind your watch" moments. Flybys of the tower and formation passes by other aircraft will not answer the "locked down" question. Insurance companies are much happier to write a relatively small check for mechanical mishaps than for policy limits in case of loss of life.

Something's Burning

(Basic maintenance procedures must be followed)

Be careful resetting the humble circuit breaker

An in-flight electrical fire is one of the worst emergencies that can confront a pilot. Fortunately, they are rare. Like many critical faults, they may start slowly at first and then build very quickly into an unrecoverable situation. The accident aircraft was a 30-year-old Cessna 310R, the only piston aircraft then operated by the National Association for Stock Car Auto Racing (NASCAR) corporate aviation division.

According to the NTSB, this accident was all but certain once the flight began. A casual approach toward a seemingly benign maintenance item turned deadly not only for the pilots but also for people on the ground. Disregarding safety procedures that may be perceived as overly conservative or irrelevant can wind up costing lives, aircraft, and tens of millions of dollars.

Cessna 310R, Cessna

Deferred Maintenance

The day before the crash, another company pilot had flown the Cessna and reported on a maintenance discrepancy sheet: "Radar went blank during cruise flight. Recycled—no response and smell of electrical components burning. Turned off unit—pulled radar [circuit breaker]—smell went away. Radar inoperative."

The pilot told investigators that upon landing, he documented and reported the discrepancy to NASCAR maintenance personnel, "leaving the white original page in the discrepancy binder in the airplane and providing the yellow copy to the director of maintenance [DOM], in accordance with the company's standard operating procedure." He also discussed the problem with the DOM and the aircraft's mechanic. The original maintenance write-up was recovered at the crash site, but the DOM was unable to find the yellow copy after the accident.

No technician inspected the aircraft, and the radar circuit breaker was not "collared" with a tie wrap to prevent it from being reset. Additionally, there was no placard or sticker placed to indicate that the equipment was inoperative. At least one or more of these actions is required under FAR Part 91.213. The NTSB

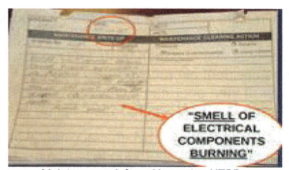

"SMELL OF ELECTRICAL COMPONENTS BURNING"

Maintenance deferred log entry, NTSB

further determined that the night before the accident flight, NASCAR's chief pilot called the ATP who was to act as the safety pilot to advise him of the radar discrepancy. Moreover, a maintenance technician informed the ATP of the issue in person on the morning of the accident flight. On both occasions, the ATP dismissed the issue as unimportant. It is unknown whether he was aware of the burning smell or just that the radar wasn't functioning..

The Flight – On July 10, 2007, around 8:22 a.m., the Cessna 310R departed Daytona Beach, Florida, on an IFR flight plan to Lakeland, Florida. The weather was good VFR. Two pilots were on board—the pilot in command, a commercial pilot with about 270 hours, and the corporate ATP (with more than 10,000 hours), who had been briefed about the discrepancy, seated in the right seat as a safety pilot.

Shortly after reaching a cruise altitude of 6,000 feet, at about 8:32 a.m., the ATP advised air traffic control that they had an emergency, stating, "Smoke in the cockpit, we need to land at Sanford." The controller cleared the flight direct to Orlando Sanford International Airport (SFB) and issued a descent to 2,000 feet.

The last radio transmission was received at 8:33 a.m. and ended mid-sentence with the phrase, "Shut off all radios, elec[trical]." This was also when the last transponder return was received, indicating that the master switch had been turned off. The Cessna was eight nautical miles northwest of SFB, turning toward the airport and rapidly descending. For the next 90 seconds, primary radar returns showed the aircraft heading toward SFB. The last radar return was recorded at 8:34:45 a.m., when the flight was about three miles northwest of SFB and descending through 1,200 feet AGL.

The Crash – The 310 crashed in a residential area about 0.7 nautical miles west of the last radar return. According to the NTSB report, witnesses said, "The airplane was traveling 'extremely fast,' was 'very low,' and its wings were 'rocking' as it descended. Just before impact, the airplane entered a 'steep bank' and made a sharp turn to the west. Several witnesses reported seeing smoke trailing from the airplane, and one witness stated, 'Smoke was trailing from the port side.'"

The Cessna first struck some trees in a right-wing-low attitude about 65 feet AGL. Then, 270 feet beyond the first tree strike, it hit a palm tree about 20 feet AGL, grazed the corner of a house, and crashed into the next two houses along the street. The massive fire that ensued resulted in the deaths of both pilots, three people on the ground, and four people suffering serious injuries.

Forensics 1.0 – The forensic analysis that follows many fatal crashes is painstaking. Not surprisingly, the findings also heavily depend on who's conducting the investigation, which we'll get to shortly. Much of the fuselage, wings, instrument panel, some avionics, seats, and the right engine were found in and around the second and third homes.

For want of a circuit breaker collar, NTSB

Although most of the airplane was destroyed during the post-impact fire, NTSB investigators noted discolorations or soot deposits on airplane parts that were not part of the ground fire. The underside of the instrument panel glare shield and deck skin, located on the roof of the first house, showed thermal damage with discolored, charred paint and soot deposits, all indicative of an in-flight fire. The unburned cabin door, found about 60 feet away from the main wreckage, had numerous soot deposits on the interior side, indicating the pilots opened the cabin door to vent smoke.

Most of the electrical system components and associated wiring were severely damaged or destroyed, and little electrical insulation remained. Examination of some small sections of recovered wiring and one partial wire bundle found among the fuselage wreckage showed characteristics of strand fusing and globules of re-solidified copper. According to the NTSB, "That may be consistent with electrical arcing and/or exposure to heat from the post-impact fire. Too little remained to positively identify which systems those wires were associated with or to determine when the observed characteristics were created."

Flight instruments, avionics, controls, switches, and circuit breakers had severe post-impact fire damage and yielded no usable information regarding their pre-accident configuration or condition. Parts of the weather radar and some attached circuit boards exhibited severe impact and thermal damage; however, there was no evidence of an electrical fault.

The aircraft was built in 1977, and at the time, the FAA allowed the use of polyvinyl chloride (PVC) insulation, but it has not been used in decades, since it gives off incapacitating fumes when burned.

Forensics 2.0 – As mentioned earlier, perception of fact is in the eye of the beholder. NASCAR, as a party to the investigation, came to a significantly different conclusion than the NTSB. In its filed comments, which are summarized for brevity, it was noted that:

- The fire did not originate in the radar unit (the NTSB concurred).

- The fire likely originated in wiring installed behind the panel by Cessna (the NTSB concurred on the likely location of the fire but could not determine which wiring was specifically at fault. The NTSB further noted that the aircraft was operated by multiple owners with numerous modifications over the years. A non-factory replacement radar unit and associated wiring was installed in 1988, so the conclusion that Cessna wiring was the source is speculative).

- The factory-installed PVC wiring on the Cessna 310 was not in compliance with the CAR 3 regulations (the NTSB did not concur, noting that the rules allowed manufacturers to continue to use PVC in aircraft in which it was already being installed).

NTSB Probable Cause – *"... were actions and decisions by NASCAR Racing Corporate Aviation Division's manager and maintenance personnel to allow the accident airplane to be released for flight with a known and unresolved discrepancy and the accident pilots' decision to operate the airplane with that known discrepancy, a discrepancy that likely resulted in an in-flight fire."*

Commentary – The NTSB found that the radar anomaly the day before could easily have developed into a fire, except that the pilot pulled the circuit breaker, or perhaps it tripped. Severe damage to the 310's wiring harness might have already occurred, and the insulation was badly compromised. The accident crew likely reset the breaker during the prestart checklist, following the pilot's operating handbook guidance that one reset is allowed.

By not following good maintenance procedures, such as possibly removing the radar, collaring the breaker, and making a log entry, the aircraft was not airworthy. If those things had been done, and the radar was not essential to safe flight (as determined by the Minimum Equipment List [MEL] and the mechanic), the aircraft could have flown legally and safely.

411

As soon as in-flight fire is evident, if it cannot be immediately extinguished, an off-airport landing is recommended because it's impossible to know how fast and how bad things are likely to get.

The pilots were aware that the radar was inoperative but may not have understood that the wiring had been compromised, as evidenced by the burning smell. When power was restored to the affected circuit, it took only a few minutes to develop into a major fire, and once adjacent wiring was involved, pulling circuit breakers and shutting off the master switch had no effect. If you notice any sort of burning smell, the aircraft is unairworthy and must be thoroughly inspected by a competent technician before flying again.

The existing guidance in POHs of the era and training of GA pilots traditionally allowed one reset of a breaker after it cooled. Current thinking, backed up by the tragic events here, is that if a breaker trips and it is not critical to flight, do not reset it—even once. More attention must be given to aging wiring by pilots, owners, and maintenance technicians.

NASCAR filed a multi-million-dollar lawsuit against Cessna, alleging improper wire was installed on the 310R. It was summarily dismissed.

Trust but Verify

(Complex machines require careful reassembly)

Aircraft have to be maintained and reassembled correctly.

The quote, "First, do no harm"—often erroneously attributed to the Hippocratic Oath—applies as much to aircraft mechanics as to doctors. After digging into the patient, remove all foreign objects and properly reconnect things. Too many emergencies, incidents, "almosts," and crashes occur coming off the shop floor. It's so obvious in hindsight.

An ASRS report on a Cessna Citation business jet shows how little things can lead to big problems. Shortly after takeoff, the right engine anti-ice failed. Cruise flight was normal, but then a cascading series of malfunctions occurred: left windshield heat, left engine control fault, followed by L fuel boost. A "large thump" ensued with the left engine failing. The crew shut it down and proceeded with a single-engine landing. Subsequent inspection revealed that a clamp on a pneumatic duct had failed, allowing hot air to fry the Full Authority Digital Engine Control (FADEC). This isn't something that would be found on a preflight. Was the proper clamp properly installed?

Critical components must be overbuilt. Barring that, a fail-safe mode is acceptable, but remember that those requirements generally add weight and cost. In piston aircraft, where the exhaust clamps were run well past normal limits, an automotive clamp that wasn't up to the rigors of the aviation environment led to an engine fire and several fatalities. It's happened more than once that substituting substandard parts is a really bad idea, even if it saved a little money on the front end. There are places, however, where requiring full FAA approval—especially when a failure will be inconsequential—is overkill. Static wicks, for example, and I can think of quite a few other things.

Misrigged – A landmark maintenance accident occurred in January 2003, but the fault didn't appear on the first several flights until the right set of circumstances created a tragedy. A Beech 1900 regional airliner, taking off from Charlotte, North Carolina, crashed when the crew was unable to maintain control. The elevator trim system was mis-rigged, and the aircraft was loaded well aft of the allowable center of gravity (CG). Multiple systemic failures were involved, including poor training of the mechanics, inadequate oversight by the FAA, and poor checklist discipline by the maintenance shop and quality control supervision.

With an aft CG, the pilots were unable to counter the massive nose-up flight characteristics. The crash was caused by several factors, including the increased weight of the traveling public—the 170-pound standard used at the time was too low. The incorrect rigging restricted the elevator's down travel to seven degrees, half of the Beech specification. Both pilots were unable to force the nose down as the 1900 pitched up to more than 50 degrees before stalling.

Thankfully, trim misrigging doesn't happen often. Beech issued a mandatory service bulletin in 1991 to prevent technicians from interchanging elevator trim mechanisms. Bonanzas, Barons, T-34s, and other aircraft with similar configurations have dual elevator trim actuators unique to the right and left sides of the elevator—if reversed, they will operate in the opposite direction of normal. Beech's solution was to color-code the actuators (left blue and right black). But why not red and green? Their best advice is to verify compliance after maintenance is complete.

If every technician and pilot did a manual ground check of flight controls—checking all controls, including trim tabs—before engine start after any flight-control maintenance, this type of incident would be greatly reduced. It should be standard procedure.

A NASA Aviation Safety Reporting System (ASRS) report outlined a near-accident involving a Cessna 182. The mechanic was highly experienced and had never mis-rigged a trim system before. But the elevator trim was reversed, so that up was down and vice versa. The two young pilots on a test flight nearly lost the aircraft before figuring out that it was "backwards day."

Complacency and Distraction – Our old nemeses, complacency and distraction, are just as prevalent on the shop floor as they are in the cockpit. Checklist discipline is necessary to ensure that no tools are left behind as stowaways. Tools have the uncanny ability to lodge in dark places that may not reveal themselves until well after the aircraft is airborne, and perhaps not even until after many flights.

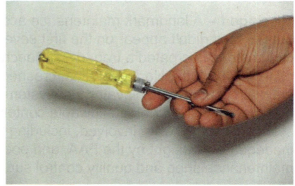

Small forgotten tools can do a lot of damage,

All tools are accounted for, all inspection plates secured, fluids topped off, clamps clamped, hoses connected, bolts and screws tightened, etc. The bigger the job, the more cautious the pilot should be. On test flights, stay close to the airport and operate in daylight visual conditions.

A U.S. Air Force F-35 fighter, valued at over $100 million per unit, sustained $4 million in damage when a flashlight was left in the engine inlet and ingested.

Technicians are often interrupted by phone calls, consultations, or to lend a helping hand to a colleague. As previously noted, humans do not divide attention well. We've learned the hard way that when a checklist is interrupted, it's best to start over from the beginning—just to be sure. The same applies on the shop floor.

Procedures Are There for a Reason – (You may have heard that before.) Following the manufacturer's procedures exactly is essential. Misplacing a washer, locknut, or using the wrong sealant or grease has led to fatal accidents. For example, a Cirrus pilot lost control when a split-flap condition developed due to missing safety wire that failed to secure a critical item. Fortunately, the parachute saved both the aircraft and the pilot, and the NTSB conducted an excellent analysis to determine exactly what happened. This failure took months and many flight hours to develop.

In 2000, Alaska Airlines lost an MD-83 when the horizontal stabilizer jammed, causing the aircraft to lose control and crash into the Pacific Ocean. According to the NTSB report: "The National Transportation Safety Board determines that the probable cause of this accident was a loss of airplane pitch control resulting from the in-flight failure of the horizontal stabilizer trim system jackscrew assembly acme nut threads. The thread failure was caused by excessive wear resulting from Alaska Airlines' insufficient lubrication of the jackscrew assembly."

MD-80, Failed Jackscrew, NTSB

"Contributing to the accident were Alaska Airlines' extended lubrication interval and the Federal Aviation Administration's (FAA) approval of that extension, which increased the likelihood that missed or inadequate lubrication would result in excessive wear of the acme nut threads. Also contributing to the accident was the absence of a fail-safe mechanism on the McDonnell Douglas MD-80 to prevent the catastrophic effects of total acme nut thread loss."

This report should cure most pilots of the desire to skimp on airframe maintenance. It should be required reading for mechanics. There are several caveats that go beyond the scope of this book, but briefly:

Be sure the technicians working on the aircraft are knowledgeable about that specific type. Inexperience with the make and model is expensive, and the shop may cause as many problems as they fix. They must also have current shop manuals.

There are components with proven life limits, and the guidance for these should be rigorously followed. It's not just about the manufacturer trying to make money. Airworthiness Directives are usually on the mark.

Service Bulletins for Part 91 operations in the United States, even when the manufacturer refers to them as "Mandatory," are optional. However, review them carefully with an experienced technician. It may be an outstanding investment in safety and aggravation avoidance—but sometimes not.

All pilots will benefit from assisting technicians working on their aircraft and learning as much as they can about what to look for during pre-and post-flight inspections. This applies to renters as well.

Assume that the machine will bite until proven that it's back to being that solid piece of hardware that you know and love. Avoid night and instrument conditions until you're certain all the gremlins have gone away.

Flameout! –

The Crew is There to Help – Listen! An overcautious approach can lead to disaster and requires a new approach to cockpit management.

> **"If you are too careful, you are so occupied in being careful that you are sure to stumble over something."**
>
> – *Gertrude Stein, Novelist, Playwright*

Distraction is typically present in almost every accident scenario, and setting priorities is essential. This landmark accident occurred decades ago, but the message remains as relevant as ever. Regardless of the number of crewmembers, some things just need to be said assertively.

On December 28, 1978, United Airlines Flight 173, a McDonnell Douglas DC-8-61, departed Denver with 181 passengers and 8 crew members, bound for Portland, Oregon. The flight-planned time was 2:26, with a scheduled arrival in Portland at 5:13 pm PST. The total amount of fuel required was 31,900 lbs, and with 46,700 lbs of fuel on board— which included the FAR requirement for fuel to destination plus 45 minutes and a company contingency of about 20 minutes—it was adequate for good VFR weather. However, the accident chain had already begun to build.

Abnormal Gear

Around 5:10 pm, as the flight was descending through about 8,000 feet, the First Officer, who was flying, asked for approach flaps and landing gear. The captain recalled, "It was noticeably unusual, and I feel it seemed to go down more rapidly. As I recall, it was a thump, thump in sound and feel. I don't recall getting the red and transient gear door light. The thump was much out of the ordinary for this airplane. It was noticeably different…"

At 5:12:20, the flight was told to contact the tower, but the pilot responded, "Negative, we'll stay with you. We'll stay at five. We'll maintain about 170 knots. We've got a gear problem. We'll let you know." The flight was cleared down to 5,000 feet and given delaying vectors.

According to the NTSB, "For the next 23 minutes, while Portland Approach was vectoring the aircraft in holding south and east of the airport, the crew discussed and accomplished all of the emergency procedures available to them to ensure the gear was down. The lead flight attendant was briefed and told to prepare the cabin for a possible emergency evacuation. About 5:38 pm, the crew contacted the United Airlines Maintenance Control Center in San Francisco, and the captain explained the situation while they continued to troubleshoot.

The captain reported 7,000 lbs of fuel on board and that he planned to hold for another 15 or 20 minutes to prepare the passengers for emergency evacuation. United maintenance concurred, but there would have been less than 15 minutes of fuel remaining.

About 5:50 pm, the captain asked the flight engineer to "Figure about another fifteen minutes." The First Officer responded, "Fifteen minutes?" To which the captain replied, "Yeah, give us three or four thousand pounds on top of zero fuel weight." The flight engineer responded, "Not enough. Fifteen minutes is gonna really run us low on fuel here." It was the strongest statement he would make.

At 6:06 pm, the first flight attendant reported the cabin was ready for an emergency landing. The aircraft was about 17 miles south of the airport, heading southwest. At this point, the captain said they would be landing in about five minutes.

Flameout!

Just then, the First Officer said, "I think you just lost number four..." At 6:06:46, the First Officer told the captain, "We're going to lose an engine..." The captain replied, "Why?" At 6:06:49, the First Officer again stated, "We're losing an engine." Again, the captain asked, "Why?" The First Officer responded, "Fuel!"

At 6:07:12, the captain requested, "Clearance for an approach into 28 Left, now!" This was the first request for an approach clearance from Flight 173 since the landing gear problem began. At 6:07:27, the Flight Engineer stated, "We're going to lose number three in a minute, too." The captain responded, "You've got a thousand pounds. You got to."

The flight was now about 18 miles out and headed toward the airport. At 6:13:21, the Flight Engineer stated, "We've lost two engines, guys." At 6:13:38, the captain said, "They're all going. We can't make Troutdale (a small GA airport nearby)." The First Officer replied, "We can't make anything." At 6:13:46, the captain told the First Officer, "Okay. Declare a Mayday." The First Officer advised, "Portland Tower, United 173 Heavy, Mayday. We're—the engines are flaming out. We're going down. We're not going to be able to make the airport."

At about 6:15 pm, the DC-8 crashed into a wooded area of suburban Portland, about 6 miles southeast of the airport. There was no fire. Two crew members (including the flight engineer) and eight passengers died. There were 23 severe injuries and 156 with little or no injuries.

The Crew

The 52-year-old captain had been employed by the airline since 1951 and upgraded to captain in 1959. At the time of the crash, he had 27,638 total flight hours, 5,517 of which were as a captain in the DC-8, and had passed his last proficiency check in September 1978.

The 45-year-old first officer had been employed by the airline since 1965 and upgraded to DC-8 first officer in 1978. He had 5,209 total flight hours but was relatively new to the DC-8, with only 247 hours in type. He passed his proficiency check in June 1978.

The 41-year-old flight engineer had been employed by the airline in December 1967 and upgraded to the DC-8 in January 1975. He had 3,895 total flight hours as a flight engineer, 2,263 of which were in the DC-8.

The Aircraft

The precipitating event was the failure of the right main landing gear retract cylinder assembly. A piston rod failed due to corrosion. McDonnell Douglas had issued a service bulletin to address corrosion, which was accomplished on this airplane, but a later service bulletin to modify the cylinder had not been done. The aircraft was 10 years old—young by GA or today's airline standards—but it highlights the importance of maintaining flight-critical systems. There was some discussion in the accident report regarding the fuel gauging, but the NTSB determined that the systems were reporting correctly.

NTSB Probable Cause

"The probable cause was the failure of the captain to properly monitor the aircraft's fuel state and to respond correctly to the low fuel state and the crew members' advisories regarding fuel state. This resulted in fuel exhaustion to all engines. His inattention resulted from preoccupation with a landing gear malfunction and preparations for a possible emergency landing. Contributing to the accident was the failure of the other two flight crew members to fully comprehend the criticality of the fuel state or to successfully communicate their concern to the captain."

The NTSB analysis noted that the captain delayed contacting maintenance for about 28 minutes from the time of the gear malfunction. After speaking with company maintenance personnel, the flight should have turned inbound to land. This would have occurred 30 to 40 minutes after the initial gear problem, and landing would have occurred with about 15 to 20 minutes of fuel remaining.

Commentary

This crash led to the formal establishment of Crew Resource Management (CRM) and helped the airline industry move beyond a mindset that was prevalent at the time:

- Rule 1: The captain is never wrong.
- Rule 2: If the captain is wrong, see Rule 1.

Good captains, however, have always understood that the other crew members are not there just for aggravation.

Had the flight engineer said something like, "Captain, we will run out of fuel in about 15 minutes and will crash—we need to land right now!" things might have turned out differently.

While the culture has generally changed, the CRM system is only as good as the willingness of crew members to apply it. As noted in the Asiana crash on a visual approach to San Francisco, respectful assertiveness is essential. Thousands of crashes have been prevented by the implementation of CRM.

Fuel: It's Our Decision Alone

First, how much fuel is actually on board the aircraft? Full tanks or using internal tank references, such as tabs, provides certainty. Without a solid physical reference, it's more complicated. Be wary of someone telling you they only flew for half an hour! No tabs? A dipstick is a great idea— calibrate it carefully. Remember that the aircraft must be level.

Fuel totalizers are extremely accurate in measuring flow, but the starting point must be known. Fuel gauges on light aircraft may be "variable," especially older ones. Flight time is an excellent measurement with a few caveats. It's dependent upon knowing the starting point, understanding the burn rate (both in climb and cruise), and leaning appropriately. Too often, in training, fuel burn is not emphasized because the aircraft remains in the pattern or practice area, and cross-country flights are typically rather short. As a primary instructor, I was told never to lean below 5,000 feet, so students only got a theoretical explanation about leaning. We were also running 80-octane fuel, which had less lead than 100 "low lead" and resulted in a bit less plug fouling.

The AOPA Air Safety Institute recommends the "Golden Hour," which means the aircraft will always be on the ground when down to one hour of fuel. If there is a slight miscalculation for any reason, it won't matter.

Some pilots felt the United captain was unfairly tagged with the responsibility, and that the gauges were faulty, but that was never proven. The January 2, 1999, *Eugene Register Guard* had a front-page story regarding the reunion of the passengers and crew of Flight 173.

"The gratitude of the survivors and their families clearly touched Capt. Malburn McBroom. The former United Airlines pilot wept Monday night during a standing ovation from nearly 200 people. The gesture celebrated the fact that 179 people aboard Flight 173 survived the 1978 crash landing... But even in the applause, the same old question haunted McBroom: ...What about the 10 who did not survive?" The article goes on to discuss the terrible burden that McBroom carried, leaving him "a broken man."

In the magazines, hangars, and coffee shops, pilots discuss and learn about crashes in clinical terms, reassuring ourselves that we would never be so stupid or careless. The reality is that in the heat of battle, distraction creeps and snares the best of us. The planned reaction and periodic refresher training are good but imperfect antidotes. When faced with an abnormal situation, fuel and flight time are finite.

Lake Placid - (Sometimes, even the best get caught - but it's rare)

We Lose a Friend and a Superb Pilot to a Worst-Case Scenario

This narrative is personal. Many of you knew Richard McSpadden, who was the Senior Vice President of the AOPA Air Safety Institute, my successor there, and a friend. I had worked with Richard on many occasions and had flown with him on a photo mission in my airplane. There was no more capable pilot. Richard retired from the Air Force after having flown F-15s in combat and F-16s as the USAF Thunderbird Commander. But he was more than a jet jock. He owned a pristine Super Cub and flew many different GA aircraft—a pilot's pilot and a wonderful person, easily approachable and open to new ideas.

On October 1, 2023, I got the phone call from then AOPA President Mark Baker that Richard and another pilot, Russ Francis, a former NFL All-Star, had just died in an attempted return to Lake Placid Airport in New York. It was a beautiful fall afternoon with benign weather when something happened to the Cessna Cardinal RG they were flying, and they immediately turned back to the airport shortly after takeoff on a photo mission. They almost made it, impacting a ravine just short of the runway. Another two seconds would have made all the difference.

At this writing, the final report has not been published. Ironically, Richard and several other AOPA pilot staff had recently completed a video on turnbacks immediately after takeoff, testing a variety of aircraft. Their conclusion: in some light aircraft, with a proficient pilot and regular training, it might be successful. In heavier, high-performance singles, it was highly unlikely.

The NTSB's preliminary report is all that's available now and is excerpted below. This information is preliminary and subject to change.

The Flight

On October 1, 2023, at 1608 Eastern Daylight Time, a Cessna 177RG, N545PZ, owned and operated by Lake Placid Air Service, crashed on a return to the airport just after takeoff. The purpose of the flight was to photograph the accident airplane while airborne for later publication in a magazine article. According to witnesses, there were two airplanes in the flight. The lead airplane was a Beech A36 Bonanza with a photographer onboard, which departed first from Runway 32 at Lake Placid Airport (LKP), Lake Placid, New York. The accident airplane took off about 700 feet behind the Bonanza. The pilot/owner was to fly the airplane during the takeoff and climb out, and after joining up in formation, the pilot-rated passenger would have taken over the controls to fly during the photo shoot.

During taxi-out, witnesses heard the engine of the Cardinal shut off, and about 10 seconds later, it restarted. During the takeoff roll, a witness described the engine as sounding as if the propeller was set for "climb" and not takeoff, then he heard the engine surge. During the initial climb, the witness further described that the engine did not sound as if it was running at full power. The accident airplane then made a gentle left turn while it was 300 to 400 feet above ground level to join up with the Beech A36. As the accident airplane closed to within about 1,000 feet of the Bonanza, it suddenly made a hard right turn back toward the airport. During the turn, the pilot of the A36 heard the passenger in the accident airplane transmit, "We have a problem and we're returning to the airport."

The Cardinal impacted a steep embankment in a right-wing, nose-low attitude about 15 feet below the top of a plateau on airport property. It then slid about 30 feet down the embankment. The initial impact point was about 440 feet from the approach end of Runway 14, and 250 feet left of the runway centerline.

The Airport & Weather

The photo flight departed from Runway 32, which was 4,196 feet long. The weather, as previously noted, was not a factor.

The Aircraft

The 1976 Cardinal RG had 5,352 hours total flight time and had received a major engine overhaul, accumulating only 36.7 hours before the crash. It was being maintained on a Part 135 certificate for commercial operations as a sightseeing aircraft.

The Owner Pilot

The pilot was no novice, with commercial single and multiengine land and instrument ratings. He held a DC-3 type rating and a private pilot helicopter certificate. He had more than 9,000 hours of flight time, as reported on his second-class medical, which was received nine months earlier. It is impossible to tell who was flying at the time of the crash.

Initial Findings

Because mechanical failure is suspected, but not confirmed at this point, a detailed description follows. Control cable continuity was established from the flight control surfaces to the cockpit controls. The flaps were extended approximately 10 degrees, and the stabilator trim was neutral.

The fuel strainer bowl was fractured during the impact, and the fuel strainer screen was clean. The fuel selector handle operated normally in all positions and positively engaged in the detents. The fuel selector valve was confirmed to be on "BOTH," and the fuel caps were secured. The position of the fuel pump switch could not be determined due to impact damage. There were no obstructions in the fuel or fuel vent system from the wing tanks to the inlet of the fuel strainer. First responders reported fuel was draining from the airplane upon arrival, and there was no water contamination.

The nose landing gear was crushed aft during the impact sequence, and the actuator was separated (its position could not be determined). The main landing gears were in an intermediate position. The single main landing gear actuator was observed attached to its frame. The sector gear teeth were intact. There was no observable damage to the main landing gear down locks or gear legs. The main landing gear wheels were observed in contact with the buckled lower fuselage and not in the wheel wells.

The cowl flap handle was in the "OPEN" position. The mixture control was set to the full-rich position, with the propeller control in the high RPM/fine pitch position. The throttle was out about two inches.

The propeller remained attached to the engine crankshaft. The propeller hub was found cracked and damaged, consistent with impact, and one blade was partially dislodged from the hub. The blade exhibited leading-edge scratches and gouges. The other blade was bent slightly aft at midspan, with an approximate two-inch curled section of the tip separated. The blade exhibited chordwise scratching and leading-edge gouging on the outer third of the blade. Chordwise scratching means the propeller was rotating. It does not indicate that the engine was producing power, but rather that it was likely windmilling. This also suggests that the engine had not seized.

The top spark plugs were all intact, undamaged, and tightly installed in each cylinder. The top spark plugs were removed, and a lighted borescope examination was conducted on each cylinder. No abnormalities were noted. The engine crankshaft was rotated by the propeller, with suction and compression noted on all cylinders through the top spark plug holes, and movement of all rocker arms observed during rotation. All eight spark plugs were removed. The coloration across the plugs ranged from normal to black carbon fouling, with normal wear to the electrodes. No mechanical electrode damage was noted or observed on any of the spark plugs. The bottom spark plugs for cylinders No. 1 and No. 3 were oil-soaked, consistent with the orientation of the engine at the accident site and oil within the cylinders.

The single-drive, dual magneto unit was found securely installed to the rear of the engine. It was removed from the engine, and the single drive was rotated using an electric drill. The magneto produced spark at all ignition leads. No damage was observed to the magneto housing, but both ignition harnesses sustained varying levels of impact damage in the form of cuts and abrasions to multiple leads.

The oil dipstick was securely installed in the filler neck and indicated 6 quarts of oil in the oil sump. Oil was found draining from various parts of the engine due to impact damage. The oil filter had sustained significant damage and had separated from its threaded base. A portion of the filter element was removed, examined, and found to be clean of metallic particles or debris. The oil suction screen plug was found to be tight and safety-wired to the oil sump. The screen was unobstructed and clear of any debris.

The fuel system, including the engine-driven fuel pump, fuel manifold, and fuel servo, were all found attached to the engine. The fuel pump's 45-degree outlet fitting was found slightly loose, with the mating hose tight to the fitting. No fuel staining was observed from the fitting or on the fuel pump housing itself. The fuel divider was tightly installed to the top of the engine, with all injection lines tight and secure to each injection nozzle. The fuel injection manifold was disassembled, and no debris or tears were noted to its diaphragm. The fuel servo was attached to the lower side of the engine with all lines tight and secure, but all four hold-down nuts were found loose when slight pressure was applied with a wrench. Torque stripes were present on the studs and nuts. The torque stripes did not appear to be disturbed or misaligned. The throttle plate was found in the closed position. When the throttle arm was actuated manually, the throttle plate moved freely within the servo, but the threaded rod for the idle thumb screw adjustment was found fractured and not connected. The fuel injection nozzles were found to be free and clear of any debris.

The propeller governor was found securely attached. Its screen was found free and clear of any debris, and oil flowed from the unit when rotated by hand.

The wreckage was retained for further examination and will be disassembled under NTSB supervision at the engine manufacturer's location. The engine remains in a sealed crate, which will not be opened until the final examination occurs at the factory, with an NTSB investigator observing.

Commentary

At this point, there are more questions than answers. As noted in the introduction to this chapter, determining the cause of a piston engine failure is a significant challenge for both the NTSB and the FAA.

Some caveats that apply to all preliminary reports:

Hindsight Bias — Of course, this was bound to end badly; anyone could see that! The Monday morning quarterback always has a better play and would have won the game if only they could have seen the outcome.

Witness Veracity — This is always a question. What we think we saw and what actually happened are often two different things. If there was an engine monitor on board, it would help tremendously in determining the engine's status.

Attribution Effect — An NTSB term that suggests you or I would never make the same mistakes as the accident pilots. Sadly, it's often not borne out by the facts.

Jumping to Conclusions — One of the hardest things to train out of new accident investigators is the snap judgment. Facts sometimes take a while to emerge, and theories are just that—hypotheses that need testing.

Time to Think — This crash occurred many months ago, and we've all had plenty of time to think about it. During the event, Richard and Russ had only a few seconds to make critical decisions.

Perhaps you'll indulge me for a little philosophy. This crash hit a lot of us hard because of the people involved and the circumstances. While it doesn't lessen the pain, the reality is that life is sometimes very unfair. This isn't just true in aviation. I'm channeling Richard now because he would be sitting here having the exact same discussion that we've been having. We wouldn't minimize the magnitude of the loss, but we would try to understand and put it into perspective. Everything would be focused on preventing the next tragedy.

Everyone knows about the "missing man" formation, where a group of aircraft flies over the airport and one aircraft soars vertically toward the heavens. In my view, Richard is not missing. He is beside me on every takeoff, reminding me about the procedures to follow at any point should the engine quit. I'd much rather he be with us, but it's comforting nonetheless. Perhaps you'll invite his spirit into your cockpit.

There are literally tens of millions of uneventful takeoffs every year. They open up the wonderful world of flight, allowing pilots to live more fully than most. You've already heard about the "free lunch"—there isn't one, and occasionally, stuff happens. If you take to heart the suggestions regarding personal skill and aircraft maintenance throughout this book, the odds shift strongly in your favor, but there's no guarantee. That's life. Richard would agree, and his signature sign-off was, and remains, "GO FLY!"

CHAPTER 7
CONCLUSIONS AND REFLECTION

"What comes down had better be able to go back up again."
Aviation Corollary to Newton's Law of Gravity

Author (Maybe)

Congratulations—you've made it to the final approach. Let's summarize and aim for that perfect landing.

Aviation is governed by the unforgiving laws of physics and aerodynamics. The FAA's rules, by comparison, are gentle. They can be overly complicated in too many places, but the intent is generally good. Importantly, there are a few critical things to get right. It doesn't require a lot of math or memorization.

- **Gravity is absolute**: It's omnipresent and never defied; we only coexist with it.

- **No two objects can occupy the same place at the same time**: This applies to other aircraft, terrain, power lines, towers, etc. Remember Pauli's Exclusion Principle.

- **You can only tie the record for low flying**: See Pauli again. Buzzing failures are catastrophic.

- **Angle of attack is more than important**: Where the nose of the aircraft is pointed isn't necessarily where it's going. This is one of the most misunderstood aspects of flight, and understanding it is essential to safety. Regardless of whether the aircraft has an AOA (angle of attack) indicator, the pilot must know what's happening with the wing.

- **Weather is infinitely variable**: Once you master the basics of operating the hardware, weather becomes the challenge. It keeps the game interesting and can substantially increase risk—if you let it. Study, learn, be suspicious, and always leave an out.

You've heard all of this before—**Pitch plus power equals performance**. On instrument approaches, be stable at the final approach fix; VFR, at about 500 feet. On takeoff, have a plan that you've verbalized in the event of a powerplant failure. In military flight training, students must memorize and recite the procedure for every flight maneuver. My memory was never that good, so when there was something I needed to get down cold, flashcards were a great way to wear the grooves into my brain. Today, we can use electronic versions to review airspace, taxiway signage, aircraft emergency checklists, and more. On long flights, when it's not so busy, review those checklists.

But there's a distinction between procedure and technique. During takeoffs and landings, everything must be within parameters. On larger aircraft, inertia makes things more critical. Just as in light aircraft, runway excursions in jets are one of the leading causes of crashes. The manufacturers' test pilots did a lot of experimentation to come up with performance numbers. We won't be able to improve on them, and probably won't match them.

Attitude – All pilots can copy the procedures that professionals use, and it will only restrict you a little. A safety management system, used by airlines and corporate pilots, is only as good as the people who participate. It's a set of guidelines where everyone actively looks for trouble and shares the insights learned in a non-punitive fashion. It's doing the right thing when no one is looking. It also means that schedule is never absolute.

> *"We are all apprentices of a craft where no one ever becomes a master."*
>
> —*Ernest Hemingway, great novelist of the 20th century*

The best pilots are constantly learning—it's the mark of a pro in any business, hobby, or sport. Amateurs think they know enough and thus remain amateurs. The next flight is the most important, and it can unravel in an instant. Then experience, occupation, or net worth doesn't matter—it's all about airmanship and judgment at that moment. When you've made a mistake, admit it. Learn from it and resolve not to do it again. Repeat until the perfect flight is flown. I'm still looking—there's always something that could have been done with a little more finesse.

> *"If you think training is expensive, try having an accident."*
>
> —*Mantra during my time at FlightSafety International*

Look for an experienced, knowledgeable, and enthusiastic instructor. The chemistry, style, integrity, and commitment of the CFI or maintenance technician must be solid. If they don't mesh with you, find someone else. Poor instruction or maintenance, like cheap clothes or tools, is never a good deal.

Less regulation is more—Conduct flights within regulatory guidelines—not because it's required, but because (usually) there's a good reason for it. Trying to justify something on nuance or at the absolute margin is gaming the system. That's neither smart nor safe.

In personal flying, there is no shortage of regulation, but we are largely masters of our own destiny. Our passengers have entrusted us with their most valuable possession—their lives. Shouldn't our decision-making be somewhat more conservative when gambling with someone else's life?

Regulators and politicians sometimes decide that since existing rules did not prevent a particular crash, there must be a loophole that must be plugged. Complexity doesn't help. Too many regulations require interpretation that even the authorities can't agree upon.

In too many cases, the rules didn't fail—the individual or organization failed to follow the rules. Was the failure systemic or individual? The latter is not the basis for good rulemaking. Occasionally, we are dealt a bad hand, and then it's best to follow the procedure that was memorized. It's your best shot, but sometimes that won't work either. Such is reality, sadly.

> **"Truth is ever to be found in simplicity, and not in the multiplicity and confusion of things."**
>
> —*Sir Isaac Newton*

Bureaucracy and lawyers promote paper and complexity, but if you follow the rules below, chances are excellent that your flights won't get tangled in the regulatory briar patch and, more importantly, will be concluded safely.

The Basic Survival Rules Haven't Changed Much in Over a Century of Flight—The Baker's Dozen (13)

- Fly VFR only when you can see well enough to avoid terrain, obstacles, or other aircraft with a comfortable margin. The faster the aircraft, the more visibility is required. VFR rules are absolute minimums. Legal and smart aren't always the same. In marginal conditions, slow down. See and avoid takes time; even then, it doesn't always work. At night or in mountainous terrain, VFR minimums are just asking for trouble.

- Be courteous in the traffic pattern. At non-towered airports, adhere to guidelines, but recognize that they are just that—not regulatory by design. Think like a controller and let faster aircraft play through, even if you're delayed slightly. An aircraft behind you is a bad deal. Fly defensively and avoid giving pattern-police lectures. Discuss it politely on the ground if you think it will make a difference.

- Fly high enough to avoid hitting anything. Allow some margin for that once-in-a-lifetime engine failure so that there's some choice in a forced landing. That's going to be at least 1,500 feet AGL, but it could be much higher depending on terrain, obstacles, and weather. New towers are going up all the time and may not be in your terrain database. It also prevents a low-altitude violation. Flat country and open fields provide many engine-out possibilities. Lots of water, night, IMC, cities, and hilly country means going higher to preserve whatever options may exist.

- Stay out of airspace that is reserved for other aircraft or high-risk activities. This means prohibited and restricted areas, TFRs, or controlled airspace where you haven't been invited in or aren't properly equipped.

- Be proficient in your aircraft and for the type of flight you're conducting. This addresses the intent of currency requirements. Current and proficient are two different things. Instrument conditions, wind, and short runways—we all have our limits—approach them carefully. VFR? Stay out of the clouds.

- Stay inside the aircraft's operating envelope. We're not trained (nor paid) to be test pilots. At the edge, the margins are gone. The numbers in the POH or flight manual reflect the best the aircraft, in the best hands, could do. We are unlikely to match them. The variance in aircraft behavior is much greater than automobiles. Some aircraft are easy to fly, others are much more demanding. Honor the difference.

- Maintain the aircraft as if life depended on it—it does. If finances don't allow proper maintenance, park the machine until they do. Get an active carbon monoxide detector. Old aircraft, just like older humans, require more maintenance. It takes time and money. No free lunch.

- Maintain yourself as if life depended on it—it does. When sick or using debilitating drugs to solve medical problems, don't fly. There's usually a reason why a drug is not approved. As a passenger, you wouldn't accept that behavior in an airline captain or a bus driver. Ditto for fatigue—humans don't perform well when tired. Oxygen is essential for brain function; many of us will begin to degrade above 8,000 feet MSL and be oblivious to it. Use a pulse oximeter.

- Carry enough fuel to complete the flight with safe reserves. One hour upon landing (not when you start to think about an alternate). IMC Rule of Thumb—If it's a big weather system, 50 percent gets you to the first fuel stop. Thirty percent gets you to a solid alternate, and 20 percent is absolute reserve.

- When flying in instrument conditions, adhere scrupulously to the charted numbers for headings and altitudes. They've been thoroughly tested, and the penalty for miscalculation, misreading, or cheating is both instant and severe. Set your own minimums with an appropriate margin. Many of us don't fly enough to be safe at Category I minimums.

- Whatever the forecast, anticipate that it will change for the worse. If it turns out better—good news. If it's worse, you're prepared. Weather is what you find, not what was forecast. Offer pireps frequently; forecasts will get better as a result and more flights will be safely completed.

- If you just think you're getting into trouble, declare an emergency. Use all resources—ATC, Flight Service, other aircraft, and passengers—to survive. Not quite an emergency? Use Pan, Pan. File a NASA (ASRS) Report. Don't worry about the legal aspects—in most cases, everyone is glad to have helped, and there is no crash investigation. In a few cases, there may be a legal slap on the wrist. You probably deserved it, won't do it again, and likely will be flying again shortly.

- YOU are Pilot-in-Command. ATC is there to help you, not the other way around. Controllers can be a tremendous help, but you are responsible for the safety of the aircraft. If vectored towards thunderstorms, high terrain, asked to remain in icing, or fly into IMC when not rated or equipped, the only answer is "unable." Weigh ATC advice, but make your own decisions. With that authority also comes accountability. Can't have one without the other, so be prepared to answer.

"They don't pay us to think much once airborne."

—*Author*

The big secret to safety is avoiding excessive risk in the first place. Then no great amount of skill or luck is needed. Paradoxically, the airlines and corporate flight departments take as much decision-making out of the cockpit as possible. Planned reactions are developed to address worst-case conditions. There's no delay, no agonizing, no heavy mental lifting at a critical time—just execute Plan B. It's simple but requires discipline and thinking about the unthinkable beforehand.

Some Examples:

- Ground speed drops below xxx knots, which means the fuel reserve will be getting into the "golden hour." Land to take on fuel.

- The weather just went below published (or personal) landing minimums, and the flight is still 10 miles out. Under FAR Part 91, an approach is allowed to take a look, but the pro will hold or divert to the alternate with much lower risk and no temptation to cheat.

- The weight and balance computations say only two passengers with a full fuel load. Anybody or anything else has to stay behind or be de-fueled.

- The ceiling is below 2,500 feet along portions of the route and/or the visibility is below four miles. Mountains and hilly country demand significantly more. Time to seriously consider IFR or postpone.

Most crashes are caused by a series of mistakes, not just one—the accident chain. Sometimes it's obvious when things start going wrong. If the approach isn't appropriately stabilized, VFR or IFR, go around. Not touching down in the first third of the runway? Go around.

Subtle warnings are tougher. The weather situation is marginal and could go either way. Perhaps there's a growing sense of unease—things don't feel quite right. That should trigger a no-nonsense assessment of what's happening and a vision of the outcome if things go bad. If the circumstances don't improve or one more domino starts to fall, then it's time for Plan B—no further assessment required and no second-guessing.

"Make sure the risk is worth the reward."

—*Business Wisdom*

Good decision-making is based on just four actions:

1. Anticipate the problem
2. Recognize it
3. Act
4. Evaluate whether the chosen path worked.

Repeat as necessary.

Do Not:

- Try to get more utility out of an aircraft than it's capable of delivering. This means trying to carry too much, flying beyond reasonable range or time, attempting to climb over terrain when the wind or temperature conditions make that impossible. Avoid icing conditions and IMC if the machine is not certificated, etc. No aircraft is certificated to fly in thunderstorms. Flying an unairworthy aircraft is not life-extending.

- Try to get more utility out of yourself than your present health, fatigue level, skill, or training allow. This applies to instrument flight, crosswinds, short runways, weather conditions, a demanding aircraft, etc. You may have heard this a few times before.

- Try to have too much fun with an aircraft. "Rapture of the Sky" has an alarmingly high fatality rate—buzzing, river running, or aerobatics in the wrong aircraft or without adequate training. Don't be an Icarus.

If it can happen to them, it can happen to me. Whenever a superstar pilot is killed, it should give us pause. Taking to the sky is never without some risk. But that is true of life in general and applies to any mode of transportation. On the highways, abject stupidity and flagrant lawbreaking kill one hundred times more people than GA every single year, with millions more suffering serious injury. It's the equivalent of three or four airliners per week crashing. In aviation, we have much more control over our environment than on the roads. Follow the guidance here and learn from the past. You'll make many fewer mistakes, and the odds for having safe flights increase significantly.

"Safe is not the equivalent of Risk Free."

—U.S. Supreme Court, Indus. Union Dept. v. Amer. Petroleum Inst., 448 U.S. 607 (1980)

In the introduction of the book, author John Shedd noted that "Ships in harbor are safe, but that is not what ships are built for." In modern, protected life, only a few choose to live in the "total reality" environment of aviation. It's completely different from "reality" TV and the smartphone virtual world that most of our fellow humans inhabit. There can be no flight without some risk, but managed intelligently, it's safer than many things that provide so much less reward. Enjoy wisely.

Freedom to Fly

The United States has spilled much blood in defense of freedom of speech, religion, political expression, and the opportunity to fly. It's an aviation freedom that much of the world does not enjoy unless flying airliners or military aircraft. U.S. general aviation pilots generally get to go where they want when they want. Many countries send their student airline/military pilots here for training. If you think our bureaucracy is bad (and in places, it is), try flying elsewhere. Protect our flying liberty by exercising piloting excellence and responsibility. We have general aviation and airline operations that are the envy of the rest of the world. Let's keep it that way.

I hope you've enjoyed the journey. It was a long road. It took me over 50 years to get to this point. More importantly, my hope is that you found a few thoughts, techniques, or procedure items to add to your aviation toolkit. I'll look forward to seeing you on the ramp or in the airport lounge for coffee or tea, and to talk flying. Safe flights.

With profound apologies to Ecclesiastes…

To Everything there is an Aviation Season

A time to takeoff, a time to land.

A time to go VFR, a time to go IFR.

A time to approach, a time to hold.

A time to climb, a time to descend.

A time to turn left, a time to turn right.

A time to transmit, a time to remain silent.

A time to add fuel, a time to reduce fuel.

A time to continue, a time to divert.

A time to look out, a time to scan within.

A time to speed up, a time to slow down.

A time to repair, a time to replace.

A time to go, a time to stay.

A time to learn, a time to apply.

A time to fly, and a time to stop

.

The Icarus List

Here's a short version for printing. It's all the "common sense" you've heard before—time-tested!

• **Departure**—For cross-country flights, watch the weather several days before a trip. Don't go, go earlier, go later, take a different route, drive, or fly commercially. Advise passengers that flexibility is needed when the trip is planned—not when they get to the airport.

• **Weight and Balance**—Stay within limits. Do you really need all that stuff? Most "four or six-passenger aircraft" aren't—especially with any appreciable fuel. If out of balance, don't even think of it! Aircraft behave quite differently and poorly when heavy or with aft CG. Fly at max weight to see—it's likely a slug!

• **Use the 50-50 Formula** to provide margin over POH performance data. The obstacles are sometimes more than 50 feet. Our performance, the aircraft, or the conditions will seldom match the factory numbers. For single-engine airplanes, it gives you a shot at accelerate-stop on the runway.

• **Conduct a Sterile Preflight**—no distractions. Loose oil caps, cowl plugs, tie-downs, and overlooked chocks can lead to embarrassment or worse. The pilot secures all doors after passengers are boarded.

• **Use a Checklist/Flow and Sterile Cockpit**—Distraction is dangerous! Do NOT take off with full flaps, on the wrong tank, or with gust locks in place while talking about lunch or politics.

• **Enroute**—Keep up with weather. Expect the worst, hope for the best, and no wishful thinking is allowed. If it's as bad or worse than expected, manage accordingly. If it's really bad and you waited too long, land off-airport. Everyone survives, although the aircraft may get dinged—that's much better than the alternative. That's been proven many times.

• **Thunder**—Early morning flights typically work better than later. Understand what's triggering the convection—be careful with datalink. Vision is the gold standard. Always have an escape route (duh!).

• **Ice**—Anticipate ice if it's close to freezing at your cruising altitude and there are clouds. It won't always be there, but at least you won't be surprised and will have an escape route. Mountains are always bad news with ice. Advise ATC immediately and work to get out of it quickly. Limit airspeed loss to 10 knots. Whatever is collected on the way in will likely double on the way out. It means extra speed and no flaps on landing—i.e., a much longer runway.

• **Give and Get Pireps**—The jungle telegraph may be primitive, but it works well to find out what's happening. If we give enough of them, forecasting will improve. Be skeptical of good reports on icing or thunderstorm turbulence. It may change when you get there. Have a plan B. Have you heard this before? Let's hope the system becomes more advanced to provide real-time weather.

• **Mountains**—Weather is more challenging. There are fewer places for a forced landing, thin air ruins performance, and fewer reporting stations to verify if the forecasts were correct. ATC and weather radar coverage may be limited. Wind that is uncomfortable in

427

the flatlands becomes dangerous in the high country. If turbulence were visible, you wouldn't go there!

• **Landing—Adequate Fuel**—The "Golden Hour" of reserve removes stress and massive amounts of expense and embarrassment. Lean appropriately and monitor groundspeed. With GPS and fuel computers, there's no reason to run short. The math is simple: Fuel on board × fuel burn per hour, allowing extra for takeoff and climb, equals time aloft.

• **Airports Are Collision Country**—Use the CTAF and be courteous. If someone's in a big hurry, let the jerk play through! Patience and courtesy go a long way. Sterile cockpit and lights on for at least 10 miles inbound or outbound. Taxiing inbound and outbound requires full attention and no programming or configuring the aircraft—sterile cockpit.

• **Checklist/Flow**—Forgetting the landing gear is expensive and embarrassing. On takeoff, aircraft don't climb well with full flaps.

• **Goldilocks Again**—Left-right, up-down, first third of the runway, not too fast or slow. Don't irritate the bears! Beware the tailwind on base leg. VFR—generally stable at 500 feet AGL, IMC—1,000 feet AGL.

• **Crosswinds**—Fly in light to moderate wind before testing the demonstrated crosswind component. More speed will be needed on landing in strong and gusty crosswinds, which means a longer runway. Landing downwind? Bad idea on a dry runway, worse on a contaminated runway.

• **Fit to Fly**—Medically, physically, chemically, energetically, mentally. Fatigue, illness, drugs, and distraction are contributing factors, if not the cause, in too many crashes. Oxygen above 8,000 feet for most of us.

• **VFR into IMC**—What part of cloud don't you understand? It's aviation's version of Russian Roulette. Not many pilots get nailed the first time or several times out, but success breeds contempt.

• **Buzz Job**—How low can you go? You can only tie the record, and the risk-vs-reward is a poor trade-off. Flying low over lakes or river-running often leads to wire strikes. All this also often leads to close encounters with the FAA.

• **Aerobatics**—Recommended if trained by a good aerobatic CFI, in an approved and well-maintained aircraft, at a safe altitude/airspace. This is not a life-prolonging activity without ALL of the aforementioned conditions. Good to get some exposure early in your flying career.

• **Proficiency**—Practice doesn't make perfect—only perfect practice makes perfect. This applies to all aspects of aviation. Good pilots don't have to prove anything and practice regularly. Invest in quality training periodically.

• **The Devil on Your Shoulder**—If you're questioning the wisdom of something, there's probably a very good reason. Reassess honestly—What could possibly go wrong? The Rapture of the Sky is deadly!

• **Passengers**—They are entrusting you with their most precious asset—their lives. They deserve the best, most conservative decision-making you can provide. Safety isn't everything—it's the only thing!

ACRONYM LIST

Absolutely not all-inclusive—there are some you'll just have to look up!

ADS-B - Automatic Dependent Surveillance Broadcast. A required system for flight into certain airspace that allows ATC to see aircraft that may be below radar coverage. It also includes weather in the cockpit and traffic information to aircraft properly equipped with ADS-B IN functionality.

AIRMET - Weather that may be hazardous, other than convective activity, to light aircraft and Visual Flight Rule (VFR) pilots. However, operators of large aircraft may also be concerned with these phenomena. Convective activity always rates a Sigmet.

AGL – Height above ground level. More is usually better, except when over a runway.

AOA – Angle of attack. The angle between airflow (relative wind) and the chord line of the wing, a critical concept that will NOT be explained in this glossary!

ASAP - Aviation Safety Awareness Program. A joint program between airlines, some large Part 135 operators, and the FAA. If pilots voluntarily reveal a problem or mistake, they generally are immune from prosecution by the FAA or retaliation by the company.

ASOS/AWOS - Automated weather systems that report weather at an airport continuously—see METAR. Mostly correct, but sometimes not.

ASRS(P) - NASA-run reporting system where pilots can report an event or situation. If no crash or criminal activity is involved, the FAA shall not prosecute the pilot.

ATC – Air Traffic Control. Usually very helpful, but they cannot fly the aircraft for you.

AWC – Aviation Weather Center, a division of the National Weather Service located in Kansas City, Missouri. Issues aviation forecasts, manages Pireps, and maintains the Aviation Weather website. A tremendous resource!

CEO – Chief Executive Officer. Someone who's supposed to know how to run a company.

CFI/CFII – Certificated Flight Instructor/Instruments. Usually helpful, and good ones are worth every penny!

CTAF – Common Traffic Advisory Frequency. The "party line" radio frequency that should be used by all aircraft when arriving or departing from airports without a control tower and ATC services. Not to be used for lectures or where to get lunch.

CVR – Cockpit Voice Recorder. Required on airliners and records both internal and external communications. Used in crash reconstruction.

DA or DENALT – Density altitude. Pressure altitude corrected for non-standard temperature. High-density altitude will result in loss of performance, sometimes catastrophically. High temperatures and/or high altitudes are problematic.

DME - Distance Measuring Equipment, used to measure the distance from a VOR.

EFB – Electronic Flight Bag. A tablet (iPad) that provides maps and instrument charts and deprives chiropractors of revenue because pilots don't need to carry 40 pounds of charts.

FAA – Federal Aviation Administration. Manager of the airspace, regulations, and aircraft and pilot certification. Pilots have a love-hate relationship with them.

FAR/CFR – Federal Air Regulations (often referred to as CFR – Code of Federal Regulations, but to me, it's a distinction without a difference). There are many of them.

FDR – Flight Data Recorder, aka "black box" (but it's always orange).

FIS-B – Flight Information System Broadcast. Part of the ADS-B system. See ADS-B.

FO – First Officer, i.e., copilot.

GA – General Aviation. All flight operations except airlines or military.

GPS – Global Positioning Satellites. A great system for area navigation. It can be jammed, which, fortunately, does not happen often in the United States.

IFR – Instrument Flight Rules. You must be instrument-rated, current (and proficient), and the aircraft must be properly equipped to operate under these rules. It also requires an ATC clearance.

IMC – Instrument Meteorological Conditions. Typically, ceiling and visibility are less than 1,000 feet and visibility is less than 3 miles—lousy weather! This generally requires an IFR flight plan.

LIFR – Low Instrument Flight Rules. Ceiling less than 500 feet and visibility less than one mile. Not a good place for occasional flyers, even with instrument ratings.

MCAS – Maneuvering Characteristics Augmentation System. A system originally introduced on the Boeing 737 MAX series to keep the aircraft from pitching up during certain flight regimes. It didn't work as intended and resulted in two fatal accidents. It was completely redesigned. See Automation, Chapter 5.

METAR – Meteorological Report. A weather report for an airport by a human or automated observer on at least an hourly basis, but sometimes more often.

MSL – Mean Sea Level. The height of an airport above sea level, or what's shown on an aircraft altimeter as corrected for atmospheric pressure.

NTSB – National Transportation Safety Board. An independent government agency that investigates accidents and incidents in all modes of transport to determine what happened, why it happened, and to make recommendations to the FAA, industry, and the public to prevent a recurrence.

NWS – National Weather Service. See also AWC.

PAPI – Precision Approach Path Indicator. A system that provides a visual glide path to runways, similar to VASI (see below). It helps pilots visually follow a proper glide path to the runway.

PIREP – Pilot Report. Used to report observed weather conditions. Extremely useful, but what you encounter may be different.

POH – Pilot Operating Handbook. Describes how the aircraft is to be flown, expected performance, and limitations, as determined by test pilots. Your performance is likely to vary!

SIGMET – Significant Meteorological Advisory. Advises of observed weather that is potentially hazardous to all aircraft. Best to avoid this airspace.

TCAS – Traffic Collision Avoidance System. A traffic detection device required on all airliners after the Cerritos midair collision. Many light aircraft are now equipped with a similar system, known as FIS-B, as part of the ADS-B system that identifies all traffic known to ATC. See FIS-B. Not as sophisticated as TCAS, but it works very well!

TFR – Temporary Flight Restriction. Restricted airspace that requires prior coordination with ATC to enter, or it may be prohibited. Typically associated with ground emergencies such as fires, floods, other disasters, or VIP presidential movements. Inadvertent penetration will often result in a close look from military or law enforcement aircraft, which are much better seen at an airshow!

UTC – Universal Time Coordinated. The time used in all official communications and reports. Also known as Greenwich Mean Time (GMT) or Zulu (Z). Requires a math correction to convert to local time, which is what most pilots relate to. This was developed to have all operations on the same confusing page, regardless of time zones and independent of daylight vs. standard time.

VASI – Vertical Approach Slope Indicator. A light system providing a visual glide slope to a runway. It functions the same as PAPI.

VFR – Visual Flight Rules. Minimums are generally at least 1,000 feet and 3 miles, but for most of us, that's a fool's game.

VOR – Very High Frequency Omni Directional Range. A ground-based radio aid that has largely been replaced by GPS. A Minimum Operational Network (MON) will be maintained in the event of a GPS outage—which means we still have to learn it!

Made in the USA
Monee, IL
26 September 2025

30458191R00247